LIFE IN THE UNIVERSE

Cellular Origin and Life in Extreme Habitats and Astrobiology

Volume 7

Series Editor:

Joseph Seckbach
Hebrew University of Jerusalem, Israel

Life in the Universe

From the Miller Experiment to the Search for Life on other Worlds

Edited by

Joseph Seckbach
Hebrew University of Jerusalem, Israel

Julian Chela-Flores
The Abdus Salam International Center for Theoretical Physics, Trieste, Italy and Instituto de Estudios Avanzados (IDEA), Caracas, Venezuela

Tobias Owen
Institute for Astronomy, Honolulu, Hawaii, U.S.A.

and

François Raulin
LISA, Universités Paris 12 & Paris 7, Faculté des Sciences et Technologie, France

KLUWER ACADEMIC PUBLISHERS
DORDRECHT / BOSTON / LONDON

A C.I.P. Catalogue record for this book is available from the Library of Congress.

ISBN 1-4020-3093-2 (PB)
ISBN 1-4020-2371-5 (HB)
ISBN 1-4020-2372-3 (e-book)

Published by Kluwer Academic Publishers,
P.O. Box 17, 3300 AA Dordrecht, The Netherlands.

Sold and distributed in North, Central and South America
by Kluwer Academic Publishers,
101 Philip Drive, Norwell, MA 02061, U.S.A.

In all other countries, sold and distributed
by Kluwer Academic Publishers,
P.O. Box 322, 3300 AH Dordrecht, The Netherlands.

Cover artwork by Malte Reimold, von Kiedrowski laboratory, Ruhr-University Bochum, Germany.

Printed on acid-free paper

Printed in the Netherlands.

DEDICATION

This book is dedicated to **Professor Stanley L. Miller** in honor of his 50th jubilee experiments in the prebiotic origin of life.

Professor Stanley Miller with the editor **Joseph Seckbach** during the **Trieste-2003** conference on ***Life in the Universe***.

TABLE OF CONTENTS

IX. On the Question of Life on Mars and on the Early Earth

X. Searching for Extraterrestrial Life, Europa, Titan and Extrasolar Planets

Biodata of **Stanley Miller**, dedicatee of the volume "***Life in the Universe***"

Professor Stanley Miller, an American chemist and biologist was born in Oakland, California. He was educated at University of California (B.Sc. in 1951), and then at University of Chicago (where he was a student of Nobelist Harold Urey, and received his Ph.D. in chemistry in 1954). He was an assistant professor (1958–1960), associate professor (1960–1968), and then full professor of chemistry at the University of California, San Diego (from 1968). His research deals primarily with the origin of life and he is considered a pioneer in the field of exobiology. He has also studied the natural occurrence of clathrate hydrates and the use of hydrogen as a commercial fuel. In the 1950's, Urey theorized that the early atmosphere of the Earth was probably like the atmosphere now present on Jupiter—i.e., rich in ammonia, methane, and hydrogen. Miller, working in his laboratory at the University of Chicago, demonstrated in 1953 that when exposed to an energy source such as spark discharges these compounds and water can react to produce amino acids essential for the formation of living matter. This watershed experiment gave the first demonstration that the building blocks of life could be produced spontaneously from a planetary environment. The current consensus is that the Earth never had an atmosphere that was so reducing, but similar results have been obtained with less reducing gas mixtures. Indeed, this experiment has been repeated tens of thousands of times with different mixtures of gases, and the basic results have been spectacularly confirmed by the discovery of amino acids in meteorites. Professor Miller is a member of the National Academy of Science and has received numerous honors and medals, including the Oparin Medal of the International Society for the Study of the Origin of Life (1983). He served as the President of the above society (ISSOL) from 1986 to 1989.

E-mail: **smiller@ucsd.edu**

Biodata of **Joseph Seckbach**, author of the chapter "***Diversity of Microbial Life on Earth and Beyond***" and editor (with co-editors Julian Chela-Flores, Tobias Owen and Francois Raulin) of this book (***Life in the Universe*** (2004).

Professor Joseph Seckbach is the initiator and chief editor of ***Cellular Origins, Life in Extreme Habitats and Astrobiology*** **(COLE)** book series, and author of several chapters in this series. He earned his Ph.D. from the University of Chicago, Chicago, IL (1965) and spent his postdoctoral years in the Division of Biology at Caltech (Pasadena, CA). Then he headed at the University of California at Los Angeles (UCLA) a team for searching of extraterrestrial life. Dr. Seckbach has been appointed to the faculty of the Hebrew University (Jerusalem, Israel) performed algal research and taught Biological courses. He spent his sabbatical periods in Tübingen (Germany), UCLA and Harvard University and served at Louisiana State University (LSU), (1997/1998) as the first selected occupant of the John P. Laborde endowed Chair for the Louisiana Sea Grant and Technology transfer, and as a visiting Professor in the Department of Life Sciences at LSU (Baton Rouge, LA).

Among his publications are books, scientific articles concerning plant ferritin (phytoferritin), cellular evolution, acidothermophilic algae, and life in extreme environments. He also edited and translated several popular books. Dr. Seckbach is the co-author (with R. Ikan) of the ***Chemistry Lexicon*** (1991, 1999) and other volumes, such as the Proceeding of **Endocytobiology VII** Conference (Freiburg, Germany, 1998) and the Proceedings of **Algae and Extreme Environments meeting (Trebon**, Czech Republic, 2000). His recent interest is in Life origins and extremophilic environments of microorganisms.

E-mail: **seckbach@huji.ac.il**

Biodata of **Julian Chela-Flores** (editor)

Professor **Chela-Flores** current positions are: Staff Associate of the Abdus Salam International Center for Theoretical Physics (ICTP), Trieste, Research Associate, Dublin Institute for Advanced Studies (DIAS) and *Professor Titular*, Institute of Advanced Studies (IDEA), Caracas. He is a Fellow of: The Latin American Academy of Sciences, The Third World Academy of Sciences, the Academy of Creative Endeavors (Moscow, Russia) and a Corresponding Member of the Academia de Fisica Matematicas y Ciencias Naturales (Caracas). Professor Julian Chela-Flores studied in the University of London, England, where he obtained his Ph.D. (1969) in quantum mechanics. He was a researcher at the Venezuelan Institute for Scientific Research (IVIC) and Professor at Simon Bolivar University (USB, Caracas) until his retirement in 1990. His particular area of expertise is astrobiology, in which he is the author of numerous publications in this new science, including some in the frontier between astrobiology and the humanities (philosophy and theology). Professor Chela-Flores has been the organizer of a series of Conferences on Chemical Evolution and the Origin of Life from 1992 till 2003 (and was the co-director from 1995 till 2003). From 1992 till 1994 he worked in collaboration with Cyril Ponnamperuma. This series continued Ponnamperuma's *College Park Colloquia on Chemical Evolution*, which had started in Maryland, USA in the 1970s. All the proceedings of the present series of conferences have been published. In 1999 Professor Chela-Flores co-directed and edited the proceedings of an Iberoamerican School of Astrobiology in Caracas at the IDEA Convention Center.

E-mail: **chelaf@ictp.trieste.it**

Biodata of **Tobias Owen** (editor)

Tobias Owen is a Professor of astronomy at the Institute for Astronomy of the University of Hawaii. He studies the Planets, satellites and comets of our solar system using the giant telescopes on Mauna Kea and by means of deep-space missions. He has participated in the Viking Lander on Mars, the Voyager, Galileo, Deep Space 1, Nozomi and Contour Missions. He is currently an Interdisciplinary Scientist and member of several experiment teams on the ESA-NASA Cassini-Huygens Mission to the Saturn System, an Associate Scientist on the Rosina team for the Rosetta Mission to Comet Churyumov-Gerasimenko, and a Science team member of the Kepler Mission to search for terrestrial planets. Professor Owen received his Ph.D. in astronomy from the University of Arizona in 1965. He has published over 250 scientific articles and has co-authored two textbooks: "The Search for Life in the Universe" (with Donald Goldsmith) and "The Planetary System" (with David Morrison). He was awarded the NASA Medal for Exceptional Scientific Achievement for his Viking study of the Martian atmosphere.

E-mail: **owen@ifa.hawaii.edu**

Biodata of **François Raulin** (Editor)

François Raulin is Full Professor at University Paris 12, and head of PCOS (Space Organic Physical Chemistry) group of LISA, Laboratoire Interuniversitaire des Systèmes Atmosphériques a joint University Paris 7, University Paris 12 and CNRS laboratory of more than 80 persons, working on terrestrial troposphere in relation to environmental problems and on extraterrestrial environments, in relation to exobiology (http://www.lisa.univ-paris12.fr/). He is Director of LISA since 1995 and , since 1999, of CNRS "Groupement de Recherche" in Exobiology (GDR Exobio), federation of laboratories working on Exobiology, affiliated to NAI (http://www.lisa.univ-paris12.fr/GDRexobio/exobio.html

He got a diploma of Engineer from the Ecole Supérieure de Physique et Chimie Industrielles de la Ville de Paris in 1969, and a Doctorat d'Etat ès Sciences Physiques (on the role of sulphur in prebiotic chemistry) from the Université Paris 6 in 1976. He).

His research fields are related to planetology and exo/astrobiology: organic chemistry in extraterrestrial environments (Titan, giant planets, comets and Mars) using laboratory experiments (experimental simulations, spectroscopy, GC techniques), theoretical modeling and observation (remove sensing and *in situ* space exploration).

F. Raulin is IDS (InterDisciplinary Scientist) of the Cassini-Huygens mission (Titan's Chemistry and Exobiology program). He is also Co-I (Co-Investigator) of the CIRS (Cassini), ACP and GC-MS (Huygens) experiments. He is Co.I of the COSAC and COSIMA experiments of the Rosetta European cometary mission. He is Chair of COSPAR Commission F (Life Sciences), Vice Chair of COSPAR Planetary Protection Panel and Irst Vice-President of ISSOL. He has been member of the Microgravity Advisory Committee of the European space Agency and of ESA Exobiology Science Group. He is currently member of the Planetary Protection Working Group of ESA.

He is the author or co-author of more than 200 scientific papers, and 9 books related to the field of the origins of Life and Exobiology. He likes classical music, swimming, tennis and mountain climbing by foot in summer and by skis in winter and spring, with his wife Florence and his three children Antoine, Stella and Nicolas.

E-mail: **raulin@lisa.univ-paris12.fr**

EDITORS' PREFACE: *LIFE IN THE UNIVERSE*

This volume is a collection of chapters that have been presented as oral presentations or posters during the Seventh Trieste Conference on Chemical Evolution and the Origin of Life: Life in the Universe—"From Miller Experiments to the Search for Life on Other Worlds." This conference took place at the International Center for Theoretical Physics (ICTP), Trieste, Italy on 15–19 September 2003 in Trieste, Italy.

We were particularly delighted that Professor Miller himself was able to be with us for this singular occasion. We all greatly appreciated his retrospective and forward-looking lecture.

The ICTP Center in collaboration with many other institutions has sponsored the previous six conferences, since we first started planning the series with Professors Abdus Salam and Cyril Ponnamperuma. We were deeply honored, privileged and grateful to have had the following 10 sponsors: on this occasion:

The Abdus Salam International ICTP, Trieste, Italy
Scuola Internazionale Superiore di Studi Avanzati (SISSA), Trieste, Italy
Consiglio Nazionale delle Ricerche (CNR), Rome, Italy
NASA Institute of Astrobiology (NAI), USA
European Space Agency (ESA), France
National Aeronautics and Space Administration (NASA) Washington, USA
University of Paris 12, France
Osservatorio Astronomico di Trieste, Italy
Laboratorio dell'Immaginario Scientifico,
and
with the collaboration of the book series of
Kluwer Academic Publishers:
"Cellular Origin, Life in Extreme Habitats and Astrobiology"

Among the many presentations that we enjoyed during the conference we had contributions from experts on questions related to the origin, evolution and cosmic distribution of life, as well as talks on cosmology and the origin of biogenic elements and even two talks about the destiny of life in the universe from the perspectives of philosophy and epistemology.

Aspects of space exploration played a prominent part in our discussions, including the Galileo, Cassini-Huygens, Mars Express and Rosetta Missions, as well as proposed missions that are currently in the planning stage for Mars and, the Jovian satellite Europa. Two additional topics were included to round out our coverage of life in the universe: The search for planets outside the solar system, and the closely related subject of the possible manifestation of intelligence in potential galactic environments.

Finally, we continued to dedicate lectures to commemorate Cyril Ponnamperuma and Abdus Salam, the two distinguished scientists who initiated this series of conferences.

We would like to express our pleasure with the wide interest that this conference attracted. We had worldwide representation from 28 nations, with attendees from: Algeria, Argentina, Austria, Brazil, Cameroon, Chile, China, Colombia, Cuba, France, Germany, Hungary, India, Iran, Israel, Italy, Mexico, Morocco, Netherlands, Nigeria, Russia, Spain, Sweden, Switzerland, Turkey, UK, USA and Venezuela,

With this conference the **Chemical Evolution and the Origin of Life** series has now brought together well over 500 scientists, philosophers and theologians, since the series began in October 1992. It seemed appropriate to remember in particular two members of our advisory board, who unfortunately were no longer with us, and whom we remember with particular affection for their many solid contributions to the previous meetings: The astrophysicist Professor **Mayo Greenberg** and the biologist Professor **Martino Rizzotti**. They shall remain perennially in our memory.

February 2004

Joseph Seckbach
Julian Chela-Flores
Tobias Owen
François Raulin

ACKNOWLEDGEMENTS

We thank all colleagues who assisted us in finishing this volume and made every effort to move things along as smoothly as possible. We acknowledge the assistance of Professor Mauro Messerotti (Trieste, Italy) for supplying a scientific document, Professor A. Lazcano (Mexico City) for assisting with some manuscripts, and Professor Andre Brack (Orleans, France) who recommended the source for the book's cover. We express our gratitude to Malte Reimold from the laboratory of Professor von Kiedrowski, Ruhr-University, Bochum, Germany, for contributing her artwork to the cover. Last but not least, we appreciate very much the great and constant interest of Ms. Claire van Heukelom and Dr. Frans van Dunne (Kluwer Academic Publishers) in the series of "Cellular Origins, Life in Extreme Habitats and Astrobiology (COLE)" books. Their faithful investment in this book is very noticeable.

February 2004

Joseph Seckbach
Julian Chela-Flores
Tobias Owen
François Raulin

GROUP PHOTOGRAPH

1 AKINDAHUNSI
2 WANG
3 RADOSIC
4 TEWARI
5 PENSADO DIAZ
6 MEJIA CARMONA
7 RAULIN
8 CHELA-FLORES
9 AFOLABI
10 MILLER, S.
11 MILLER, D.
12 MILLER, Miss
13 OWEN
14 CHADHA, (Mrs)
15 SECKBACH
16 BALTSCHEFFSKY, H.
17 BALTSCHEFFSKY, M.
18 CHAKRABARTI
19 BHATTACHARJEE
20 CIPRIANI FITA
21 PEREZ DE VLADAR
22 GUIMARAES
23 LAZCANO
24 GRYMES
25 COSMOVICI
26 RAMOS-BERNAL
27 Participant
28 CHADHA
29 NEGRON-MENDOZA
30 WESTALL
31 DRAKE
32 STAN-LOTTER
33 SHAH
34 SCAPPINI
35 LANCET
36 FRASER
37 GATTA
38 MINNITI, Dante
39 MINNITI NOGUERAS, Alicia
40 COYNE
41 MAYOR
42 FANI
43 BRILLI
44 FUSI
45 BIONDI
46 GALLORI
47 TURNBULL
48 ROEDERER
49 GALLARDO
50 ACHARYYA
51 JOHNSON
52 PAPPELIS
53 KRITSKIY
54 VIDYASAGAR
55 MESSEROTTI
56 DICK
57 GENTA
58 MUSSO
59 PAGAN
60 MICHEAU
61 Participant
62 MEIERHENRICH
63 Participant
64 Participant
65 KAMALUDDIN
66 Participant
67 RAMIREZ {JIMENEZ}
68 CARNERUP
68a SHIL
69 MOORBATH
70 MEYER
71 Participant
72 VLADILO
73 TANCREDI-BARONE
74 SIMON
74a BUSHE
75 Participant
76 Participant
77 PLATTS
78 Participant
79 SI LAKHAL
80 Participant
81 VN DUNNE
82 SIMAKOV
83 COLLIS
84 SCHWEHM
85 Participant
86 BRACK
87 PUY
88 DELAYE
89 Participant
90 Participant
91 UNAK

“Young Miller” in University of Chicago (1953)

Photos by **NINA ALEXANDRINA** (BRAZIL) from "Life in The Universe-2003" Meeting

PHOTOS BY **JOSEPH SECKBACH** FROM THE LIFE IN THE UNIVERSE (2003) MEETING

STANLEY

I. Opening

INTRODUCTION TO LIFE IN THE UNIVERSE
"Are we alone?"

TORRENCE V. JOHNSON
Jet Propulsion Laboratory, California Institute of Technology, Pasadena, CA. USA

Fifty years ago, the Miller experiment took the search for the chemical origins of life to a new level with the laboratory synthesis of compounds required for life under conditions that resembled the early environment of the Earth. Since that time, scientists from around the world have built on these seminal results, developing a multi-disciplinary field that explores the fundamental questions of life's origin and emergence from the outer reaches of the universe to the finest details of microfossils seen in an electron microscope.

At the Seventh Trieste Conference on the *Chemical Evolution and the Origin of Life*, over one hundred scientists from many countries gathered to discuss and debate the current state of knowledge in the field. Highlights included the Cyril Ponnamperuma Lecture by George Coyne on Cosmic Evolution, Frank Drake's pre-dinner talk on the future of SETI, and the Abdus Salam Lecture by Stanley Miller himself, reflecting on the history and current state of chemical evolution experiments. This volume contains papers describing the research and results presented during these deliberations. These papers clearly demonstrate the health and vibrancy of the field seen fifty years after the Miller experiment.

In those fifty years, great advances have been made in many disciplines affecting the chemical evolution and origin life. Advances in biology, biochemistry and related fields have allowed the exploration of the mechanisms and processes of life at the molecular and atomic scale. Astronomical explorations have revealed a universe rich in the building blocks of life. Exploration of our own planet has demonstrated that primitive life emerged very early in Earth's history, and we have found life flourishing today under extreme environmental conditions previously believed to be completely inimical to any biologic activity. Beyond the Earth, reconnaissance of most of the solar system has revealed a remarkable diversity of planetary environments, greatly expanding our concept of the 'habitable zone' in the solar system. Finally, as other planetary systems around distant stars are being discovered at an ever-increasing rate, we find ourselves asking the question "Are we alone?" in the context of new, quantitative information in all these fields.

The program of the conference featured reviews and new research contributions related to all of the above. Looking at the various contributions and listening to the debates from my perspective as a planetary scientist, I was struck by several general currents running through the proceedings:

First, the conclusions and insights from the original Miller experiment have proved to be remarkably robust. Laboratory equipment and measurement capabilities have vastly

J. Seckbach et al. (eds.), Life in the Universe, 3–5.

improved in the last fifty years. Ideas about the chemistry of the early Earth's atmosphere and the physics and chemistry of the solar nebula and planet formation have also evolved greatly. Nevertheless, the Miller experiment has remained the underpinning for understanding the chemical processes leading to the production of biologically interesting compounds, to the extent that modern variants of the experiment are still being performed routinely to provide samples for comparison with comet surfaces, Titan's aerosols and interstellar material.

Second, the search for the earliest evidence for life on Earth illustrates both the tremendous sensitivity and sophistication of our current technical capabilities but also the frustrations of trying to wrest unambiguous results from the fragmentary records available to us from that early era. Interpretations of claims made on the basis of the fossil record, micro- and "nano-" fossils, and bio-markers of various kinds are still being advanced, challenged and hotly debated. It is notable that many of these issues arise also in one form or another with respect to interpretation of the Martian meteorite results. Clearly, this on-going debate should make us appropriately cautious and inform our approach to evaluating future searches for evidence of extra-terrestrial life.

Third, the convergence of research on terrestrial life in extreme environments and exploration of the diverse worlds of our solar system is expanding the range of investigations into sites for pre-biotic chemistry and possible habitats for extra-terrestrial life. The following are just a sampling of the exciting research areas related to our expanding exploration of our local cosmic neighborhood:

Interstellar dust grains and cometary material have long been regarded as potential windows to the original material from which the planets formed and reservoirs of pre-biotic organic compounds. The prospects of sample return from the NASA Stardust mission and the approaching launch of ESA's Rosetta promise an exciting new chapter in our understanding of these materials.

Evidence for liquid water flowing on the surface of Mars in past epochs has driven a new wave of Mars exploration. Results from NASA's Odyssey mission have provided maps of likely subsurface ice deposits even at low latitudes. During the meeting, the results from past missions were reviewed and prospects for future exploration explored. At the time of the meeting two NASA rovers and the ESA Mars Express mission with the Beagle 2 lander were headed toward Mars. As these proceedings go to press, the Beagle 2 appears to be lost, but both Spirit and Opportunity rovers and the Mars Express Orbiter are returning exciting new data, beginning a new phase of Mars exploration.

Other planetary environments of intense interest to the conferees are possible subsurface oceans on the icy satellites of the outer solar system. Raised as a theoretical possibility over twenty-five years ago, oceans under the icy crusts of Europa, Ganymede and Callisto are strongly suggested by a number of lines of evidence from the Galileo mission. Magnetic field data taken during close fly-bys of these satellites showed perturbations from induction magnetic fields created by the time-varying Jupiter magnetic field (Ganymede also has a large, permanent dipole field presumably generated in its core). The best current explanation for this behavior is the presence of a global electrically conducting layer near the surface. Rock and ice are poor electrical conductors, but saline ocean water would be consistent with the data.

Europa is particularly interesting because of its geologically young, lightly cratered surface. Tidally produced fracture patterns and chaotically disrupted regions reminiscent of melting and/or solid-state convective activity in the ice crust suggest that the ocean has been in communication with the surface in recent geologic times. Gravitational data suggest a total water plus ice thickness of about one hundred kilometers, raising the possibility that this satellite has a volume of liquid water in its oceans over twice that in all of Earth's oceans. NASA is studying a future mission, the Jupiter Icy Moons Orbiter, which would return to the Jupiter system and, using nuclear ion-drive propulsion, orbit each of the icy satellites and study them in detail.

Saturn's large icy satellite Titan is another fascinating world for study of natural organic synthesis. With a massive nitrogen/methane atmosphere (surface pressure $\sim$1.5 Earth's despite its low gravity), Titan represents an extraordinary natural laboratory where versions of the Miller experiment have been conducted in its atmosphere for perhaps billions of years. Its atmosphere rich in hydrocarbons and aerosols and its surface probably a mixture of water ice and liquid hydrocarbons, Titan is believed to be an ideal place to study pre-biotic processes and compounds. The NASA Cassini mission carrying ESA's Huygens probe to Titan will arrive at Saturn in July 2004, and Titan will one of the major targets of its projected four-year mission of exploration.

Finally, the discovery of extra-solar planetary systems around other stars has added a new dimension to the discussions of life in the universe. Discussions at the conference ranged from the nature of these systems and the types of environments for life they might harbor to methods of detecting the presence of life by remote observations of these systems. It is clear that we are now only looking at the first hints of what will be in the future a large, exciting field of study.

The discussions of this Trieste Conference and the papers in this volume represent a "snap-shot" of the status of our understanding of the chemical evolution of life and the current issues related to life in the universe – fifty years beyond the Miller experiment and at the opening of the next century of exploration. We can only speculate about and eagerly anticipate the advances the next fifty years may bring.

THE ABDUS SALAM LECTURE
Introduction

H. BALTSCHEFFSKY Chairman
Department of Biochemistry and Biophysics, Arrhenius Laboratories, Stockholm University, S-106 91 Stockholm, Sweden.

Professor Stanley Miller, Professor K. R. Srinivasan, Organizers and Sponsors of this Conference, Ladies and Gentlemen;

We are getting ready for the Abdus Salam Lecture, honoring two most distinguished scientists. Both have very significantly contributed to the rapid growth of the sphere of fundamental knowledge in the second half of the twentieth century.

Abdus Salam, theoretical physicist, Nobel Prize winner, creator and long time leader of The Abdus Salam Center of Theoretical Physics. With his active interest in the origin of life he played a leading role in instigating these conferences on Chemical Evolution and the Origin of Life here in Trieste, which still are of such primary importance in this field. He left this world in 1996.

And **Stanley Miller**, who most generously, as the Abdus Salam Lecturer, is going to give us his "Recollections of the beginning of chemical evolution experiments":

Dear Stanley, it is a great privilege, and indeed a pleasure to introduce you. This is in a way a quite easy task, because we all already know that "the Miller experiment", which is most appropriately placed in the title of this conference, in 1953, exactly 50 years ago, was a major breakthrough, opening up a new research field with, and for, rational and advanced chemical experimentation on the molecular origin of life.

It would take too much time to try to describe here your scientific carrier, your prices, your Presidency of ISSOL and your many other successes. So I rather will end this introduction with a couple of personal recollections.

First I would like to combine something of Abdus Salam and Stanley Miller. Abdus Salam gave the very first invited lecture of the University of Stockholm International Lectures on Human, Global and Universal Problems, in 1975. And 10 years later, at Lidingö close to Stockholm, Stanley Miller gave the opening lecture of a conference on the Molecular Evolution of Life. On a picture I took, as a co-arranger of these events, Stanley is seen approaching in his usual, modest way, more focussed on scientific discussion than on the camera.

Last but not least, I shall tell you the true story about when we learned that Stanley is an enthusiastic environmentalist, in the best sense of the word. About 25 years ago, in Stockholm, Stanley, my wife and I strolled in the King's Garden. Its elmtrees were full

J. Seckbach et al. (eds.), Life in the Universe, 7–8.

of young people who, some even spending nights in the trees, prevented the authorities from removing the elmtrees, by ax and saw. Also Stanley signed a petition to save the elmtrees—and they were saved!

Stanley, I believe that your greatness as a scientist and as a friend must be linked to the many facets of your wonderful personality. We much look forward to your lecture.

THE BEGINNING OF CHEMICAL EVOLUTION EXPERIMENTS
Recollections and Perspectives

S.L. MILLER[1], J.L. BADA[2], and A. LAZCANO[3]
[1]Department of Chemistry and Biochemistry University of California, San Diego 9500 Gilman Dr La Jolla, CA, USA, [2]Scripps Institution of Oceanography University of California at San Diego La Jolla, CA 92093-0212, USA, [3]Facultad de Ciencias, UNAM Apdo. Postal 70-407 Cd. Universitaria, 04510 Mexico D.F. México.

1. Introduction

In spite of the ongoing debate about the oldest morphological evidence of life, it is generally accepted by scientists that the first living beings emerged on Earth early in the history of the planet. However, our understanding of the processes that led to the emergence of life is also hindered by the lack of geological evidence of the prebiotic environment, i.e., we have no direct information on the chemical composition of the primitive atmosphere, the temperature of the planet, the pH of the primitive hydrosphere, and other conditions which may have been important for the origin of life. Hence, it is not surprising that this has led to intense debates and the formulation of different and even contradictory explanations of how life came into being.

This situation is not new. Since the late 19th century, the belief in a natural origin of life had become widespread, and few years afterwards many attempts had been made by prominent scientists to explain the origin of life. As reviewed elsewhere (Miller et al., 1997), the list covers a rather wide range of explanations that include Pflügger's ideas on the role of HCN, to those of Svante Arrhenius on panspermia, and include Leonard Troland's hypothesis of a primordial enzyme formed by chance events in the early oceans, the sulfocyanic theory of the autotrophic origin of the protoplasm developed by Alfonso L. Herrera, and Herman Muller's proposal of the abrupt formation of a mutable gene endowed with hetero- and autocatalytic properties. In spite of the diversity of these concepts, most of them went unnoticed, in part because each was incomplete and speculative, unsubstantiated by direct evidence, and not subject to empirical investigation.

Although it is true that many scientists favored the idea of primordial beings endowed with a plant-like, autotrophic metabolism that could allow them to use CO_2 as their source of cellular carbon, others like A. I. Oparin, J. B. S. Haldane, C. B. Lipman, and R. B. Harvey proposed independently an heterotrophic origin of life, required the synthesis of simple organic compounds by various processes. The most successful and best-known proposal was that by Oparin, who, from a Darwinian analysis, proposed a series of events from the synthesis and accumulation of organic compounds to primordial

J. Seckbach et al. (eds.), Life in the Universe, 9–16.

life forms whose maintenance and reproduction depended on external sources of reduced carbon.

The assumption of an abiotic origin of organic compounds rested on firm grounds. In 1824 Wöhler has achieved the chemical synthesis of urea and oxalic acid from inorganic starting materials, and in 1828 he was able to confirm the identification of urea formed from silver cyanide and ammonium chloride. In fact, many of the reactions previously studied by organic chemists have turned out to be important primitive Earth synthesis, including the laboratory synthesis of alanine achieved by Strecker in 1850 from a mixture of acetaldehyde, ammonia, and hydrogen cyanide, and the formation of sugars from formaldehyde under strong alkaline conditions reported by Butlerow in 1861 (Bada and Lazcano, 2003). These efforts in fact heralded the era of organic chemistry but, as discussed elsewhere (Lazcano and Bada, 2003), the motivation of such studies was to synthesize organic compounds and not to understand what happened on the primitive Earth. The first successful prebiotic experiments in support of Oparin's ideas came first from the laboratory of Harold C. Urey's in the University of Chicago in 1953, and here we discuss come these experiments came about.

2. A Laboratory Model of the Primitive Earth

From the 1950s, chemists were drawn toward the origin of life. Driven by his interest in evolutionary biology, Melvin Calvin and his collaborators at the University of California, Berkeley, attempted to study the possibility of organic compounds under primitive Earth conditions with high-energy radiation sources. They irradiated a gas mixture of CO_2, H_2O, H_2 and a solution of Fe^{2+} with 40-meV helium ions using the Crocker Laboratory's 60-inch cyclotron (Garrison et al., 1951). The results, however, were not encouraging: only small amounts of formic acid and formaldehyde were obtained, which is similar to results obtained in experiments done since the 1920s by several other researchers attempting to understand the chemical mechanisms underlying photosynthetic processes (Rabinowith, 1945).

In 1950 the Nobel laureate Harold C. Urey, who had been involved with the study of the origin of the solar system and the chemical events associated with this process, began to consider the origin of life in the context of his proposal of a highly reducing terrestrial atmosphere. Urey presented his ideas during a seminar at the University of Chicago in October of 1951. In his lecture Urey pointed out that the solar system is reducing (that is, there is an excess of molecular hydrogen), except for the Earth and the minor planets (Mars, Venus and Mercury), which are more oxidized, with the terrestrial atmosphere being highly oxidized. According to Urey, a reducing atmosphere would contain methane, ammonia, water, and molecular hydrogen, as found on Jupiter and Saturn (except that the water has been frozen out) and on Uranus and Neptune, where both the water and ammonia have been frozen out. Urey thought that such type of reducing atmosphere would be a favorable place to synthesize organic compounds, which would form the basis to make the first living organism on Earth. It was only after this seminar that a biochemjist brought Oparin's book to Urey attention, pointing out how Oparin had discussed the origin of life and the possibility of synthesis of organic compounds in a reducing atmosphere. Urey of course acknowledged Oparin's contributions in a paper he published one year later detailing his model of the primitive atmosphere (Urey, 1952).

The electric discharge experiment followed shortly after (Miller, 1953). In September 1952, almost a year and a half after attending Urey's seminar, one of us (S. L. M.) approached

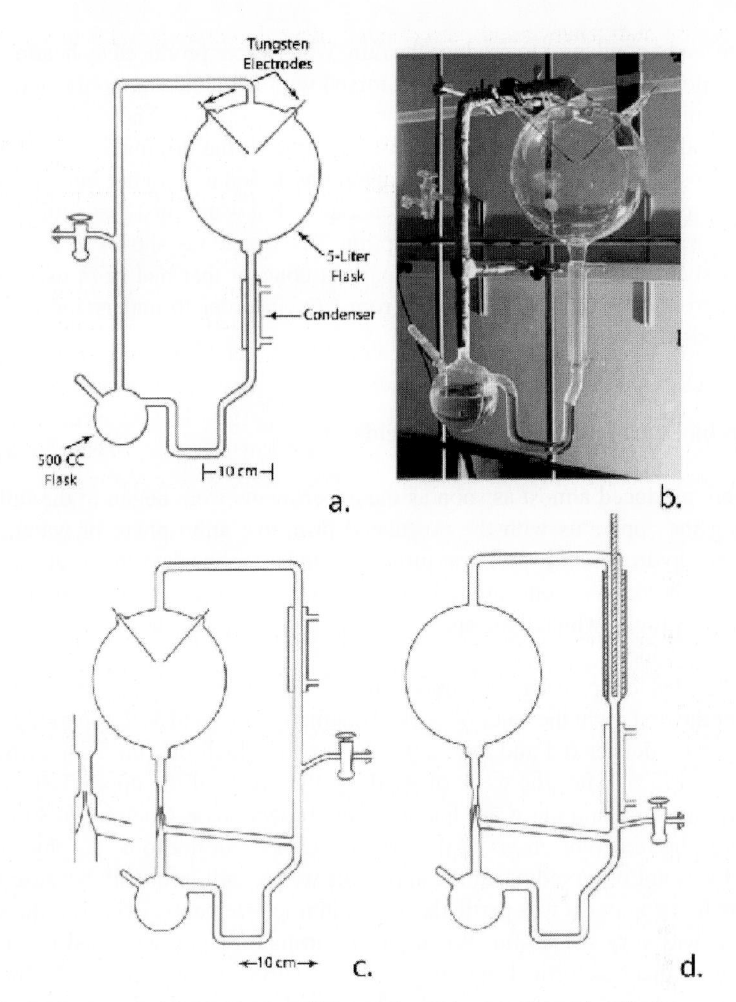

Figure 1. The various apparatus used in the Miller experiment (see text for explanation).

him about the possibility of doing a prebiotic synthesis experiment as a doctoral thesis topic. The issue was not just how to synthesize organic molecules under primitive terrestrial conditions, it was the first step in understanding how life started on our planet. Urey's initial resistance to this project was outdone after agreeing to try the organic compound synthesis for six months or a year. If nothing came of the project by this time, then a different, more conventional thesis project would be undertaken.

After realizing that ultraviolet light and electric discharges must have been the most abundant energy sources on the primitive Earth, and that very little work had been done on the effect of electric discharges on mixtures of methane and nitrogen or oxygen-bearing compounds, it was decided that amino acids were the best group of compounds to look for first, since they are the building blocks of proteins and since the analytical methods for their detection were at that time relatively well developed. Three spark discharge different apparatus to be used in the experiment where then designed (Figure 1). They were meant to simulate the ocean-atmosphere system on the primitive Earth, containing a model ocean, an atmosphere, and a condenser to produce the rain. Water vapor produced by heating would be like evaporation from the oceans, and as it mixed with methane, ammonia, and hydrogen, it would mimic a water vapor-saturated primitive atmosphere. The apparatus shown in Figures 1a and 1b was the one most extensively used in the original experiments, and is the design most widely known today. The apparatus 1c led to a higher inner pressure, and an important aspect of its design is that it generates a hot water mist that could be consider similar to a water vapor-rich volcanic eruption. The apparatus shown in Figure 1d used a so-called silent discharge instead of a spark, a concept that had been used previously in attempts to make organic compounds from CO_2 in order to understand the nature of photosynthesis (Lazcano and Bada, 1953).

3. The Prebiotic Chemistry of Amino Acids

Results were produced almost as soon as the experiments were begun in the fall of 1952. After filling the apparatus with the postulated primitive atmosphere of water, methane, ammonia, and hydrogen, the spark was turned on, and after two days the solution was a pale yellow. The solution was concentrated, and after running a paper chromatogram, a small purple spot was found which upon spraying with ninhydrin was shown to be moving at the same rate as glycine.

It was decided to repeat the experiment one more time, sparking the mixture for a whole week. After the first night the solution looked distinctly pink, and as the sparking continued it became first a deeper red and then a yellow-brown which obscured somewhat the red color (Miller, 1953). After the week of sparking the inside of the upper flask was coated with an oily material and the water had a yellow-brown color. Contrary to some original speculations, the red color observed the first day turned out not to be porphyrins (Miller, 1974), and was not observed when the apparatus was rebuilt, probably because the Pyrex glass of the first apparatus was particularly rich in trace elements. However, the search for amino acids was very successful. When paper chromatography was used to characterize the compounds that had formed, spraying with ninhydrin showed that the glycine spot was much more intense and spots corresponding to several other amino acids were also detected (Miller, 1974). Three of these amino acids were strong enough and in the correct position to be identified as glycine, α-alanine, and β-alanine. Two spots were considerable weaker in color but corresponded to aspartic acid and α-amino-n-butyric acid. The two remaining spots did not correspond to any of the amino acids that occur in proteins or any known amino acids then available, so they were simply labeled as A and B (Figure 2). It was estimated that at least 10 mg of amino acids had been formed, and it was concluded that the total yield was in the milligram range (Miller, 1953). Experiments with the apparatus in Figure 1a and 1b, as well as the one in Figure 1c produced in general a similar distribution and quantities of amino acids and other organic compounds. In contrast, experiments with the apparatus in Figure 1d led to lower overall yields and a much more limited suite of amino acids: essentially only sarcosine and glycine were produced (Miller, 1955).

After discussing the outcome of the experiment with Urey, it was decided to have its results published as soon as possible. A short manuscript was thus prepared, but in a very generous way Urey decided not to be included as a coauthor, in recognition of the work done largely by S. L Miller. The paper was first sent to *Science* on February 1952, but after some

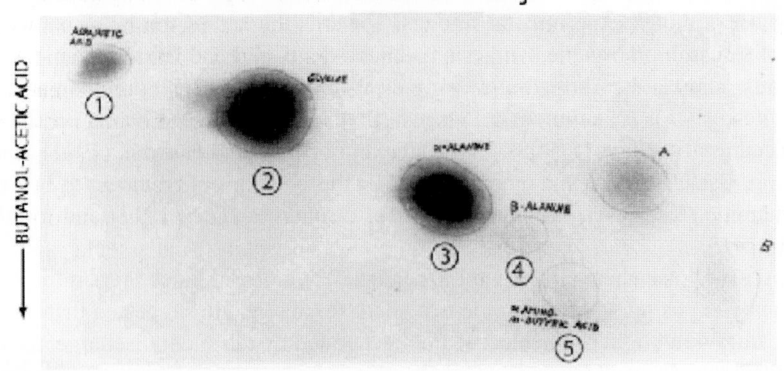

Figure 2. The two dimensional paper chromatogram of the amino acids produced from the sparking experiment (Miller, 1953).

delay and confusion it was retired and submitted to the *Journal of the American Chemical Society*, only to be retired and resubmitted to *Science*, where it was published on May 1953 (Lazcano and Bada, 2003).

Although by comparison with contemporary analytical tools the methods available in the early 1950s were crude, the search for amino acids was very thorough in terms of the techniques then available. These included the determination of the melting points of the amino acids that had been synthesized, which was at the time the usual method of positively identifying organic compounds. This represented several months work, but the identifications were then firmly established and have not been disputed ever since. Equally important, the possibility of biological contamination was completely ruled out by autoclaving the apparatus, which had been filled with the mixture of reduced gases, for 18 hours. The experiment was repeated and the yield of amino acids was the same as the runs where the apparatus was not autoclaved (Miller, 1974).

4. The Dawn of Organic Chemistry and Prebiotic Syntheses

It is not strictly correct to say that the 1953 Miller experiments were the first organic compound synthesis under primitive Earth conditions, since many of the reactions previously studied by organic chemists have turned out to be important primitive Earth synthetic reactions (Miller, 1974). In 1824 Friedrich Wohler reported that the reaction of cyanogen with ammonia solution led, in addition to several other products, to oxalic acid and "a crystalline, white substance" which was properly identified as urea four years later (Wohler, 1828). As reviewed elsewhere (Bada and Lazcano, 2003), although it was not immediately recognized as such, a new era in chemical research had begun: in 1850 Adolph Strecker achieved the laboratory synthesis of alanine from a mixture of acetaldehyde, ammonia, and hydrogen cyanide, and in 1861 Alexandr M. Butlerov showed that the treatment of formaldehyde with strong alkaline catalysts, such as sodium hydroxide (NaOH), leads to the synthesis of a wide array of sugars.

Research on the laboratory synthesis of biochemical compounds was soon extended to include more complex experimental settings. Towards the end of the 19th century a large amount of research on organic synthesis had been done, and had led to the abiotic formation of fatty acids and carbohydrates using electric discharges with various gas mixtures (Rabinovitch, 1945). As summarized elsewhere, this line of research was continued into the 20th century by Klages (1903), Ling and Nanji (1922), and Herrera. (1942) Moreover, Walter Lob, Oskar Baudish and others worked on the synthesis of amino acids by exposing wet formamide ($CHO\text{-}NH_2$) to silent electrical discharges (Lob, 1913) and to UV light (Baudish, 1913).

As discussed elsewhere (Bada and Lazcano, 2002), Lob did indeed report the synthesis of glycine by exposing wet formamide to a silent discharge. He suggested that because of either the ultraviolet light or the electric field generated by the silent discharge, formamide is first converted to oxamic acid, which in turn is reduced to glycine. He also claimed that glycine is produced when wet carbon monoxide and ammonia are subject to the silent discharge, and proposed formamide as the intermediate in this synthesis. Lob also theorized that glycine might also be produced from wet carbon dioxide and ammonia in a pathway wherein formamide was again the intermediate, but he did not demonstrate this directly. Although Lob apparently did produce glycine from formamide, this cannot be considered a prebiotic reaction because formamide would not have been present on the primitive Earth in any significant concentrations. It is also possible that the wet carbon monoxide and ammonia led to the synthesis of HCN, which would have produced glycine on polymerization and hydrolysis (Lazcano and Bada, 2003).

In retrospect, the efforts by Wohler, Strecker, Butlerov, Klages, Lob and others to produce simple organic compounds from simple reagents heralded the dawn of prebiotic chemistry, and have some bearing on our understanding of prebiological evolution. However, there is no indication that Lob and the others who carried out these studies were interested in how life began in Earth, or in the synthesis of organic compounds under possible prebiotic conditions. This is not surprising; since it was generally assumed that the first living beings had been autotrophic, plant-like microbes, the abiotic synthesis and accumulation of organic compounds was not considered to be a necessary prerequisite for the emergence of life. In fact, these experiments were not conceived as laboratory simulations of Darwin's warm little pond, but rather as chemical models designed to understand the autotrophic mechanisms of nitrogen assimilation and CO_2 fixation in green plants (Bada and Lazcano, 2002).

5. Afterword

The lack of direct geological evidence of the environmental conditions of the primitive Earth at the time life emerged always raises the question of whether the spark discharge experiments and other laboratory simulations are an adequate model of the primitive Earth. Contemporary geoscientists tend to tend to doubt that the primitive atmosphere had the highly reducing atmosphere modeled in the original Miller experiment. One additional objection would be that the input of electrical energy was far higher than possible in the primitive atmosphere. However, a striking confirmation of the general validity of this experimental simulations came from the analysis of the Murchison meteorite, which fell in Australia on September 29, 1969. This meteorite was shown to be a carbonaceous chondrite that was particularly rich in organic compounds, including a wide array of indigenous amino acids. The amino acids first reported in the Murchison meteorite were glycine, alanine, sarcosine, glutamic acid, and α-aminoisobutyric (Ring et al., 1972), all of which had been found in the original electric discharge experiment (Miller, 1953), as well as valine and proline. A second paper on amino acids in Murchison reported the presence of N-methyl alanine, β-alanine, aspartic acid, and α-amino-n-butyric acid (Wolamn et al., 1972). Although a number of amino acids have been reported in the Murchison had not been found in the electric discharge experiment, the meteorite contained most of those formed in the 1953 Miller experiment. Thus, although there is no geological evidence for the existence of Oparin's postulated prebiotic soup, the occurrence of amino acids, as well as purines, pyrimidines, sugar derivatives, and many other compounds in the 4.6×10^9 years-old Murchison meteorite makes it plausible (but does not prove) that similar abiotic synthesis took place on the primitive Earth.

Many have suggested that the organic compounds needed for the appearance of life may have originated from extraterrestrials sources such as meteorites. It is of course unlikely that a single mechanism can account for the wide range of organic molecules that may have been present in the prebiotic Earth. Rather, the primitive soup was almost certainly formed by contributions from endogenous and exogenous sources—from local synthesis in reducing environments, and on metal sulfides at deep-sea vents, as well as from comets, meteorites and interplanetary dust. This eclectic view does not beg the issue of the relative significance of the different contributors of organic compounds: it simply recognizes the wide variety of potential sources of the raw material required for the origin of life.

The existence of diverse non-biological mechanisms by which biochemically relevant monomers can be synthesized under plausible prebiotic conditions is now established. Of course, not all prebiotic pathways are equally efficient, but the wide range of experimental conditions under which organic compounds can be synthesized demonstrates that prebiotic syntheses of the building blocks of life are robust. The abiotic reactions producing such compounds span a broad range of seetings, and are not limited to a narrow spectrum of highly selective reaction conditions. Our ideas on the prebiotic formation of organic compounds are largely based on experiments in model systems. How life arose on Earth is of course unknown, but it is certainly encouraging to see that we have reached this level of understanding of the chemistry of the prebiotic Earth 50 years after the electric discharge experiments simulating the primitive environment were first performed.

6. Acknowledgements

For grant support, we thank the NASA Specialized Center of Research and Training (NSCORT, UCSD) (S.L.M, J.L.B, A.L.) and UNAM-DGAPA Proyecto PAPIIT IN 111003-3 (A.L.).

7. References

Bada, J. L. and Lazcano, A. (2002) Miller revealed new ways to study the origin of life. *Nature* **416**, 475.
Bada, J. L. and Lazcano, A. (2003) Prebiotic soup—revisiting the Miller experiment. *Science* **300**, 745–746.

Baudish, O. (1913) Ueber das CO_2 fixation. *Angew. Chem.* **2**, 612–616.

Garrison, W. M., Morrison, D. C., Hamilton, J. G., Benson, A. A., and Calvin, M. (1951) The reduction of carbon dioxide in aqueous solutions by ionizing radiation. *Science* **114**, 416–418.

Herrera, A. L. (1942) A new theory of the origin and nature of life. *Science* **96**, 14.

Klages, A. (1903) Ueber das methilamino-acetonitril. *Berichte der Deutschen Chemischen Gesellschaft* **36**, 1506.

Lazcano, A. and Bada, J. L. (2003) The 1953 Stanley L. Miller experiment: fifty years of prebiotic organic chemistry. *Origins Life Evol. Biosph.* **33**, 235–242.

Ling, A. R. and Nanji, D. R. (1922) The synthesis of glycine from formaldehyde. *Biochem. J.* **16**, 702–705.

Lob, W. (1913) Uber das Verhalten des Formamids unter der Wirkung des stillen Entladung. Ein Beilrag zur Frage der Stickstoff-Assimilation. *Bercfihte der Deutschen Chemischen Gesellschaft* **46**, 684–697.

Miller, S. L. (1953) A production of amino acids under possible primitive Earth conditions. *Science* **117**, 528–529.

Miller, S. L. (1955) Production of some organic compounds under possible primitive Earth conditions. *J. Am. Chem. Soc.* **77**, 2351–2361.

Miller, S. L. (1974) The first laboratory synthesis of organic compounds under primitive Earth conditions, In Jerzy Neyman (ed.), The Heritage of Copernicus: theories "pleasing to the mind" (MIT Press, Cambridge), 228–242.

Miller, S. L., Schopf, J. W., and Lazcano, A. (1997) Oparin's "Origin of Life": sixty years later. *J. Mol. Evol.* **44**, 351–353.

Rabinovich, E. I. (1945) Photosynthesis. Vol I (Interscience, New York), pp. 61–98.

Ring, D., Wolman, Y., Friedmann, N., and Miller, S. L. (1972) Prebiotic synthesis of hydrophobic and protein amino acids. *Proc. Natl. Acad. Sci. USA* **69**, 765–768.

Urey, H. C. (1952) On the early chemical history of the Earth and the origin of life. *Proc. Natl. Acad. Sci. USA* **38**, 351–363.

Wohler, F. (1828) Ueber das organische synthese. *Ann. Physik* **12**, 253.

Wolman, Y., Haverland, W. J., and Miller, S. L. (1972) Nonprotein amino acids from spark discharges and their comparison with the Murchison meteorite amino acids. *Proc. Natl. Acad. Sci. USA* **69**, 923–926.

AN OVERVIEW OF COSMIC EVOLUTION

GEORGE V. COYNE, S.J.
Specola Vaticana, V-00120 Città del Vaticano Rome, Italy

Abstract. Since chemical complexity and then life itself came about as part of the evolutionary history of the cosmos, it would be helpful to have an overview of that history. After an introduction to a few key concepts, several relevant topics in modern cosmology are reviewed. These include: the Hubble Law and the age of the universe; large scale structuring; galaxy formation and evolution; and cosmic chemistry.

1. Key Concepts

A preliminary time scale for the evolution of the universe is basic to our discussion. If we take the age of the universe as of the order of 10^{18} secs (we will refine this shortly) then the universe became transparent, the decoupling of radiation from matter, at 10^{13} secs. The Planck time is measured as 10^{-43} secs and inflation, of which we will speak shortly, occurred shortly after the Planck time. In the intervening period, from inflation to the present, stars and galaxies formed, structure in the radiating and non-radiating material developed and we came to be.

Cosmological distances are typically expressed as a measure of the redshift, z. The relationship between distance and redshift requires a cosmological model which gives the fundamental constants of the expanding universe. It is usual to adopt the Standard Model (Longair, 1996). Thus, at $z = 0.1$ the distance is approximately 1.5×10^9 light years, at $z = 1.0$, about 9×10^9 light years and at $z = 10$, about 13×10^9 light years. When cosmologists refer to "high redshift galaxies" they mean those with redshifts greater than $z = 0.1$. The most distant galaxies yet detected have redshifts of about $z = 6.5$.

When cosmologists say that the universe is "flat", they mean that in its expansion it is just on the edge of expanding forever or collapsing in upon itself. Among all the possibilities for the expansion rate of the universe, it is remarkable that it is "flat" and this requires explanation. In 1980 Alan Guth (Guth and Tye, 1980) first proposed that the universe inflated at many times the velocity of light very shortly after the Planck time. The energy source is found in quantum mechanical phase transitions in the early universe before there was matter, so that velocities exceeding that of light could occur. A schematic presentation is given in Fig. 1. In a non-inflationary universe, depicted at the top, when the universe was at $t = 10^{-35}$ sec it had a radius of 1 mm, which is much larger than the horizon distance of the universe at that time. With expansion the horizon of the visible universe is at a distance of 10^{28} cm (based on an age for the universe of 13.7×10^9 years). In an inflationary universe, depicted at the bottom, the universe inflated so that at $t = 10^{-35}$ sec it had a radius of

J. Seckbach et al. (eds.), Life in the Universe, 17–26.

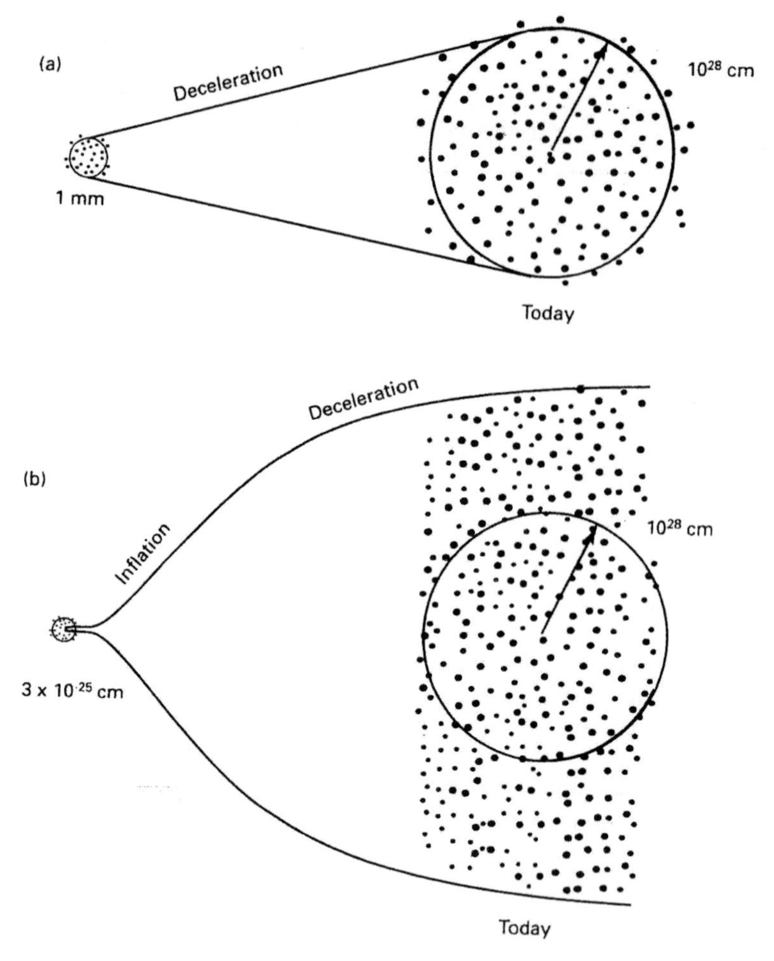

Figure 1. A schema of a non-inflationary (a) and an inflationary universe (b)

3×10^{25} cm.Thus today it has dimensions which are much larger than the horizon distance of 10^{28} cm, the distance which light can travel in the total age of the universe. Our visible universe is, therefore, inserted in a "multi-verse" invisible to us. This, of course, raises the question of the verifiability or falsifiability of such a multi-verse. However, the first predictions of observational tests of inflationary cosmologies are now being made (Hogan, 2002).

2. The Hubble Law Revisited and the Age of the Universe

One of the most fundamental ways of measuring the age of the universe is to determine its expansion rate, the velocity of expansion with distance, and thus calculate when the expansion began. This is called the "Hubble Law" and it is illustrated in Fig. 2 where the velocity in kilometers per second is plotted against distance measured in millions of

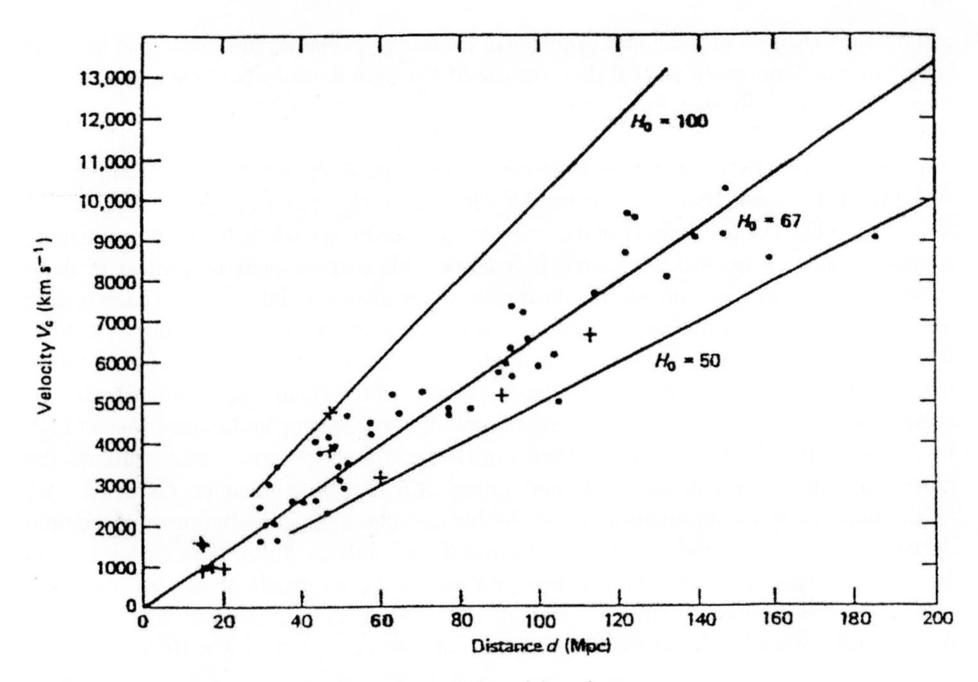

Figure 2. The Hubble Law interpreted as the expansion of the universe.

parsecs (one parsec equals 3.26 light years). The slope of this relationship, the Hubble constant H_0, is inversely proportional to the time when the expansion began and thus to the age of the universe. The best fit to the observations (dots) is $H_0 = 67$ km/sec/megaparsec; extreme values of H_0 are also shown. The scatter in the observations, which increases with distance, reveals serious difficulties involved in obtaining accurate measurements both of the velocities and of the distances. In fact, measuring velocities is quite straight forward since it simply involves measuring the shift in wavelength of spectral lines in celestial objects, Doppler shift or redshift. But it is the selection of the objects and the interpretation of the measurements that create difficulties. What one wishes to measure is the expansion in the space-time structure of the universe, the so-called "Hubble flow." To do this one must select objects that are participating in that expansion. This typically means measuring the redshifts of galaxies and clusters of galaxies. However, galaxies have their own peculiar motions in addition to their motion with the Hubble flow and clusters of galaxies have the peculiar motions of their member galaxies and the peculiar motion of the cluster as a whole. These difficulties are best overcome by measurements made for objects at very large distances where the expansion velocities are very large, so that peculiar velocities are an insignificant fraction of the total velocities measured. But then the difficulty arises in the measurement of distances, since the error in the distance measurement increases with distance. In fact, the whole case for getting an accurate measurement of the Hubble constant comes down to measuring astronomical distances. Even with telescopes in space, outside the perturbations of the earth's atmosphere, geometrical determinations of distances are accurate to only about 3,000 light years, three percent of the diameter of the Milky Way. The usual way to proceed beyond is to use the so-called photometric distance measurements, whereby we assume the existence of "standard candles" in the universe, classes of objects that have the same intrinsic luminosity, so that their measured apparent luminosity is the result of only the inverse square distance for light propagation. Hubble, for instance, first measured the distance to the Andromeda nebula by using Cepheid variable stars as a standard candle. In order, however, to measure to very large distances we need objects which are intrinsically very bright. In recent years use has been made of type Ia supernovae (Saha et al., 2001; Sandage, 2002). The progenitors of these supernovae are binary white dwarfs wherein mass exchange causes a nuclear explosion. It is remarkable that the peak brightness in these systems is quite constant considering the origins of the energy and the fact that each system results in the nucleosynthesis of different amounts of certain isotopes, for instance Ni^{56}. Recently three other non-photometric methods have been used to measure large distances: measurements of the size and optical depth of holes in the cosmic background radiation (CBR) due to the Sunyaev-Zeldovich effect, gravitational lensing and fluctuations at high resolution in the CBR. Gravitational lensing is of particular interest since it allows the rather accurate determination to a lensed galaxy at a very large distance. Cardone et al. (2002) have made a determination of the Hubble constant by the measurement of the time delays in the arrival of the four images from a distant galaxy. Since there are more than 50 multiply imaged systems known this promises to be an excellent way to obtain an independent measure of the Hubble constant. The result of all of the tedious measurements described above is a Hubble constant which translates into an age of 13.7×10^9 years for the universe.

3. Large Scale Structure. Clustering and Voids

Why is it important to know the large scale distribution of matter in the universe? The distribution of matter today is the inheritance of the fluctuations in the early universe, propagated to the current epoch according to a given world model, i.e. the fundamental parameters of the universe: deceleration or acceleration, matter and energy densities, including dark matter and dark energy, etc. So the large scale distribution of matter will be a test of how well we know those fundamental parameters. Since galaxies are the luminous tracers of the way matter is distributed, it is important to map the distribution of galaxies to large distances. Hubble Space Telescope has observed rich clusters of galaxies, containing thousands of galaxies, out to distances of 8×10^9 light years. The first large galaxy redshift survey was conducted at the Harvard-Smithsonian Center for Astrophysics in the 1980s (Geller, 1990). That survey plotted 1,065 galaxies in a slice of the universe whose outer edge was at 450 million light years. Later about 10,000 additional galaxies were measured. This survey already began to reveal the so-called "soap bubble" structure with large clumps of galaxies and large voids. In recent years large automated sky surveys have already measured redshifts for hundreds of thousands of galaxies. At the Anglo-Australian Observatory the Two-Degree Field (2dF) Galaxy Redshift Survey was completed in April 2002 (Colless, 2003). It plotted more than 200,000 galaxies out to distances of 3×10^9 light years and confirmed the soap bubble structure of the universe to larger distances. A detailed comparison of the galaxy distribution with the small scale temperature fluctuations in the CBR showed that the distribution of visible mass (galaxies) is an excellent tracer of the over all mass of the universe (including dark matter). It also confirmed that some unknown form of dark energy is the dominant constituent of the universe and is causing the universe to accelerate in its expansion. An even more ambitious survey, The Sloan Digital Sky Survey (SDSS), an international collaboration, is under way and will be completed in June 2005. It will have measured redshifts for 600,000 galaxies in addition to positions and 5-color photometry for 100 million celestial objects (mostly galaxies) over 6,600 square degrees or one-sixth of the sky (Frieman and SubbaRao, 2003).

What is the largest scale on which clustering of galaxies and voids occur? How does one determine this? The apparent sizes must be corrected for inaccuracies in distance determinations. For instance, galaxies that appear to be at the edge of a void may be foreground or background objects to the actual void. Statistical methods have been developed for "walking around" the edges of an apparent void and determining its actual diameter (Hoyle and Vogeley, 2002). The result is that the largest voids measure about 150 million light years in diameter, about 1500 times the diameter of the Milky Way. This result is confirmed by a detailed study of the 2dF Survey mentioned above (Roukema, Mamon and Bajtlik, 2002). A surprising result has been that intergalactic clouds have been found statistically to be as common within voids as elsewhere (Manning 2002). These clouds are discovered by the so-called "Lyman-alpha forest," a series of Lyman-alpha absorption lines of the light from distant quasars shifted into the visual spectral regions by the Hubble flow. These clouds may be either proto-galactic, galaxies in formation, or simply the dissipation of a gas cloud which failed to form a galaxy. At any rate voids are apparently not altogether empty!

An important issue is the change in structure with distance and thus with time since the Big Bang. The problem here is that there are two effects which tend to cancel out each other. Any clustering that begins will generally increase with time. That is the way gravity works. So we should see more clustering nearby. But as we observe to greater distances we preferentially see the more clustered arrays, since they are the brightest and more obvious. The surveys mentioned have not yet succeeded in separating out these two effects and providing an answer to the development of structure with time in the expanding universe.

4. Galaxy Formation and Evolution. The Rate of Star Formation

Did the nuclei of galaxies form first and then by their gravity assemble stars already formed elsewhere? Or did stars form within a galaxy as the nucleus and other galactic features were also forming. The most likely answer is both with one or other scenario dominating in certain types of galaxies. One model of interest starts with the collapse of a 10^{10} solar mass cloud at about 12×10^9 years ago, just 1.7×10^9 years after the Big Bang (Mangalam, 2001). As a core forms the first generation of massive stars also forms. Thus enriched material is supplied to a rotational disk which forms after 0.5×10^9 years from the beginning of the cloud collapse. Stars continue to form in this disk, a second generation from material enriched by supernovae from the first generation, and the core continues to collapse to from a black hole of 10^7 solar masses. The only reason that star formation can continue is that the supernovae heat energy is less than the binding energy or that the cooling time of the hot gas is less than the time to escape the halo. At the end of this process we have a black hole nucleus, a bulge, a halo and a disk in which star formation continues. These structures are typical for galaxies as we observe them, varying in how much one or other structure dominates in different galaxy types: elliptical, spiral, irregular, etc.

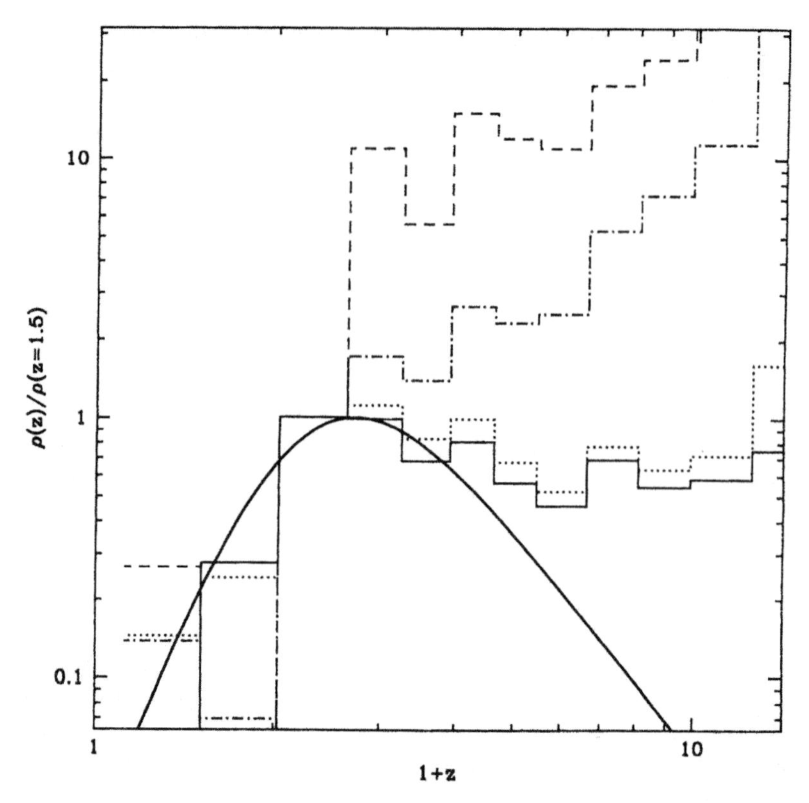

Figure 3. The rate of star formation is plotted as a function of "look back time" for: observations in the ultraviolet to submillimeter wavelength range (solid line); for gamma-ray bursters interpreted as "collapsers" (histograms to the lower right); for gamma-ray bursters interpreted as "mergers" (histograms to the upper right). See text for further explanation.

Since one of the principal reasons for the difference in galaxy types is the rate of star formation, it is important for an understanding of galaxy evolution to study star formation rates as a function of redshift or "look back time" in the expanding universe. A best representation using data from the ultraviolet to submillimeter wavelengths is shown by the solid line in Fig. 3 (Blaine 2001) where the star formation rate, $\rho(z)$, normalized to $z = 1.5$, is plotted against $1 + z$. The maximum rate occurs at a look back time of about 10×10^9 years, 3.7×10^9 years after the Big Bang. However, there is almost certainly an undersampling at large distances where only very intrinsically bright objects can be seen at the wavelengths covered. For this reason an interesting attempt has been made to sample star formation by using gamma-ray bursts, since these are very energetic events both in gamma-rays and in the associated optical transient emission and since their progenitors are thought to be short lived massive stars (Lloyd-Ronning et al., 2002). The results of this study of 220 gamma-ray bursts is shown by the four histograms in Fig. 3. It is thought that gamma-ray bursts arise from either the mergers of massive stellar remnants (neutron stars and black holes; Paczyński 1986) or from the collapse of massive stellar cores (hypernovae, Paczyński 1998). In either case they apparently help us to sample massive star formation at early epochs. In Fig. 3 the two histograms extending to the bottom right are derived by considering the collapse model and those rising to the right top by considering the merger model. In each case the difference between the two histograms is of no significance to our discussion, since it considers the flatness of the initial mass function which is not of concern to us now. It is clear, however, that, whatever the interpretation of the massive progenitors of gamma-ray bursts, massive stars formed in the very early epochs of the universe, as early as 0.5×10^9 years after the Big Bang. This could not be determined by observations in the uv to submillimeter wave range (solid curve in Fig. 3). The universe was developing structure: stars, galaxies, even clustering, at a much earlier epoch than had been thought possible a few years ago.

5. Cosmic Chemistry

All chemicals, except for the lightest elements, come from nucleosynthesis in stars. In the very early hot universe hydrogen, deuterium, helium and a few other light elements were made, but, as the universe expanded and cooled down, hot spots which housed thermonuclear furnaces were required to generate the heavier elements. With the passage of each generation of stars the universe became more metal abundant. The heaviest elements come from supernovae explosions, the death of massive stars. In cool, dark clouds in the interstellar medium molecules, even somewhat complex ones, can form sometimes with the help of interstellar grains on whose surfaces they find fertile grounds for development. These grains are, of course, themselves assemblages of somewhat complex molecules. A great deal of terrestrial biochemistry takes place in liquid water. This cannot be the usual case in the interstellar medium. Table 1, taken from Ehrenfreund and Charnley (2000), lists organic molecules discovered thus far in the interstellar and circumstellar environments. Of particular interest are the studies of infrared absorption spectra to young stellar objects embedded in cold dense envelopes. The line of sight typically passes through a large column density of interstellar ices. An example is seen in the spectrum of the star W33A, taken by the Infrared Space Observatory (Gibb et al. 2000) and shown in Fig. 4 where the absorption features of several organic molecules are identified. In recent years there has been an ongoing debate about the possible biotic origin of the 3.4 micron absorption signature of C-H stretching modes found in an amazing variety of objects, as shown in Fig. 5 taken from Ehrenfreund and Charnley (2000): the solid line is for an infrared galaxy at a distance of about 9×10^9 light years; the points are for the center of the Milky Way; the dashed line is for an organic residue of the Murchison meteorite. It appears that certain activities in carbon chemistry occur throughout the universe and there may be a rather universal reservoir of prebiotic organic carbon. However, from a study of the whole spectral region from 2 to 20 microns Pendleton and Allamandola (2002) conclude that there is no evidence for a biological origin to the 3.4 micron feature.

6. Summary

Since the physical evolution of the universe has led through increasing chemical complexity to the origins of biotic systems, an overview of recent results in cosmology is proposed as a background to discussions of the chemistry which led to life. Various independent methods

TABLE 1. Interstellar and circumstellar molecules

Number of Atoms										
2	3	4	5	6	7	8	9	10	11	13
H_2	C_3	c-C_3H	C_5	C_5H	C_6H	CH_3C_3N	CH_3C_4H	CH_3C_5N?	HC_9N	$HC_{11}N$
AlF	C_2H	l-C_3H	C_4H	l-H_2C_4	CH_2CHCN	$HCOOCH_3$	CH_3CH_2CN	$(CH_3)_2CO$		
AlCl	C_2O	C_3N	C_4Si	C_2H_4	CH_3C_2H	CH_3COOH?	$(CH_3)_2O$	NH_2CH_2COOH?		
C_2	C_2S	C_3O	l-C_3H_2	CH_3CN	HC_5N	C_7H	CH_3CH_2OH			
CH	CH_2	C_3S	c-C_3H_2	CH_3NC	$HCOCH_3$	H_2C_6	HC_7N			
CH^+	HCN	C_2H_2	CH_2CN	CH_3OH	NH_2CH_3		C_8H			
CN	HCO	CH_2D^+?	CH_4	CH_3SH	c-C_2H_4O					
CO	HCO^+	HCCN	HC_3N	HC_3NH^+						
CO^+	HCS^+	$HCNH^+$	HC_2NC	HC_2CHO						
CP	HOC^+	HNCO	HCOOH	NH_2CHO						
CSi	H_2O	HNCS	H_2CHN	C_5N						
HCl	H_2S	$HOCO^+$	H_2C_2O							
KCl	HNC	H_2CO	H_2NCN							
NH	HNO	H_2CN	HNC_3							
NO	MgCN	H_2CS	SiH_4							
NS	MgNC	H_3O^+	H_2COH^+							
NaCl	N_2H^+	NH_3								
OH	N_2O	SiC_3								
PN	NaCN									
SO	OCS									
SO^+	SO_2									
SiN	c-SiC_2									
SiO	CO_2									
SiS	NH_2									
CS	H_3^+									
HF										

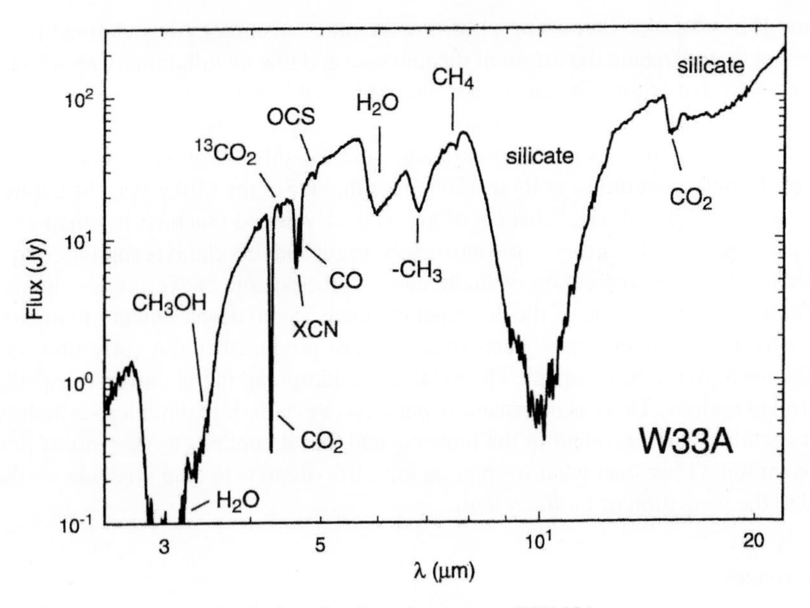

Figure 4. Absorption spectrum in the line of sight to the protostar WW 33A.

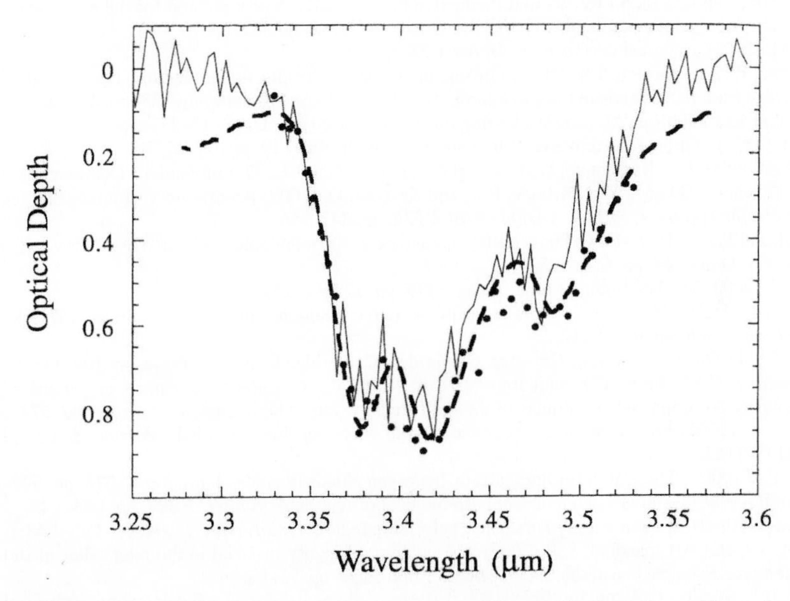

Figure 5. The 3.4 micron absorption feature due to C-H stretching modes in various objects. See text for further discussion.

conclude that the age of the universe is $13.7 \pm 0.2 \times 10^9$ years. For the first time in the history of these determinations a decimal point has been placed on the year in which we celebrate the birthday of the universe. This is due in no small part to the refinements in old methods and to the development of new methods for determining large distances in the universe. It has become increasing clear that quantum cosmology has claimed priority in our attempt to understand the origin of the universe and that an inflationary epoch occurred very soon after that origin. The large scale distribution of luminous matter has become much better known with ambitious redshift surveys of galaxies and it appears that the distribution of all gravitating matter, including dark matter, follows this distribution. The largest scale on which the universe clumps is about 1500 times the size of the Milky Way. Structure as we know it today: stars, galaxies, clusters of galaxies, developed much earlier than expected, within perhaps 0.5×10^9 years of the universe's origin and this claim is supported by recent measurements at high resolution of fluctuations in the cosmic background radiation. The rate of star formation is one of the principal elements in the development of structure and it now appears that many more massive stars were produced in the early universe than we had known previously. Organic chemicals are widespread in the universe, especially in star forming regions. There is increasingly persuasive evidence that numerous pathways of carbon chemistry are prevalent in the universe and that it contains a reservoir of prebiotic organic carbon. Other than what we find on the earth, there is to date no evidence that this has led to the formation of biotic systems.

7. References

Blaine, A.W. (2001) *Starburst Galaxies*, Berlin, Springer, p. 303.

Cardone, V.F., Capozzielo, S., Re, V., and Piedipalumbo, E. (2002) A new method for the estimate of H_0 from quadruply imaged gravitational lens systems, *Astron. & Astrophys.* **382**, pp. 792–803.

Colless, M. (2003) The great cosmic map, *Mercury* **32**, pp. 30–36.

Ehrenfreund, P. and Charnley, S.B. (2000) Organic molecules in the interstellar medium, comets and meteorites: A voyage from dark clouds to the early earth, *Ann. Rev. of Astron. & Astrophys.* **38**, pp. 427–483.

Frieman, J.A. and SubbRao, M. (2003) Charting the heavens, *Mercury* **32**, pp. 13–21.

Geller, M. (1990) Mapping the universe: Slices and bubbles, *Mercury* **19**, pp. 66–76.

Gibb, E.L., Whittet, D.C.B., Schutte, W.A., Boogert, A.C.A., Chiar, J.E., Ehrenfreund, P., Gerakines, P.A., Keane, J.V., Tielens, A.G.G.M., Van Dishoek, E.F., and Kerkhof, O. (2000) An inventory of interstellar ices toward the embedded protostar W33A, *Astrophysical J.* **536**, pp. 347–356.

Guth, A.H. and Tye, S.-H.H. (1980) Phase transitions and magnetic monopole production in the very early universe, *Phys. Rev. Letters* **44**, pp. 631–635.

Hogan, C.J. (2002) The beginning of time, *Science* **295**, pp. 2222–2225.

Hoyle, F. and Vogeley, M.S. (2002) Voids in the point source catalogue survey and the updated Zwicky catalog, *Astrophys. J.* **566**, pp. 641–651.

Longair, M.S. (1996) *Our Evolving Universe*, Cambridge, Cambridge University Press, pp. 109–113.

Lloyd-Ronning, N.M., Fryer, C.L. and Ramirez-Ruiz, E. (2002) Cosmological aspects of gamma-ray bursts: Luminosity evolution and an estimate of the star formation rate at high redshifts, *Astrophys. J.* **574**, 554–565.

Mangalam, A. (2001) Formation of a proto-quasar from accretion flows in a halo, *Astron. & Astrophys.* **379**, pp. 1138–1152.

Manning, C.V. (2002) The search for intergalactic hydrogen clouds in voids, *Astrophys. J.* **574**, pp. 599–622.

Paczyński, B. (1986) Gamma-ray bursters at cosmological distances, *Astrophys. J.* **308**, pp. L43–L46.

Paczyński, B. (1998) Are gamma-ray bursts in star-forming regions?, *Astrophys. J.* **494**, pp. L45–L48.

Pendleton, Y.J. and Allamandola, L.J. (2002) The organic refractory material in the interstellar medium: Mid-infrared spectroscopic constraints, *Astrophys. J. Suppl.* **138**, pp. 75–98.

Roukema, B.F., Mamon, G.A. and Bajtlik, S. (2002) The cosmological constant and quintessence from a correlation function comoving fine feature in the 2dF quasar redshift survey, *Astron. & Astrophys.* **382**, pp. 397–411.

Saha, A., Sandage, A., Tammann, G.A., Dolphin, A.E., Christensen, J., Panagia, N. and Macchetto, F.D. (2001) Cepheid calibration of the peak brightness of type Ia supernovae, *Astrophys. J.* **562**, pp.314–336.

Sandage, A. (2002) Bias properties of extragalactic distance indicators, *Astron. J.* **123**, pp. 1179–1187.

PHYSICAL PHENOMENA UNDERLYING THE ORIGIN OF LIFE*†

JUAN PÉREZ–MERCADER
Centro de Astrobiología (CSIC–INTA)
Associated to the NASA Astrobiology Institute
Carretera de Ajalvir, km 4
28850 Torrejón de Ardoz, Madrid

Living systems are examples of emergent behavior whereby complex chemicals assemble into hierarchical structures and systems with a collective set of properties that characterize them. We start by highlighting the basic properties of living systems in order that one can adopt an operational description (not really a definition) of living systems. This allows one to actually attempt a description of the properties of living systems and from there an unified description in terms of physics. Such a unified description can indeed be found by abandoning the traditional reductionist approach and using the notion of emergence. We give an analytical description of how emergence takes place. We will learn that the physics associated with the emergence of Life points towards the notion that life is a consequence of the evolution of the Universe.

1. Introduction

Today we recognize that the Origin of Life[1] on planet Earth must have involved at least the following phases: (1) synthesis of the basic chemical components and molecules; (2) transition to a proto-biochemistry; that is, the formation of complex molecules capable of handling information, controlling self-reproduction, etc.; (3) RNA-to-DNA transition; (4) generation of a Proto-cell; (5) evolution into a cell. See for example Reference [1] for a description of the currently favorite "scheme".

This sequence points to an evolutionary pattern where certain out of equilibrium physico/chemical aggregates evolve from disordered configurations into ordered systems and where, *in general*, there are transitions from the "simpler" into the "more complex". (Note that these stages can be viewed as a time-dependent "hierarchical" organization.)

We will address the question of which fundamental physical principles must be at work. To use an analogy, what we will describe here puts together recent advances which can help

---

* Inaugural Lecture delivered at the Symposium Celebrating the 50th Anniversary of Miller's Experiment. Trieste, September 2003.
† In honor of Professor Stanley Miller.
[1] When we refer to life on planet Earth, the only one we know so far, we capitalize and write "Life". For the generic phenomenon we will use lower caseand write "life".

J. Seckbach et al. (eds.), Life in the Universe, 27–51.

in establishing for the cell what the Carnot cycle is to the internal combustion, steam or engines in general.

The material presented here has features of both, the bottom-up and top-down approaches to the study of Life. From knowledge of the characteristics of living systems we can extract a generic set of properties that characterize living systems; from the current knowledge of the evolution of general out-of-equilibrium systems we can infer the properties that the system must have in order to match the properties of living systems. These approaches meet at the Origin of Life and make its study an excellent scenario for the application and testing of ideas on the emergence and evolution of complexity. Furthermore, the ideas presented here have the advantage of being explicitly analytical and quantitative, and therefore are useful both in helping us understand the results of experiments and for proposing new experiments to understand the Origin of Life. Our aim is to provide a phenomenological framework for the study of these problems and contribute in helping to overcome the "intuitive understanding" that we now have. This time we find more than just "encouragement" as was the situation [2] ten years after Stanley Miller's experiment, when J. Lederberg wrote that *"... a consideration of contemporary theory on the origin of life is justified for two reasons: (1) exobiological research gives us a unique, fresh approach to this problem; and (2) we can find some encouragement that the recurrent evolution of life is more probable than was once believed."*

For useful discussions on these topics, see for example References [3] and [4].

2. Some Generic (Basic) Properties of Living Systems

To identify what physical phenomena underlie the origin of Life, we must first have a notion of "What is Life" [3]. This has been attempted many times (see for example [5] for a recent survey on this problem) and it may even be that it is not possible, so we must be for now contempt with just giving a "proxy description": only in terms of its properties.

We will describe the top-level properties of living systems. They will be used to give a list of the minimal set of properties that a system must have in order that it can qualify as a "living system". In this way we will be able to infer the essential physics of such systems and, by exploring the physics, we will be able to say a lot about living systems and their potential origin.

Somewhat vaguely, living systems are (1) *open*, (2) *out-of-equilibrium*[2] (3) *chemical systems* with (4) *limited available resources* which (5) are subject to *noise and fluctuations*.

The properties of living systems can be roughly classified into three categories: (i) intrinsic, (ii) operational (functional) and (c) historical (diachronic). The *intrinsic* properties are the ones with a phenomenology associated with the more essential or generic features of a living system; the *operational* properties are the ones that manifest by the expression or workings of the system.

---

[2] These systems can support long-range correlations generated out of shorter range interactions via the collective involvement of all the parts of the system. By short range we mean scales very small compared with typical system sizes.

TABLE 1. Properties of Living Systems tabulated according as to whether they are Intrinsic, Operational or Historical in nature. We have (i) *Replication/reproduction*, i.e., the system is capable of giving rise to a new system, either by itself or by associating with other systems; (ii) *Increasing free-energy gradient*, the system is capable of changing its own free-energy by interacting with its environment; (iii) *Robustness*, meaning that the system maintains its integrity and operation under considerable changes in external conditions; (iv) *Hierarchical/Modular (Scaffolding) organization*, the system organizes itself into interconnected subsystems relevant for the stable operation of the full system. The next property we list is (v) *Self-organization* which means that the system organizes at all scales, both in space and in time, without any external intervention. Under (vi) *Information* we imply that the system is capable of both generating and handling external information in ways essential to being alive. They also display what is known as (vii) *Collective/cooperative behavior*. In these systems there is not only a free-energy flow but because of inhomogeneities and fluctuations in the system there are (viii) *Fluctuations in free-energy* which play a decisive rôle in the collective behavior. Next we list some basic Operational properties and finally the Historical properties of living systems. The (Darwinian) (i) *evolution* of living systems manifests through events that take place during a finite period of time and depends both on the set of circumstances intrinsic to the living system and those of the environment into which it is inserted. This is "historical" in nature: it manifests only after we compare records which are separate in time. There is an accumulation of change in the form of a more or less pronounced (ii) *continuing novelty* that eventually leads to (iii) *transitions* in the system. Sometimes these transitions involve such important quantitative and qualitative changes that they manifest in the form of *major events* in the historical process. Occasionally, these do not seem to be continuous, and instead manifest only after a period of subtle accumulated change; this is known as (iv) *punctuated equilibrium*, meaning that the changes that accumulate over time in the populations are so subtle that they seem to be in "equilibrium" until a certain threshold is reached and a new form of the system emerges.

Properties of Living Systems	
Intrinsic	Replication/Reproduction
	Increasing free-energy gradient
	Robustness
	Hierarchy/Modular organization
	Self-organization
	Information generation and Management
	Collective behavior
	Fluctuations in free-energy
Operational	Interaction with the environment
	Metabolic networks
	Variation in reproduction
	Functional patterns
	Compartmentalized
Historical	Evolution
	Continuing novelty
	Transitions/Major Events
	Punctuated Equilibrium

Finally, by *historical* properties we mean the global manifestations of the living system as time elapses. Living systems change with time both at the level of the individual system and at the population or species levels. These properties are summarized in Tables 1 and 2.

TABLE 2. The above characteristics of Living Systems lead to establishing an analogy with stochastic non-linear out of equilibrium systems with many chemical species in interaction.

Features of Life as an Assembled Chemically Operated System
a Self-Reproduction
b Dynamical stability when subject to fluctuations: chemical/Quantum-Thermal
c Hierarchical Organization
d Chemical Patterns in space-time
e Metabolism
f Information storage and processing
g Adaptation/Functional interactions with environment
h Living/lineage evolve

2.1. OPERATIONAL DEFINITION OF LIFE AND SYNTHETIC SUMMARY OF PROPERTIES

With this we can begin to draft a definition of a living system. It can be done in at least two ways: via a "synthetic" definition and via an "operational" definition. We will not pursue synthetic definitions here[3].

An operational definition allows one to capture the properties of living systems by making direct reference to their general properties. The most complete[4] is that Life is *a self-reproducing, dynamically stable to environmental fluctuations, hierarchically organized set of chemical space-time patterns, which collectively are capable of metabolizing, storing and processing information, adapting to the environment and whose assembled results are evolving lineages.*

3. The Physics of Generic Systems with the Basic Properties of Living Systems

The above signatures and properties are known from work carried out over the last two decades, that they are present in condensed many-body systems with dynamics involving both random and regular attributes.

These systems can be modeled in terms of sets of Stochastic, PArabolic, NOn-linear, Differential EquationS (SPANODES) [7]. As a particular application of these equations we have the spatio-temporal evolution of a chemical species in interaction with itself, other species and a fluctuating environment. We will see that this provides a rich scenario for the study of the Origin of Life.

The parabolic character of the equation ensures that the system is out of equilibrium (as in diffusion) and that there are next-to-nearest-neighbor interactions in the species'

[3] An example of a "synthetic" definition is furnished by the following: *"life is a self-sustained chemical system capable of undergoing Darwinian evolution"*.

[4] Except for the inclusion of fluctuations and the specificity to chemical processes, this definition of Life is due to Farmer and Belin [6]. The fluctuations could be quantum, thermal or chemical fluctuations due to the presence of composition inhomogeneities in the system or in the environment.

concentration. This is so because parabolic partial differential equations involve the first derivative with respect to time of the unknown and therefore imply that there is the time asymmetry present in out-of-equilibrium systems. Parabolic partial differential equations involve the laplacian of the unknown quantity (say the concentration of a chemical species); the laplacian being a second derivative involves comparisons of the unknown at three consecutive space points. *In this part* of the collective dynamics there are contributions from both, the regular and the random characters. The regular come from the nature of the derivatives (differences between values of the unknown function at different points in space and time). The random character is present because the parabolic character implies the existence of irreversibility and this irreversibility ultimately can be traced to the presence of thermal and quantum fluctuations [8].

When there are many species involved in the dynamics many possibilities open up. From the existence of chaotic behavior as in the Rössler oscillator to the existence of synchronous patterns, the appearance of small worlds or the emergence of the various collective behaviors [9]. This is mostly regular in its character.

Non-linearity ensures the presence of reaction terms: the species in the system will mix with each other and/or themselves. If the mixing term is non-linear then it can generate new species out of the original species and, simultaneously, do it in a way that goes beyond simple additivity. Again this is mostly regular in nature.

The stochastic feature models the presence of noise or any other fluctuation in the system. Noise is nothing but a means of (a) summarizing the contribution to the dynamics of degrees of freedom whose typical scales are *smaller* than the scales at which we study the system or (b) modeling the contribution to the dynamics of non-deterministic[5] processes which can only be statistically characterized through some probability distribution.

These systems model at a very basic level a time-dependent flow of free-energy which, due to the presence of reaction-diffusion and a random component, becomes scale dependent both in space and in time.

Next we briefly discuss some of their phenomenological features.

3.1. A QUICK SURVEY OF PHENOMENOLOGICAL FEATURES

These systems display critical behavior. That is, for certain values of the parameters in the system such as reaction or diffusion constants, the short range forces acting among the components of the system "align", and the correlation between components does not decay exponentially. The system enters into a regime with pure power law correlations. This has very deep and important consequences. Exponential decay is associated with some dominant length scale and the system is not scale invariant. When the system enters into a power-law correlation regime it becomes scale invariant and very long range correlations between parts of the system are established: there is collective behavior! The transition into these regimes is of course what happens when a system experiences a phase transition (see Figure 1).

---

[5] Note the implication that this has for "reductionist" views. The presence of non-deterministic processes, even if they can be statistically represented by a noise term, implies that one is giving up the standard reductionist views.

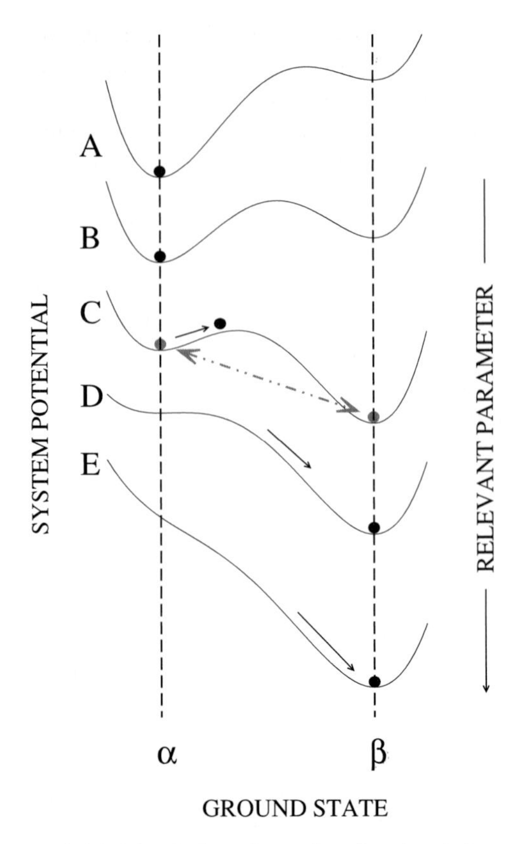

Figure 1. Change in the ground state (phase) of a system as the relevant control parameter evolves with scale. The ground state is the most stable state and corresponds to the lowest potential energy. In C, we indicate a situation where a "tunneling" phenomenon could happen: the system tunnels from state α to β.

The systems we are considering are out of equilibrium; the phase transitions are said to be dynamical [10]. These phase transitions in chemical systems are known to lead not only to long range correlations, but also to self-organization and hierarchical organization [9] and [11]. Which type of organization the system follows depends, as we will see later, on the nature of the scaling exponent in the power law: for imaginary exponents the system organizes into hierarchies. We will also see that the values of the exponents, and hence the nature of the critical point, are related to how the physical parameters change with length/time scales, a process known as "coarse graining" of the system. As in the better known equilibrium phase transitions the system goes into very different states depending on the side of the critical point where it sits. For example, below the critical point the system may be in its solid state whereas above the critical point it could be in a liquid or gaseous state.

In addition, the states into which these systems transit can be very *robust*, implying that the system will have a tendency to organize itself into such states and to *remain* in them. This is often seen in living systems, where a particular pattern such as DNA chemistry, embryonic development or eyes, once they emerged, are used over and over again in the evolution of Life. They are states that organize themselves into spatial "clusters" [12] or into temporally synchronized phenomena, very much like the songs of crickets or the croaking of frogs in the night of a tropical forest. Additionally, these systems display geometrical patterns or even complex patterns that move together through space as time elapses. Some of them can even "spontaneously" "arise" and "die"! or evolve through variation [12] induced by fluctuations. Their potential for complexity generation is huge.

What we have just described are precisely the properties of what are now called "Emergent Phenomena". These phenomena and an analytical framework to describe them will be presented in some detail in the next section. For now, we simply remember that emergent behavior is usually associated with the simultaneous presence of (a) Collective Phenomena, (b) Evolution, (c) Self-organization, (d) Generation of Patterns, (e) Scaling behavior (not geometrical or engineering scaling!), (f) Hierarchical organization and (g) the Assembly of the "Probable into the Little-Probable"[6]. Hence Emergence provides a unifying theme from Cosmological and Astrophysical phenomena into Biological and Ecological phenomena where all of the above are known to take place.

4. What Is Emergence? How Can It Be Described?

Emergence can be defined as what occurs in phenomena where "the whole is more than the simple sum of its parts". This happens, for example, in an ant hole where we see that the capabilities of the ant colony go far beyond the ability of a single ant. This is because the whole includes not only the individual ants, but also their social relations, communication abilities, etc. [14].

In prebiotic chemistry we can think of a "chemical soup" with many chemical species present in an environment where the chemicals are "stirred" by some noise. The chemicals interact among themselves as well as with the environment. The chemical system is open, i. e., exchanges energy and matter with the environment. The combined effect of non-linear interaction, spread in space, temporal and spatial evolution, as well as stochastic effects, are various forms of self-organized behavior: the "soup" goes into different basins of attraction for the parameters, each of these basins characterize the "emergent" behavior of the system. Emergence is what takes place when the system goes from one state into another.

In chemistry we can think of the aromatic properties of benzene as an emergent property of aromatic rings. Or we can think of emergence in more complex systems like reaction networks, metabolism or living cells. Another example is provided by the processes from musical tones to notes to the emotion generated by listening to the music.

Emergence is (1) associated with the spatio-temporal evolution of the system and (2) related to scale (and complexity) changes in the system.

Properties (1) and (2) above imply that emergence arises when *coarse-graining* an *open* system. Coarse-graining [15] a system is the process of studying the system at different spatial and/or temporal scales. Or the process of considering various parts of the system

[6] Holland (page 231 of Ref. [13]) puts this as "Transformation from the extremely unlikely to the likely", meaning that once a little-probable configurations have emerged, they become very common and are used as "parts" of larger, more complex, systems. For example, once eyes emerged, Life has used them over and over.

by averaging out subsets of the system's degrees of freedom. In other words, as we coarse grain a system and take into consideration more and more of the details of its components, we also perceive an increase of the number and classes of states to which the system has access. The result is emergent behavior.

5. "Predicting" Emergent Behavior. The Dynamical Renormalization Group and Emergent Behavior

Let us consider a generic many body system. Let us for the sake of the argument assume that its components are very diluted and that, in addition, they interact very weakly with each other. Under these conditions the whole is simply the sum of its parts and one does not expect any cooperative phenomena to arise: additivity is all we will see.

Next, imagine that we substantially increase the number of components to the point where the interactions between them become more significant. This has the implication that new states become available to the system as the relevance of the interactions is enhanced. Due to the non-linear character of the interactions the principle of superposition does no longer apply, but conservation of energy or maximization of available entropy may force the interacting parts of the system into assuming new configurations. Let us also have the system exchange energy and/or matter with its environment, i.e., let the system be *open*. Then the system may fragment or coagulate into new systems which, *effectively*, maintain some of the properties that made the original system stable. Not just any energy/matter transfer rate will lead to stable configurations. There will have to be, at least, some balance between the strength of the interactions and their associated rates, the energy/matter transfer and the relaxation times of the new states assumed by the evolving system. Many of these phenomena could be due to processes taking place at scales smaller than the one at which we are observing the system: to what we have called "noise".

5.1. COARSE GRAINING A MANY-BODY SYSTEM

The scale at which we study the system is thus basic to understand the phenomena that we observe in a many-body system. To implement "coarse graining" requires that we introduce a description of the physical variables (such as concentrations) in terms of fields, space-time dependent quantities which evolve according to some equations. We study these systems using what is known as Statistical Field Theory (see [10] and references therein), one example of which is provided by the previously mentioned "spanodes".

Observing the system at different scales and connecting the results of these observations to a phenomenological description means that we average the quantities in the system over some (graining) scale. In particular this implies that parameters such as viscosities, reaction "constants", masses, noise amplitudes, etc., become scale dependent. The scale dependent parameters are called "effective parameters" and they obey some ordinary differential equations which describe how they change with scale. For example, in chemical reactions the reaction rates become scale-dependent and make the dynamics scale-dependent.

In general these differential equations have "fixed points". That is values of the parameters which remain independent of scale changes. When the parameters have these values, the system becomes scale independent and, therefore, all scales in the system participate in its dynamics. The system is scale independent and the correlations must become power law correlations (cf. Appendix A). These are "critical states" and the transition from one state to another as the graining scale changes is called a "phase transition" (see Figure 1). The power laws are characterized by some (calculable) exponent which depends on the values at the fixed point of the effective parameters. For out of equilibrium systems, these states have all the properties described earlier for "emergent" states and we can calculate many of their properties even if we cannot predict in complete detail all of them. The system is *not* fully reducible.

There exists a quantitative and qualitative tool that allows us to carry out such calculations. It is known as the Renormalization Group (RG) and has been extensively used in many body phenomena[7] as well as in particle physics. It is now finding application in many other realms of science, including Biology and Ecology. We explore how this happens.

5.1.1. Combining regular and random dynamics. The tools. (i) The effective action and (ii) the RGEs

In statistical mechanics, the full description of a system is given in terms of a partition function. In a statistical field theory there is an analogue of the partition function that can be calculated and from which all the physical parameters can be derived: the Generating Functional, $Z[J; \{\alpha_i\}]$. Here $J(\vec{x}, t)$ represents the source for the field $\phi(\vec{x}, t)$ and the α_i represent the couplings in the system, i. e., reaction constants, masses, etc.

The generating functional comes with a problem which turns out to be a boon: it is plagued by divergences. But removing them is precisely what allows one to coarse grain the system! The divergences imply the existence of a scale that is either very large or very small compared with the length/time scales typical of the system. In some cases the divergences can be systematically removed by introducing an *arbitrary* length scale λ. The scale λ is called[8] the "sliding scale" and can be identified with the "graining" scale. The mathematical procedure for removing the divergences is called "renormalization". Systems for which this is possible are called "renormalizable".

This procedure introduces a λ-dependence in $Z[J]$ which now becomes $Z[J; \{\alpha_i\}; \lambda]$. Since the scale λ was arbitrary, we must have that

$$\lambda \frac{dZ[J; \{\alpha_i\}; \lambda]}{d\lambda} = 0. \tag{1}$$

The only way this can be satisfied is if the couplings themselves become scale dependent and satisfy equations equivalent to Eq. (1). These equations are the Renormalization Group Equations RGEs mentioned above. They are of the form,

$$\lambda \frac{d\alpha_i}{d\lambda} = -\beta_{\alpha_i}(\{\alpha_j\}), \tag{2}$$

---

[7] The RG was introduced in the early fifties by Peterman and Stuckelberg and, independently by Gell-Mann and Low [16]. In the 60's and 70's it was extended to many-body systems. In fact, the 1982 Nobel Prize in Physics was awarded to K. Wilson for his application of the RG to the study of equilibrium phase transitions, and in 1991 to P. G. de Gennes for his application to the study of polymer dynamics.

[8] Because the calculations are usually performed in momentum space, instead of configuration space, the sliding scale is usually taken in the literature to be its inverse, a momentum scale μ. Here we will always refer to the length scale λ as the sliding scale.

where the (non-linear) function $\beta_{\alpha_i}(\{\alpha_j\})$ can be explicitly calculated and contains the combined effect of the regular and fluctuating parts of the dynamics.

5.1.2. Power laws. (i) Real and (ii) complex exponents. Some examples

Since the generating functional satisfies Equation (1) and the derivatives of the generating functional with respect to the source J give the correlation functions for the field[9] ϕ, $G_N(r_1, \ldots, r_N; \{\alpha_j(\lambda)\}; \lambda)$, it follows that the correlation functions also satisfy RGEs. These are partial differential equations whose coefficients are functions of the various coupling parameters and which, as was the case for the $\beta_{\alpha_i}(\{\alpha_j\})$, can in general be calculated using perturbation theory. The $\gamma_\phi(\{\alpha_j\})$ are the anomalous dimensions of the field ϕ, and they are to ϕ what the β_{α_i} were to the α_i. They are of the form,

$$\left[-\lambda\frac{\partial}{\partial\lambda} + \sum_i \beta_{\alpha_i}\frac{\partial}{\partial\alpha_i} + \frac{N}{2}\gamma_\phi(\{\alpha_j\})\right] G_N(r_1, \ldots, r_N; \{\alpha_j(\lambda)\}; \lambda) = 0 \tag{3}$$

and can be immediately solved using the method of characteristics. Their solution is

$$\begin{aligned} &G_N(r_1, r_2, \ldots, r_N; \{\alpha_j(\lambda)\}; \lambda) \\ &\quad = G_N(r_1, r_2, \ldots, r_N; \{\alpha_j(\lambda_0)\}; \lambda_0) \times \exp\left(-\frac{N}{2}\int_1^s \frac{d\lambda'}{\lambda'}\,\gamma_\phi[\alpha_i(\lambda')]\right) \end{aligned} \tag{4}$$

Here the subindex zero refers to a reference value: the length scale corresponding to the system configuration that we have observed and where the parameter had *measured* values $\alpha_j(\lambda_0)$.

Near a fixed point of the couplings, i. e., for those values $\alpha_i = \alpha_i^*$ where $\beta_{\alpha_i}(\{\alpha_j^*\}) = 0$, the couplings go to their *constant* fixed point value and the integral in the above N-point correlation function becomes proportional to $\log\frac{\lambda}{\lambda_0}$. In other words,

$$\begin{aligned} \exp\left(-\frac{N}{2}\int_1^s \frac{d\lambda'}{\lambda'}\,\gamma_\phi(\lambda')\right) &\longrightarrow \exp\left(-\frac{N}{2}\gamma_\phi^*\int_1^s \frac{d\lambda'}{\lambda'}\right) \\ &= \exp\left(-\frac{N}{2}\gamma_\phi^*\ln s\right) = \left(\frac{\lambda}{\lambda_0}\right)^{-\frac{N}{2}\gamma_\phi^*}. \end{aligned} \tag{5}$$

Here γ_ϕ^* is the value of the anomalous dimension when the couplings are at their fixed point values. Note from the last line that at the critical point the correlation functions follow a power law and, as advertised earlier, the system is scale invariant and therefore *all* its components are involved in the dynamics: there *emerges* a cooperative phenomenon that involves the full system and which self-organizes all of its parts.

In the case of the two-point correlation function (2PCF) Equation (5) becomes

$$G_2^{Imp}(r_1, t_1; r_2, t_2) \propto |r_1 - r_2|^{2\chi}\, F\left(\frac{|t_1 - t_2|}{|r_1 - r_2|^z}\right) \tag{6}$$

[9] In the following, and for the sake of brevity, the r_k represent with a single symbol the space-time coordinates $\vec{x}_k$ and t_k.

with χ and z related to various suitably chosen (for convenience) combinations of anomalous, engineering and space dimensions. The scaling function $F(u)$ can be shown to have the limiting behaviors

$$\lim_{u \to 0} F(u) \propto constant \tag{7}$$

for small argument values, and

$$\lim_{u \to \infty} F(u) \propto u^{2\chi/z} \tag{8}$$

for large argument values.

There are many examples in the literature of the above. A particularly useful one is provided by the evolution of an incompressible fluid subject to a stirring force which is random in both space and time [17]. This system is very generic in its properties and may be used to model many different physical situations from the evolution of the Universe at large scales, to self-reacting chemical systems to hypercycles. It has fixed points for the noise couplings which are both real and complex. The real fixed points lead to the standard power laws familiar from chemical kinetics [11]. The complex fixed points lead instead to hierarchical behavior [18]. One can also treat the case of several coupled chemical species [12], but the currently available calculational techniques escalate very quickly in complexity with the number of species and the only way to proceed is by numerical calculation where, unfortunately, there can be a loss of intuition.

5.2. COARSE-GRAINING OUT-OF-EQUILIBRIUM SYSTEMS AND EMERGENCE

The deep reason why there are complex fixed points of the RG is as follows: because the system is not isolated and exchanges energy and matter with its environment, it does not have symmetry under time reversal and conservation of probability no longer holds. This opens up the possibility for fragmentation and coagulation phenomena together with self-organized behavior in these systems. Coarse graining out of equilibrium systems is the tool to describe the large class of phenomenological behaviors observed in these systems. Each set of fixed points defining an "emergent state". The global properties of these states are determined by the stability properties of the fixed points.

To understand this, we need to imagine the system at some fixed scale. At this scale reaction *constants*, diffusion *constants*, etc., will have some fixed values. As we change scale these couplings evolve according to their RGEs and "move" into different values. Towards which states the system evolves depends on where in the "basins of attraction" of the various fixed points are the initial values of the couplings. The system will "transit" to the new state corresponding to the couplings towards which it is being attracted (see Figure 1).

We can now describe in some useful detail how the phenomenon of "emergence" takes place in these systems.

In the system of ordinary differential equations Eq. (2) we can have various stability situations depending on the nature of the eigenvalues of the matrices characterizing the linearized version of the system. We can have improper nodes, saddle points, proper nodes, degenerate improper modes spiral points and centers. They can be attractive or repulsive at either short or long length scales. Each of these stability classes gives rise to a particular phenomenology and associated pattern. In going from one size scale to another the state of the system changes and its global stability properties also do change.

Once a fixed point is reached, the properties and patterns corresponding to this particular state become manifest. If the new state is stable, the emergent state will be "persistent". On the other hand, if the state is only metastable, the system will display emergence into a new ephemeral state[10]. However all parts of the system are fully involved in its dynamics and this dynamics, in turn, drives the system into the emergence of a self-organized state.

The n-point correlation functions are the coefficients in the functional expansion of the probability distribution function (PDF) [19]. In fact, the PDF can be shown to satisfy an RGE [20] which is a generalization of Eq. (3). The various emergent behaviors are characterized by the properties of the PDF in the basins of attraction of the respective fixed points. Of course these properties mirror and refine the results that are derived from the simpler 2-point correlation functions.

Thus the emergence of complexity is to be studied by analyzing the scale-evolution of the PDF. The consequences of this are very deep: emergent behavior cannot be fully predicted although it can be characterized in terms of some basic *effective parameters*. The most one can make are probabilistic statements about the accessible states. These states can be identified for each level of description or of the hierarchy in the system. Full reducibility is not possible and we enter into a new realm where the binomial complexity/simplicity is harmonized through probability.

5.3. EXPONENTS: REAL AND COMPLEX. SCALING AND HIERARCHIES. SOME EXAMPLES

It can be seen (cf. Appendix A) that because of the implied scale invariance, the real exponents are indicators of cooperative behavior. These exponents can be measured by studying log-log graphs of the correlation functions which in the scaling (power-law) regime will show up in this graph as a straight line with slope given by the exponent. The more negative an exponent is the tighter the correlation between components located at the various points to which the n-correlation function refers.

Let us restrict ourselves to the two-point correlation function (2PCF), which together with the one-point correlation function determines the PDF for gaussian processes. The 2PCF is *the joint probability of finding two objects located in two independent volume elements*. Since the 2PCF is a probability, it must be a real quantity. This has very important implications in the physical interpretation of complex exponents.

We now examine the behavior of the 2PCF in the limit of large spatial separation. As we saw in Eqs. (6) and (7) the two point correlation function for the fluctuation in concentration of a chemical species, $\delta\rho(r)$, measured at points r and r' has in the critical regime the form

$$\langle \delta\rho(r')\delta\rho(r)\rangle \propto |r - r'|^{2\chi} \tag{9}$$

---

[10] The state could also be an ephemeral but synchronous state.

where the angle-brackets denote that an ensemble average has been taken. Let us assume that χ is complex and given by $\chi = \chi_R + i\chi_I$. Then, since the 2PCF is real, Eq. (9) becomes

$$\langle \delta\rho(r')\,\delta\rho(r)\rangle = \mathrm{Re}\,\tilde{c}\,e^{i\beta}\left|\frac{r-r'}{r_0}\right|^{2(\chi_R+i\chi_I)}$$

$$= c\cdot|r-r'|^{2\chi_R}\cdot\cos\left[\beta+2\chi_I\log\left(\frac{|r-r'|}{r_0}\right)\right]. \tag{10}$$

Here the quantities c, r_0, and β are constants. They are obtained by direct comparison with the experimental results. The form of Eq. (10) in a log-log plot is that of a log-periodic function modulated by a power law. The periodicity shows up in the form of "wiggles" in the graph of the 2PCF. The period of the logarithmic periodicity is controlled by χ_I while the slope of the modulating power law is controlled by χ_R.

From Eq. (10) one obtains that *maximal correlation* occurs for finite-size regions which are quantized by an integer n and whose size is given by

$$|r_{(n)}-r'| = r_0\cdot\exp\left\{\frac{1}{2\chi_I}\left[\tan^{-1}(\chi_R/\chi_I)-\beta\right]\right\}\cdot(e^{\frac{\pi}{2\chi_I}})^n \equiv A\cdot b^n \tag{11}$$

with $n = 0, \pm1, \pm2, \ldots$. We conclude that

(i) there is a *hierarchy* of regions of average linear sizes $R_{(n)} \equiv |r_{(n)} - r'|$ such that correlation on those scales is favored,

(ii) the hierarchy can be *classified* according to an *integer n*, and

(iii) note also that *within* each level of the hierarchy the correlation is a power law.

In other words, for systems away from equilibrium, the collective organization of the system leads to self-organized cooperative behavior and, in some instances, the system self-organizes into interrelated regions whose average finite sizes can be classified according to integers. These regions are nested within each other and have the property that at a given level they all seem similar, however their specific *small* scale properties can be quite different. This is what is understood by a hierarchy (see for example [21] and [22]).

Examples of the phenomenology just mentioned are abundant. They range all the way from the large scale structure of the Universe, to planetary systems, to geological faults, to chemical networks and cellular automata, evolutionary games and replicators, fragmentation and coagulation phenomena, proteins, allometries in living systems and ecologies. They also include self-replicating networks, catalytic hypercycles and the emergence of environmentally selected evolutionary pathways. All these are examples of "systems", as discussed by Miller [23] more than two decades ago, but in a new incarnation and with a vengeance: we can describe them numerically (through algorithmic simulations) and are well on our way to describe them analytically, from first principles.

6. The Emergence of Life and the Evolution of the Universe

The history of the Universe started about 13.7 Gyr ago, when the period known as the Big-bang took place. It is a story progressing in space and time where at each stage the dominant

Figure 2. An iconic representation of the evolution of the Universe and Life.

forces have generated characteristic structures and patterns. The evolutionary histories of the Universe and of Life are iconically depicted in Figure 2.

These patterns have to do with properties such as "synchronization", "networks", "emergence", "scaling" and "hierarchies". All of them are part of what is now known as "complexity science" or "complexity theory". Complexity theory tries to identify non-obvious patterns of self-organization in Nature that occur in complex systems, i.e., in systems where "the whole is more that the sum of the parts". Thus, trying to understand if "Life is a cosmic imperative" [24] or, more precisely, if "Life is a consequence of the evolution of the Universe" [25], i.e., that it emerges every place in the Universe where there is an opportunity for chemical evolution, requires that we frame these questions within the context of "complexity theory". Complexity theory becomes the basic analytical tool for describing, understanding and unifying astrobiological phenomena.

The presence of scale invariance with real and/or complex exponents is one of the hallmarks of self-organized behavior. As is the generation of complexity in open out of equilibrium systems. The existence of power laws in the galaxy-to-galaxy correlation function has been known for a long time as well as a similar relation for clusters of galaxies and other structures. This can be understood as a consequence of the evolution of matter in gravitational interaction in the noisy environment provided by the Universe itself [26]. The fragmentation of the initial structures into their various hierarchical fragments (top-down components such as clusters of galaxies, galaxies of the various types, clusters of stars and the various types of molecular clouds) can also be described in terms of this evolution in an amazingly precise way [27]. At these scales the dominant few body force is the gravitational force. At smaller scales the dominant interaction is the electromagnetic force which dominates through chemistry and its many allied macroscopic processes.

6.1. SCALING OF FREE-ENERGY-RATE DENSITY

In correspondence with the above one can estimate the free-energy-rate density[11] involved in these processes to find [30] that, considered as a function of the structure size-scale, it follows a power law! Therefore, these structures and their emergence fit very well into the notion we have developed here of a hierarchical and self-organized emergent system. One can extend this calculation to living systems and find that indeed they *also* follow the same pattern of power laws with a common exponent: all the way from large clusters of galaxies to ecosystems there is a power law dependence of the free-energy-rate density. This is an excellent token connecting the evolution of structure in the Universe with the evolution of living systems. It should be telling us something about the period when the transition between them took place: the Origin of Life (at least!) on Earth. It is furthermore telling us that the Origin of Life can be considered as yet another (emergent) manifestation of the evolution of the Universe.

6.2. HIERARCHIES IN THE UNIVERSE AND IN LIFE

Since the Universe is out of equilibrium, one *expects* to have complex exponents in the correlation functions. It therefore comes as no surprise that the Universe, at large scales, self-organizes into hierarchies. At smaller scales it is also expected that hierarchical organization will emerge, because the non-equilibrium of the Universe is true at all scales. This hierarchical organization can be appreciated in the following example, which is a direct application of the out-of-equilibrium hierarchy theory deduced from Eq. (11).

If we study the two-point correlation function for matter density fluctuations from the average density, $\delta\rho(\vec{x}, t)$, corresponding to the large scale structures in the Universe, all the way from the known horizon to the various classes of molecular clouds, one *predicts* [18] that each structure type j has a typical mass $M_{(j)}^{Pred}$ contained within a region of size $R_{(j)}$, and that they are related by $\Delta M_{(j)}^{Pred} \sim R_{(j)}^{2.175\pm0.075}$. The *observational* data for all the structures is best fit by $\Delta M_{(j)}^{Obs} \sim R_{(j)}^{2.10\pm0.07}$, which agrees, within the estimated uncertainties, with what is predicted. Predictions and observations are plotted in Figure 3.

The data for all the structures can be fit by the power law of Eq. (10). From this one *predicts* the existence of a mass-size hierarchy, and the size for each structure-type is predicted to be given by $R_{(n)} \equiv |r_{(n)} - r'|$ as in Eq. (11), with $A = (3.92 \pm 0.67) \times 10^{16}$ cm and $b = 9.02 \pm 0.24$. This is represented in Figure 4. Similarly, the mass is *predicted* to be $M_{(n)}^{Pred} = (123 \pm 27)^n \times [(2.46 \pm 0.62) \times 10^{30}]$ g. In Figure 5, we plot these predictions

[11] The free-energy density can be obtained from the partition function by using the standard procedures of statistical mechanics. It follows once the n-point correlation functions are known. See for example [28] and for a more advanced treatment [29]. What follows can therefore be interpreted as a manifestation of the power-law behavior for correlation functions that has been described above.

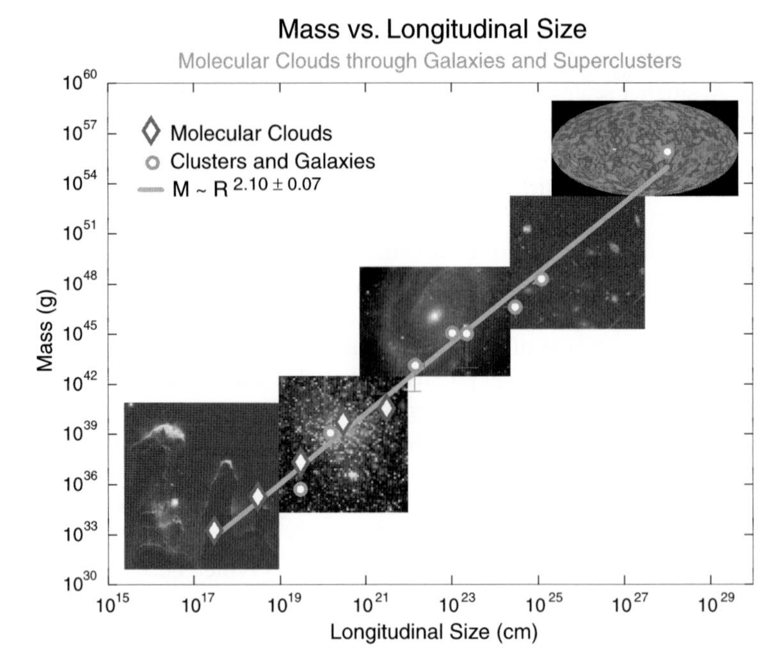

Figure 3. Mass vs. longitudinal size distribution of discrete structures in Universe from molecular clouds through galaxies to superclusters of galaxies and power law fit (slope $= 2.10 \pm 0.07$).

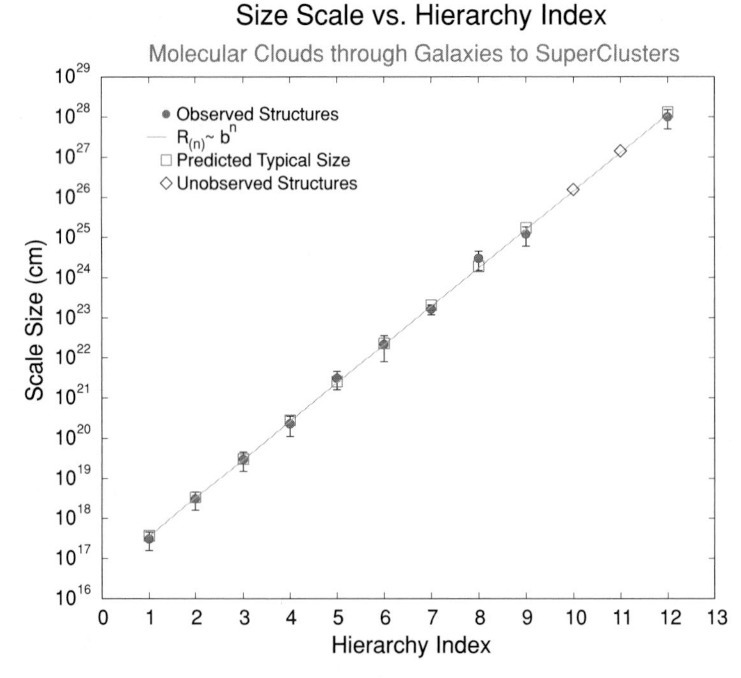

Figure 4. Longitudinal size scale vs. hierarchy index n for discrete structures in Universe from molecular clouds through galaxies to superclusters of galaxies and predictions from Eq. (11). The value of b is 9.02 ± 0.24.

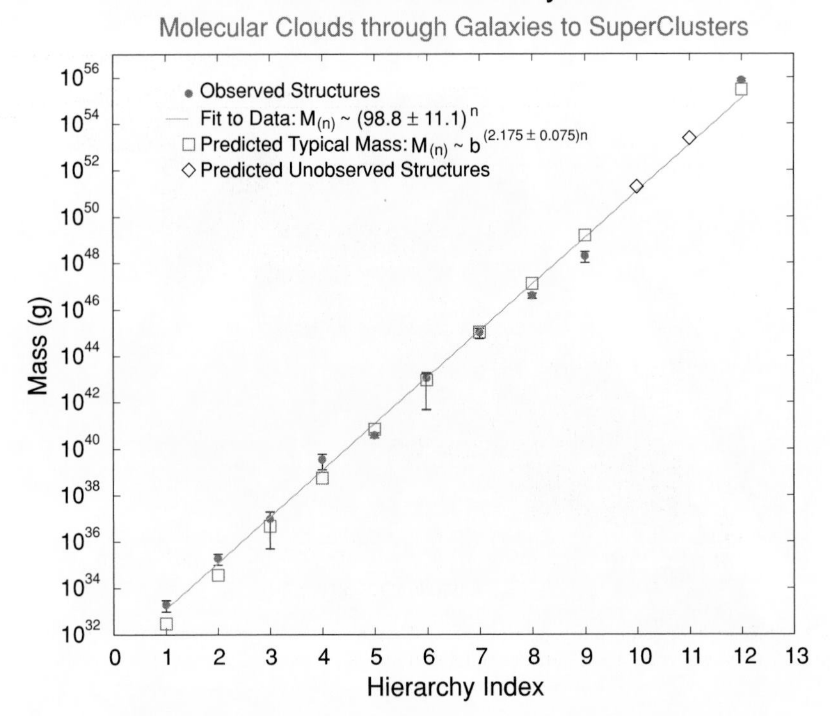

Figure 5. Mass vs. hierarchy index n for discrete structures in Universe from molecular clouds through galaxies to superclusters of galaxies and theoretical predictions from hierarchy theory. The value of b is 9.02 ± 0.24.

for the masses together with the observed values. The observed data is best fit by $M_{(n)}^{Obs} = (98.8 \pm 11.1)^n \times [(7.64 \pm 4.82) \times 10^{30}]$ g. The agreement between this quantity and the prediction is also well within the observational uncertainties.

These hierarchies can be *iconically* represented like a pyramid on whose vertex is the most encompassing class of structures. Together with power-law correlations at the individual levels, they imply the presence of a network (in this case a "scale-free" and "hierarchical" network [31]). For gravity-dominated structures we display this in Figure 6 which is to be understood only as a metaphor for Figures 3 to 5. See also [32].

A similar situation is known for living systems (see for example Refs. [33], [34] and references cited). There is ample evidence of power-law behavior for metabolic rate vs. mass all the way from bacteria to elephants and whales, and an analogous relationship is known to hold for plants over several orders of magnitude. In addition, there is evidence that the traditional organizational levels in the cell, the genome, transcriptome, proteome and metabolome are associated with networks governed by principles *similar* to the ones just described except that they are chemically operated (see Figure 7). As will be argued below, their generated complexity is therefore much greater.

Note also that each level in the hierarchy has a *characteristic* mass/size relationship. And that this is true for both gravity and chemistry dominated structures. For example, galaxies have a typical sizes and the same, in general, is true for cells.

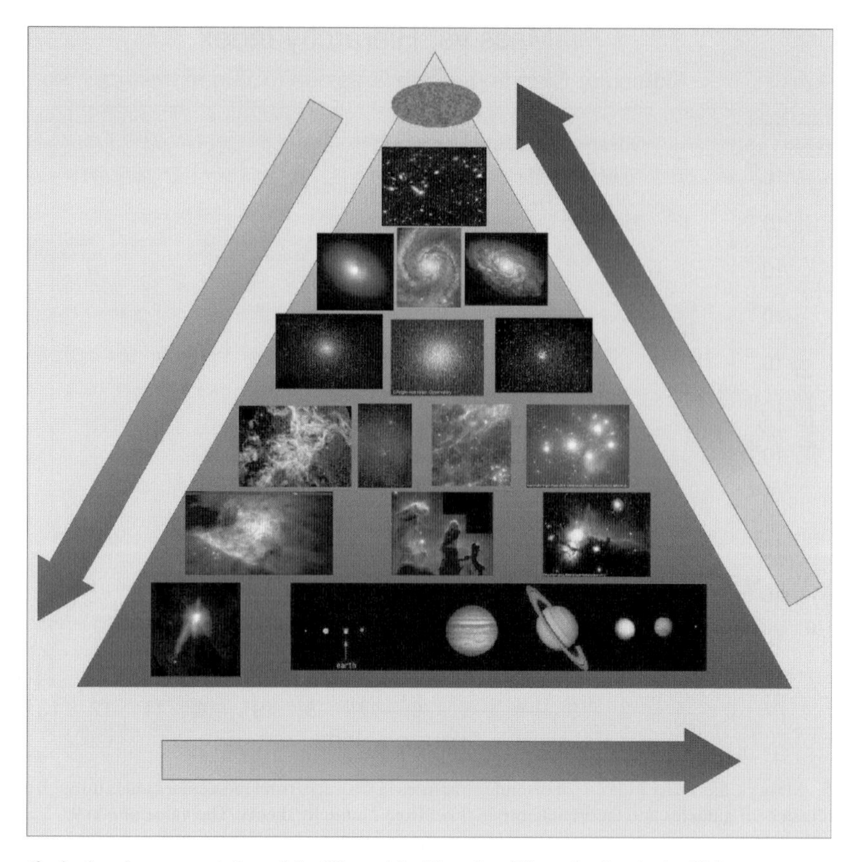

Figure 6. An iconic representation of the Hierarchical Levels of Organization in the Universe.

For gravity induced structures, one can construct arguments which fall under the (rather unfortunate!) name of the Anthropic Principle and which shed lots of light on their nature and how they are formed [35]. The features shared with living systems seem to suggest the existence of an analogous principle for living objects: the chemically dominated structures. But these relations still await discovery and much experimental as well as phenomenological work needs still be done in this area.

6.3. THE CHEMISTRY-DOMINATED UNIVERSE. THE TRANSITION INTO LIVING MATTER

We have argued that Life is not only chemically based, but that it also fulfills all the basic properties of emergent, complexity generating and out-of-equilibrium phenomena. The complexity of a system is related (in ways we still do not completely understand) to the number of states accessible to the system. Of the two long-range fundamental interactions, the gravitational and the electromagnetic forces, the strongest and the one with more potential for generating complexity is the electromagnetic interaction when acting at the atomic/molecular and mesoscopic levels. That is, when the conditions are such that chemical

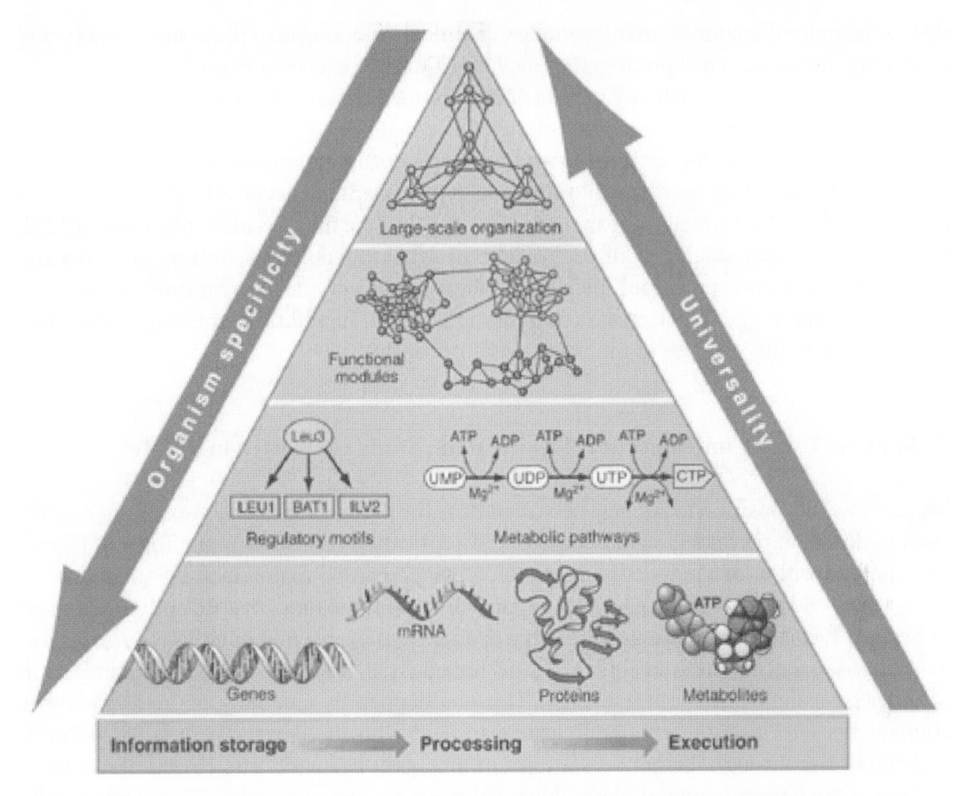

Figure 7. The main hierarchical levels in present day living systems. Taken from reference [34].

processes can happen. We then expect to have emergence of life in those out-of-equilibrium environments where the conditions for the emergence of some complex chemistry with some sufficient degree of stability (but still out-of-equilibrium) are met, as was discussed in the introduction to Section 5. The transition from simpler to more complex chemistry must have taken place as the environmental and kinetic conditions for Life became compatible. Once the conditions for a sufficiently complex chemistry were present, then the laws of physics as discussed here *inevitably* lead to systems with the properties of Table (2): the combined action of the electromagnetic force and quantum mechanics (chemistry) leads to the assemblage of supramolecular structures with the basic properties of the living systems that we see today. Properties such as hierarchical organization, metabolic networks, etc. are the indicators of what must have been at work during the Origin of Life. This provides us with useful guidance to reconstruct or try to reproduce the steps listed in Section 1. Something that may have happened at many locations in the Universe where the conditions for this class of universal phenomena were possible.

The fact that we know of common regimes as described above for both living and non-living organized matter, already gives us clues to use in understanding the transition. We observe that there is a *larger* change in the free-energy-rate density for living things than for non-living matter. It is conceivable that such changes are favored because of the existence of a hierarchical organization. This is, for example, the case in metabolic activity where due to the help of enzymes, many complex chemical reactions take place individually and reversibly; however, when put together they constitute an irreversible process.

With the hierarchical self-organizing principles that we have discussed here we begin to envisage scenarios for the sequential origin of Life mentioned in Section 1 and in Ref. [1]. Indeed, the presence of hierarchical (or modular) organization is an excellent vehicle for the use of the available free energy, while allowing synchronicity to occur. At a global level, this is facilitated by the formation of membranes, itself a dynamical critical phenomena [25], [36]. Note also that staging is of importance in synchronization as well as in increasing free-energy gradients: processes that would otherwise be very fast can be slowed down and thus lead to more stable collective configuration which, in addition, are more robust and complex due to the increase in the interaction probability.

7. Reprise: Using Contemporary Physics in the Study of the Origin of Life

We are now in a position to suggest new *scenarios* for prebiotic chemistry experiments. We need to design experiments where there is self-replication leading to some form of hierarchical organization of the reaction byproducts which can be maintained by an in-flow of energy. The chemistry does not have to happen all at the same time. Instead and as described in Section 1, it is conceivable that there is a sequence of events that manifest as some form of a time-sequence. These events could be linked by a tendency to favor maximization of order by using available free-energy. This can happen by assembly of smaller/simpler components into a more "efficient" system to deal with the environment surrounding it and giving rise through this process a form of emergent behavior. The emergent chemical dynamics can have a temporal hierarchical behavior, i.e., be sequential. There can even be the combined effect of space and time hierarchical behaviors!

Thus we need to design experiments with reaction kinetics that leads to complex exponents (not only real exponents) in the correlation functions. This can be accomplished by making appropriate choices of environmental parameters such as pressures, temperatures, fluctuations of various types ("noise") or the chemical substrate on which the chemical reactions takes place, just to mention a *few* possibilities. Under these conditions and based on the guidance that physics gives us, one can expect the spontaneous emergence[12] of states where the available free-energy-rate density jumps by amounts large enough to saturate the available complexity gaps.

It has already been mentioned that stirring or random-forcing of many-component chemical reactions can lead to collective self-organized behavior, including hierarchies. This seems to be a very basic feature of complex systems with a high potential to reproduce the phenomenology of living systems [12]. This spontaneous emergence of "living-like" order reminds one of what happens in some cases of structure formation in the Universe. For example, in the formation of stars there are regions of space-time which for some random reason have a higher concentration of matter/energy; this concentration, with the help of self-gravitation, can reach the levels required by the nuclear force to initiate a fusion

[12] Technically, what can happen is that, due to the imaginary component of the exponent, there is "tunneling" between states.

reaction and start the burning into a star. This process is described (except for the origin of the primeval fluctuation) by the Oppenheimer-Volkoff theory of star formation, and is very much at the basis of what we know today about star formation. It is conceivable that an *analogous* process of "concentration" (enrichment) of complexity potential and free-energy-rate density gradients could do it for Life. Of course one cannot simply go and take *any* arbitrary chemicals: many must probably be involved, but a judicious choice of compounds is needed. The choice could be informed by involving chemicals that lead to both a large number of reaction by-products which are sufficiently stable so as to increase the probability of further reaction or chemical-node formation in the emergent chemical networks.

There are several strategies that may be used to implement the ideas discussed here. They must involve theoretical and phenomenological exploration of the principles just discussed combined with the power of (numerical) molecular dynamics simulations. From these one can design actual chemical experiments and even attempt biological synthesis experiments. Crucial to these experiments is the assembly or integration of the various components *at each* hierarchy level, both in length (spatial hierarchy) and in duration (temporal hierarchy). The integration could be self-organized if the conditions are such that the free-energy-rate density has the levels required by emergent hierarchical behavior as described here: then the system maximizes correlation for a given available energy. Given a self-replicating heteropolymer, not every sequence-length will work: only those that follow hierarchical behavior and overall power-law in the dependence of the free-energy-rate density with size (and/or time) should work.

These considerations can be of use in some of the "Life in a Test Tube" experiments being performed now [37] where a phage is constructed from synthetic oligonucleotides. One of the main difficulties encountered in these experiments lies in the "error-free" assembly of long-enough genome segments of the known sequence using synthetic oligonucleotides. The removal of this and allied problems can benefit from the hindsight gathered by analyzing the assembly with the techniques discussed here and in [12] and [20].

8. Conclusions

We have described the basic properties of living systems and shown how they are unified by current notions of emergent phenomena. These can be analytically studied through a class of stochastic systems which (1) can model systems of many-chemical species in reaction-diffusion configurations inserted in a noisy environment; (2) lead to the notion that Life is a consequence of the evolution of the Universe and (3) can be effectively used to guide in the design of experiments and the proposal of scenarios to understand, albeit in a non-reducible and to a certain extent probabilistic way, the Origin of Life.

The Origin of Life is an extraordinary field to experimentally test ideas on emergence. Emergence becomes an overarching theme that connects the evolution of Life with the evolution of the Universe, and allows the application of the scientific method to these problems. Phenomenological studies indicate that Life can be integrated with the evolution of the Universe which, for chemically dominated matter, naturally leads via a "bottom-up" scenario to the notion that the Universe must be teaming with life.

Many outstanding problems can benefit from the application of the guidance provided by the methods and results discussed here. They range from understanding in detail the chemical reaction dynamics of complex chemical systems with hierarchical correlations, to the bottom-up approach to the synthesis of Life, to the top-down synthesis of living organisms in the test tube. It is of special interest to combine the methods described here with the top-down synthesis of living systems, where they can for example, help in providing effective methods for the synthesis of the target genome.

But even the successful generation of quasi-living or living systems does not solve the problem of the Origin of Life, since we would then need to correlate the synthetic genome with the current notion about LUCA (Last Universal Common Ancestor). The physics we have at our disposal today can greatly help in these issues, and we are rightfully hopeful, but lots of creative work from people from many disciplines still needs to be done, even if we now have much more guidance than 50 years ago.

Acknowledgments

The author thanks Herrick Baltscheffsky, Baruch Blumberg, Esteban Domingo, Murray Gell-Mann, Alvaro Giménez, Terry Goldman, David Hochberg, Andy Knoll, Bernd-Olaf Küppers, Jonathan Lunine, Martin Rees, Bruce Runnegar, Jack Szostak, Gunther von Kiedrowsky and Geoffrey West for many discussions on the many issues discussed in this paper. Finally, I wish to thank Julian Chela-Flores for his vision and Joseph Seckbach for his encouragement, patience and professionality.

References

1. Lehninger, A. L., Nelson, D. L. and Cox, N. M., *Principles of Biochemistry*, 2nd edition, Worth Publishers. See for example page 73 and Figure 3–18. Also Küppers, B.-O., *Information and the Origin of Life*, MIT Press, Cambridge, Massachusetts, 1990. For a text on the origin of Life, see Zubay, G., *Origins of Life on the Earth and in the Cosmos*, 2nd edition, Academic Press, San Diego, 2000.
2. Lederberg, J., Exobiology: *Experiments, Approaches to Life beyond the Earth*, in p. 412 of *Science in Space*, edited by L. V. Berkner and H. Odishaw, McGraw-Hill Book Company, New York, 1961.
3. Popper, K. R., *Objective Knowledge. An Evolutionary Approach*, Revised Edition, Oxford University Press, 1979. Also, *The Poverty of Historicism*, Routledge Classics, London 2002.
4. Cavalli-Sforza, L. L., *Genes, Peoples and languages*, North Point Press, New York, 2000.
5. Cleland, C. and Chyba, C., *Origins of Life and Evolution of the Biosphere 32 (2002): 387–393*.
6. Farmer, J. D. and Belin, A., in *Artificial life II*, C. Langton, C. Taylor, J. Doyne Farmer and S. Rasmussen editors, Addison-Wesley, Redwood City, Ca. 1992.
7. Gell-Mann, M. and Pérez-Mercader, J., *unpublished work on the Statistical Mechanics of Power-law Probability Distributions and on Stochastic Non-linear Processes*, 1998–2004.
8. See for example, Halliwell, J., Pérez-Mercader, J. and Zurek, W., *The Physical Origins of Time Asymmetry*, Cambridge University Press, Cambridge, UK, 1994.
9. Kuramoto, Y., *Chemical Oscillations, Waves, and Turbulence*, Dover Publications; Inc., New York, 2003.
10. Onuki, A., *Dynamical Phase Transitions*, Cambridge University Press, New York, 2002. See also Horsthemke, W. and Lefever, R., *Noise-Induced Transitions*, Springer Verlag, New York 1984 and the more modern Garcia-Ojalvo, J. and Sancho, J. M., *Noise in Spatially Extended Systems*, Springer Verlag, New York 1999. Many useful ideas are contained in Sornette, D., *Critical Phenomena in Natural Sciences: Chaos, Fractals, Self-Organization and Disorder: Concepts and Tools*, Springer-Verlag, New York, 2000.
11. See for example, House, J. E., *Principles of Chemical Kinetics*, Wm. C. Brown Publishers, Dubuque, Iowa, 1997. See also Epstein, I. R. and Pojman, J. A., *An Introduction to Non-linear Chemical Dynamics*, Oxford University Press, New York 1998.
12. Lesmes, F., Hochberg, D., Morán, F. and Pérez-Mercader, J., *Phys. Rev. Lett. 91 (2003) 238301-4*. See also Hochberg, D., Lesmes, F., Morán, F. and Pérez-Mercader, J. in *Phys. Rev. E, (2004)* to appear.
13. Holland, J. H., *Emergence. From Chaos to Order*, Perseus Books, Reading, Massachusetts, 1998.

14. Morowitz, H., *The Emergence of Everything. How the World became Complex*, Oxford University Press, New York, 2002.
15. Creswick, R. J., Farach, H. A. and Poole Jr., C. P., *Renormalization Group Methods in Physics*, John Wiley and Sons, New York, 1992.
16. Gell–Mann, M. and Low, F., *Phys. Rev.* **75** (1954) 1024.
17. Martin, P. C., Siggia, E. D. and Rose, H. A., *Phys. Rev. A8 (1973) 423.* [27].
18. Pérez-Mercader, J., *Coarse Graining, Scaling, and Hierarchies* in *CoarseNonextensive Entropy – Interdisciplinary Applications* edited by M. Gell-Mann and C. Tsallis, Oxford University Press, New York, 2004.
19. Feller, W., *An Introduction to Probability Theory and Its Applications*, Volume I, 3rd edition, John Wiley and Sons, New York, 1968 and Volume II, 2nd edition, John Wiley and Sons, New York, 1971.
20. Hochberg, D. H. and Pérez-Mercader, J., *Phys. Lett. A 296 (2002) 272–279.*
21. For a classical description of what is understood by a "hierarchy"see Simon, H., *The Sciences of the Artificial*, 3rd edition, MIT Press, 1996.
22. Pattee, H. H., *Hierarchy Theory. The Challenge of Complex Systems*, George Braziller, New York, 1978.
23. Miller, J. G., *Living Systems*, McGraw-Hill Book Company, New York, 1978. For some up to date reviews and a source of many references see the collection of papers on Systems Theory in *Science 295 (2002) 1661 et ff.* and the excellent perspective article by Pennisi, E., *Science 302 (2003) 1649.*
24. de Duve, C., *Vital Dust. Life as a Cosmic Imperative*, Basic Books, New York, 1995 and *Life evolving. Molecules, Mind and Meaning*, Oxford University Press, New York, 2002.
25. Pérez-Mercader, J., *"Scaling and the the Emergence of Complexity in the Universe"* in *Astrobiology: The quest for Life in the Universe*, edited by G. Horneck and C. Baum-Starck, Springer-Verlag, 2002.
26. Berera, A. and Fang, L.-Z., *Phys. Rev. Lett. 72 (1994) 458.* Barbero, J. F., Domínguez, A., Goldman, T. and Pérez-Mercader, J., *Europhys. Lett. 38 (1997) 637–642.*
27. Domínguez, A., Hochberg, D., Martín-García, J. M., Pérez-Mercader, J. and Schulman, L., *Astron. and Astrophys. 344 (1999) 27–25* and *ibid. 363 (2000) 373–374.* See also [18].
28. Kittel, C. and Kroemer, H., *Thermal Physics*, 2nd edition, W. H. Freeman and Company, 22nd printing, 2002.
29. Itzykson, C. and Zuber, B., *Statistical Field Theory*, Volumes I and II, Cambridge University Press, Cambridge, Massachusetts, 1989.
30. Chaisson, E., *Cosmic Evolution. An Evolutionary Perspective*, Harvard University Press, Cambridge, Massachusetts, 2001.
31. An excellent introduction to networks and relevant bibliography is given in Watts, D. J., *Six Degrees: the Science of a Connected Age*, W. W. Norton and Co., New York, 2003.
32. de Vaucouleurs, G., *Science 167 (1970) 1203–1213.*
33. West, G. B., Brown, J. H. and Enquist, B. J., *Science 276 (1997) 122–126.* For more references and general comments, cf. Whittfield, J. in *Nature 413 (2001) 342–344.*
34. Barabási, A.-L., *Linked. The New Science of Networks*, Perseus Publishing, Cambridge, Massachusetts, 2002.
35. Carr, B. and Rees, M., *Nature 278 (1979) 605–612.*
36. Bucksnall, D. G. and Anderson, H. L., *Science 302 (2003) 1904.*
37. Smith, H. O., Hutchison III, C. A., Pfannkoch, C. and Venter, J. C., *Proc. Nat. Acad. Sciences USA, 100 (2003) 15440–15445* and references therein.
38. Douçot, B., W. Wang, J. Chaussy, B. Pannetier, R. Rammal, A. Vareille and D. Henry, *Phys. Rev. Lett.* **57** (1986) 1235.

Appendix A: Scale Invariance and Power Laws

Let us quickly show how the connection between scale invariance (hence collective behavior and a basic ingredient of self-organization) arises mathematically. For this we consider a function $f(x)$ of a variable x and subject the independent variable x to a change of scale as

$$x \rightarrow \lambda x \tag{A1}$$

Here λ is the scale factor, which for infinitesimal changes is convenient to take as $\lambda \approx 1 + \epsilon$, with ϵ much smaller than 1. Scale invariance of the function $f(x)$ means that under the scale change in Eq. (A1) the function behaves as

$$f(x) = \frac{1}{a} f(\lambda x) \tag{A2}$$

where the *affine* parameter a must be not equal to one. Expanding in a Taylor series we find that for λ close to 1, the above equation turns into an easily solved ordinary differential equation:

$$\frac{df}{f} = \frac{a-1}{\epsilon} \frac{dx}{x} \tag{A3}$$

whose general solution is

$$f(x) = \text{Constant} \cdot x^{\chi}; \quad \chi = \frac{a-1}{\epsilon} \tag{A4}$$

i.e. a power law with exponent χ. We see at once that scaling invariance of a function Eq. (A2) implies that its functional form must be the one of a power law. And because scale invariance means that *all* the scales in the system are involved in its dynamics, we have the statement that scale invariance is the harbinger of collective behavior. But the solution of an ordinary differential equation requires that we impose some boundary or initial condition. The condition here is that Eq. (A4) must reproduce Eq. (A2), and this leads to [38] the conclusion that in general χ is of the form

$$\chi = \frac{\ln a}{\ln \lambda} + i \frac{2\pi}{\ln \lambda} \cdot n \equiv \chi_R + i\chi_I \tag{A5}$$

with $n = 0, 1, 2, \ldots$ i.e., the scaling exponent χ can be both real or complex. Furthermore, the imaginary part is discrete and quantized by the natural[13] numbers.

Appendix B: Synopsis

We have seen that for many body systems where one can write evolution equations of the form

$$\text{Flow with Time} = \text{Spatial Diffusion} + \text{Reaction} + \text{Noise} \tag{B1}$$

for the various species making up the system, there is a redistribution of the available free-energy; and with this a flow of free-energy. Given some initial and boundary conditions for the fields and parameters, the system goes into configurations where a stable state is reached at a rate faster than the "thermal" approach and there are catalysis and"Darwinian evolution". In addition persistent patterns as well as hierarchical organization and synchronization emerge by coupling enhancement [12].

The above suggest a "picture" of life fitting in the evolution of the Universe as one more phenomenon that takes place as soon as the conditions for the emergence of complex

[13] There are not many generalizations of the proportionality factors. By studying the statistical mechanics of power laws in a general framework one finds a straightforward connection with the Riemann Zeta function, ζ. In this case the complex exponents are related to the zeroes of ζ: each zero leads to a full *family* of complex exponents! [7]

chemistry are present. The phenomenon of Life on Earth would be just one example of a Universal phenomenon, very much as our galaxy or our Solar System are only *one* instance among many in the Universe.

There is, however, a very basic difference: due to the interplay between the presence of noise, the large number of complex available states, the possibility of reaction between many different components and the scale dependence of the final states, one cannot give a fully deterministic description and has, instead, to resort to a description in terms of probabilities. The result is a non-reductionist approach to the Origin of Life and its Evolution.

II. Where did the Chemical Elements Come From and When did Life Begin?

THE ORIGIN OF BIOGENIC ELEMENTS

F. MATTEUCCI[1] and C. CHIAPPINI[2]
[1]Department of Astronomy, University of Trieste, Via G.B. Tiepolo 11, 34131 Trieste, Italy and [2]Astronomical Observatory of Trieste (INAF), Via G.B. Tiepolo 11, 34131 Trieste, Italy

1. Introduction

We discuss the origin of the cosmic abundances of the chemical elements and in particular of biogenic elements such as H, C, N, O and Fe. Interesting enough, many of these elements are among the most abundant in the Universe.

The solar system abundances of chemical elements measured in the Sun photosphere and in meteorites represent the chemical composition of the gas at time of formation of the Solar System (4.5Gyr ago). The solar chemical abundances are called *Cosmic Abundances*. The cosmic chemical composition can be described by three quantities: $X = 0.71$, $Y = 0.27$, $Z = 0.02$, representing the abundance by mass of H, He and metals (all the elements heavier than He). In fact, when considering elements other than H and He, we find that C, O and N are among the most abundant metals.

By means of a chemical evolution model for the Galaxy together with our knowledge on stellar evolution it is possible not only to reproduce the observed solar abundances but also predict the evolution of chemical elements in time. Chemical evolution models also predict abundance variations inside the galaxy as by instance, abundance gradients along the thin-disk. This can be interesting in view of some recent results (Santos et al. 2000) suggesting that a higher than solar Fe abundance favors the formation of stars hosting planets. In fact, due to abundance gradients, we would expect a larger fraction of stars hosting planets towards the inner disk where the Fe abundance is larger.

2. The Production of the Light Elements

While all the metals have originated in stars, the light elements (H, D, He and ^{7}Li were created during the Big Bang and some of them (He, Li) also in stars. The main phases of the Big Bang nucleosynthesis can be summarized as follows: at $T = 10^{12}$K only weak interactions causing conversions between protons and neutrons occurred, whereas the nucleosynthesis started when $T = 10^9$K and lasted until $T = 10^8$K. At that point the deuterium followed by Helium and lithium were formed. Then also very small fractions of ^{7}Li (10^{-9} by mass) were produced.

One of the major achievements in cosmology is that the Big Bang Nucleosynthesis (BBN) can account simultaneously for the primordial abundances of H, D, ^{3}He, ^{4}He and

J. Seckbach et al. (eds.), Life in the Universe, 55–58.

^{7}Li. The BBN and CBR (Cosmic Microwave Background, WMAP experiment) measures agree on the same value for the baryon/photon ratio $\eta = 6.1 \times 10^{-10}$ which implies a baryonic density for the universe $\Omega_b\, h^2 \sim 0.0224$ (Bennet et al. 2003; Spergel et al. 2003). The primordial chemical composition was then X = 0.76, Y = 0.24, Z = 0.

3. The Production of "Metals"

All the elements contained in the 2% of the cosmic chemical composition have then been manufactured in stars. Elements with mass number A from 12 to 60 have been formed in stars during the quiescent burnings during stellar evolution. Stars transform H into He and then He into heavier until the Fe-peak elements, where the binding energy per nucleon reaches a maximum ($A \sim 56$).

Hydrogen is transformed into He through the proton-proton chain or the CNO-cycle, then ^{4}He is transformed into ^{12}C through the triple-α reaction:

$$^4\mathrm{He} + {}^4\mathrm{He} \rightarrow {}^8\mathrm{Be}$$
$$^8\mathrm{Be} + {}^4\mathrm{He} \rightarrow {}^{12}\mathrm{C}^* + \gamma + \gamma$$

The triple-α reaction is a resonant reaction and creates C in an excited state. This is probably the reason why carbon has survived instead of reacting with α-particles and give rise to ^{16}O. The above reaction (triple-α) is very important for the terrestrial life which is a carbon-based chemistry.

Oxygen and α-elements (O, Ne, Mg, Si, S, Ca) originate from the capture of α-particles and the chain arrives until the formation of ^{28}Si. The last main burning in stars is the ^{28}Si-burning which produces ^{56}Ni which then decays into ^{56}Co and ^{56}Fe. Si-burning can be quiescent or explosive (depending on the temperature) and it always produces Fe. Explosive nucleosynthesis occurs during Super Nova (SN) explosions and mainly produces Fe-peak elements. Finally, s- and r-process elements (elements with $A > 60$ up to Th and U) are formed by means of slow or rapid (relative to the β-decay) neutron capture by Fe seed nuclei and are produced during He-burning and SN explosions, respectively.

4. The Dispersal of Elements from Stars

Supernovae and stellar winds restore the newly synthesized and the old elements into the interstellar medium (ISM). Type II supernovae or core-collapse supernovae originate from the explosion of stars with $M > 8\, M_{SUN}$. They contribute mainly to Oxygen and other α-elements (Ne, Mg, Si, S, Ca) plus some Fe and Fe-peak elements. Type Ia SNe are believed to originate from white dwarfs in binary systems exploding by C-deflagration. They produce mainly Fe. Carbon and Nitrogen are mainly produced in intermediate and low mass stars ($0.8 \leq M/M_{SUN} \leq 8$). In particular, carbon is produced via the triple-α reaction which acts in stars of masses larger than $0.5 M_{SUN}$, whereas nitrogen is a product of the H-burning via the CN and CNO cycle and generally is a secondary element (produced from C and O already present in the stars). Possible mechanisms such as dredge-up and rotation can however produce primary nitrogen.

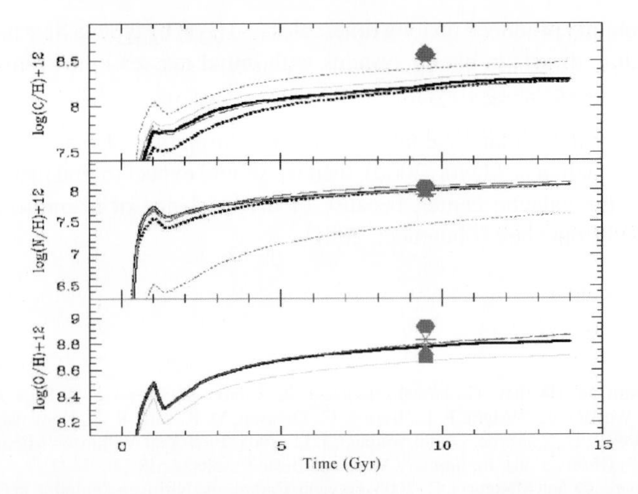

Figure 1. The predicted temporal evolution of the abundances of C, N, O in the gas in the solar neighbourhood, from the model of Chiappini et al. (2002). The various curves refer to different nucleosynthesis prescriptions. The large symbols indicate the solar abundance measurements according different authors (for more details see Chiappini et al. 2002 and references therein).

5. Stars as Probes of Chemical Evolution

Four main stellar populations inhabit the Galaxy: halo, thick-disk, thin-disk and bulge. Galactic chemical evolution models take into account the star formation history in galaxies plus the nucleosynthesis and predict the evolution of the abundances of the most common chemical elements in the gas in galaxies. In particular, a model of the chemical evolution of the galactic disk should reproduce the solar and present time abundances observed.

The abundances of C, N, O as all the abundances of the elements heavier than ^{4}He increase with the cosmic time whereas H is consumed but only in a negligible amount. In Figure 1 we show an example of predicted temporal evolution of the abundances of the C, N, O elements in the solar neighbourhood by the model of Chiappini et al. (2002).

6. Conclusions

By comparing the predictions from chemical evolution models and observational data we can conclude that:

- Carbon is mainly produced in stars with masses in the range 2–8 M_{SUN}, and partly by massive stars. The bulk of this element is produced on timescales $\geq$250Myr.
- Nitrogen is also mainly produced in stars with masses in the range 2–8 M_{SUN} on relatively long timescales but its production is complicated by the secondary/primary processes in stars. The bulk of this element is also produced on timescales $\sim$250Myr.
- Oxygen is mainly produced by massive stars on very short timescales (from few Myr to 30Myr).

- Iron is mainly produced on long timescales ($\geq$1Gyr) by type Ia SNe, which originate from white dwarfs in binary systems with initial masses in the range 0.8–8 M_{SUN} on long timescales ($\geq$1Gyr).
- If a higher than solar Fe content favors the formation of planetary systems (e.g. Santos, Israelian & Mayor, 2000), then we should expect to find them preferentially towards the galactic center, because of the existence of a gradient in [Fe/H] as measured from stars (Matteucci, 2001).

7. References

Bennet, C. L. Halpern, M., Hinshaw, G., Jarosik, N., Kogut, A., Limon, M., Meyer, S. S., Page, L., Spergel, D. N., Tucker, G. S., Wollack, E., Wright, E. L., Barnes, C., Greason, M. R., Hill, R. S., Komatsu, E., Nolta, M. R., Odegard, N., Peiris, H. V., Verde, L. and Weiland, J. L. (2003) First-Year Wilkinson Microwave Anisotropy Probe (WMAP) Observations: Preliminary Maps and Basic Results. ApJS **148**, 1–27.

Chiappini, C., Romano, D. and Matteucci, F. (2003) Oxygen, Carbon and Nitrogen Evolution in Galaxies. MNRAS **339**, 63–81.

Matteucci, F. (2001) *The Chemical Evolution of the Galaxy*, ASSL, Vol. 253, Kluwer Academic Publishers, Dordrecht, The Netherlands.

Santos, N. C., Israelian, G. and Mayor, M. (2000) Chemical Analysis of 8 Recently Discovered Extra-solar Planet Host Stars. Astron. Astrophys. **363**, 228–238.

Spergel, D. N., Verde, L., Peiris, H. V., Komatsu, E., Nolta, M. R., Bennett, C. L., Halpern, M., Hinshaw, G., Jarosik, N., Kogut, A., Limon, M., Meyer, S. S., Page, L., Tucker, G. S., Weiland, J. L., Wollack, E. and Wright, E. L. (2003) First-Year Wilkinson Microwave Anisotropy Probe (WMAP) Observations: Determination of Cosmological Parameters. ApJS **148**, 175–194.

THERMOCHEMISTRY OF THE DARK AGE

DENIS PUY
University of Geneva – Observatory of Geneva
Chemin des Maillettes 51, 1290 Sauverny-Switzerland

1. Introduction

In the conventional view the Universe began in a hot Big Bang some 15 billions years ago, and has been expanding ever since. The dark age of the Universe is pointed out as the period between the hydrogen recombination epoch and the horizon of current astrophysical observations. At very early stage in the expansion, when the temperature of the Universe was still 10^9–10^{10} K, collisions between subatomic particles created hydrogen and helium nuclei with very minor traces of deuterium and lithium nuclei. The chemistry of the early Universe is the chemistry of these light elements. This chemistry (Standard Big Bang Chemistry or SBBC) has been source of large studies. One of the most important consequences, of the existence of a significant abundance of molecules, is the crucial role played in the dynamical evolution of the first collapsing structures. The arrow of time in the cosmic history describes the progression from simplicity to complexity, because the present Universe is clumpy and complicated unlike the homogeneous early Universe. Thus it is crucial to know the nature of the constituents, in order to understand the conditions of the formation of the first bound objects. In this paper we analyse the chemical history of this Dark Age and the consequences on the birth of the first astrophysical objects. Thus in section 2 we describe the chemical evolution of the Universe, then in section 3 we analyze the implications on the formation of first stars. In section 4 some possible outlooks are pointed out.

2. Primordial Chemistry

During the recombination period the ions became progressively neutralized, which led to molecular formation. Nevertheless at early epochs where a total absence of dust grains appears justified, the chemistry is different from the typical interstellar medium astrochemistry. Some groups proposed an assembled comprehensive set of reactions for the early Universe, such as Lepp and Shull (1984), Puy et al. (1993), Galli and Palla (1998), Lepp et al. (2002) and Puy and Pfenniger (2003). The chemical network is coupled with the temperatures (radiation and matter) and the matter density. The evolution of baryons depends on the expansion and on the chemical kinetic, when the radiation only depends on the expansion. Matter temperature depends on different factors such as expansion, coupling between free electrons and background photons, and thermochemistry (see Puy et al. 1993,

J. Seckbach et al. (eds.), Life in the Universe, 59–62.

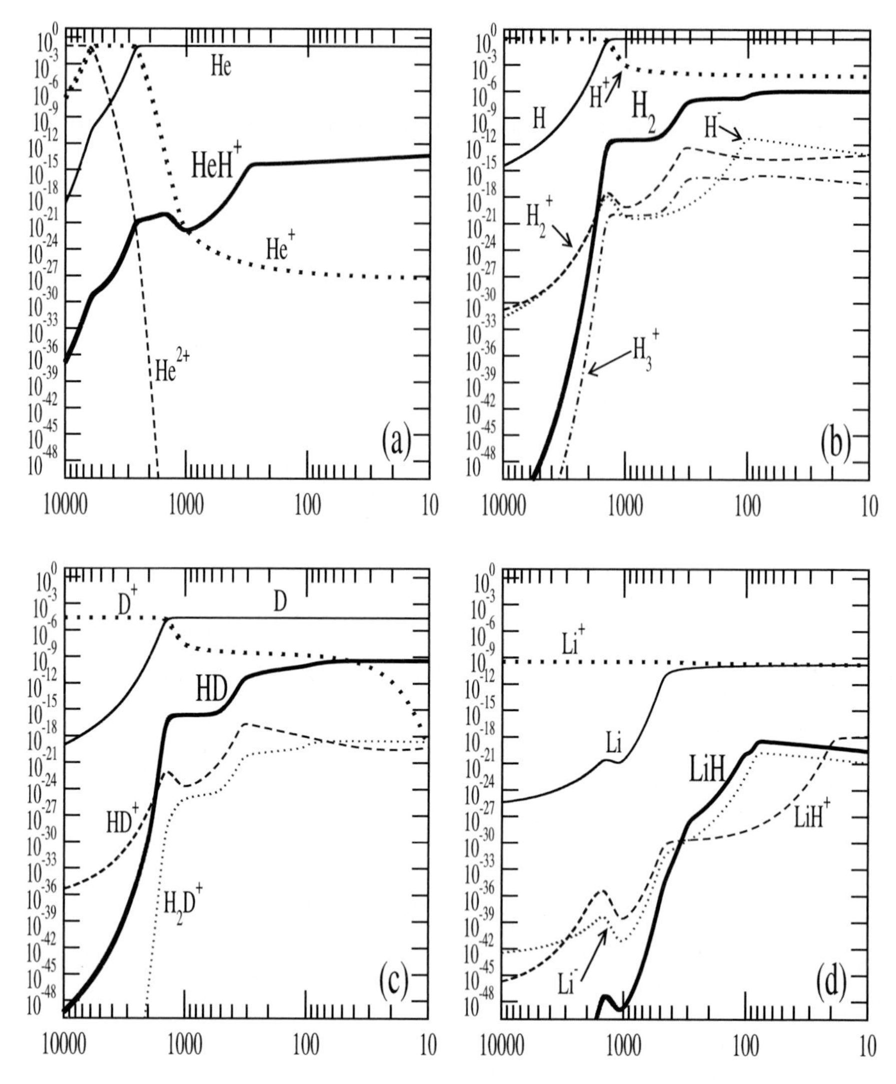

Figure 1. Evolution of abundance in the SBBC. Vertical axis characterizes the abundances when the horizontal axis is relative to the redshift. (a) describes the He-chemistry, (b) the H-chemistry, (c) the D-chemistry and (d) the Li-chemistry, see Puy and Pfenniger 2003.

Puy and Pfenniger 2003). We started our calculation at the redshift $z_i = 10^4$ where the Universe is full ionized, and stop the integration at $z_f = 10$ where the process of reionization is supposed to act up (see Bennett et al. 2003). We pull the initial abundances of the standard big bang nucleosynthesis (see Cyburt et al. 2003, Puy and Pfenniger 2003).

Helium is the first neutral atom which appeared in the Universe, then charge transfer initiate neutral formation (see Fig 1a). Despite the absence of any surfaces of dust grains, it is possible to form neutral molecules through the radiative association between two neutral atoms. Nevertheless H_2 cannot form by this radiative process, charge transfer from ${H_2}^+$

become an alternative or associative detachment with H^-, see Fig 1(b). HD has permanent dipole moments which lead to the capacity to be formed by radiative association. Nevertheless HD formation is more significative, when H_2 appeared, through the dissociative collision with D^+, see Fig 1(c). Besides the radiative association, LiH can be formed from Li by exchange reaction with H_2, by associative detachment with H^- or by mutual neutralization process ($Li^+ \cdot H^-$)or ($Li^- \cdot H$). see Fig 1(d).

3. Formation of the First Structures

In the post recombination context, H_2 HD and LiH have a negligible role on the thermal balance of early Universe (see Puy et al. 1993, Pfenniger and Puy 2003), but these primordial molecules play a role in the dynamics of the first objects in the primordial gas (Puy and Signore 1996). HD and H_2 are important coolant agents and initiate, in numerous cases, the process of thermal instability then the mechanism of fragmentation. The study of forming structure is crucial, in the bottom-up hierarchical picture of structure formation; earliest baryonic objects provide the elementary building blocks for larger mass objects that form later. First massive objects could be early stars, which contaminate the medium with metals, through supernova-driven winds, then lead to a chemistry of heavy elements (see Harwitt and Spaans 2003). The recent discovery of a high optical depth Thomson scattering from the WMAP date implies that significant reionization took place, see Bennett et al. (2003). Reionization theory can be supported by a scenario of early star formation, see Loeb and Barkana 2001.

4. Outlooks

Many important questions still await answers. SBBC is still in a nascent stage and the development of computing techniques and software for the quantum mechanics will open large possibilities. Pfenniger and Puy (2003) suggested that primordial molecules H_2 and HD can freeze out and lead to the growth of flakes of solid H_2 and HD at $z \sim 6$–12 in the unperturbated medium and under-dense regions. Thus in the current understanding of the formation of the first bound structures during the dark age, the possibility that solid hydrogen flakes exist and modify the subsequent evolution must be considered. More recently Puy and Pfenniger (2003) showed that SBBC is sensitive to density perturbations, and lead to relative abundance variations of several percents. Larger variations are expected in the non-linear phases such as collapses. A definitive understanding of the collapse of the first structures is still lacking, because the coupling of gravity, chemistry, and radiation constitutes a formidable non-linear system for which not all equations are presently well known. This question is central because provide strong indications on the mass of the first formed stars, and ultimately knowledge on the contamination in heavier elements at high redshifts. Chakrabarti and Chakrabarti (2000) suggested that adenine formation is possible by successive addition of HCNs in collapsing clouds. Thus as soon as protostellar processes occur, carbon chemistry can be developed then lead to different ways of early pre-biotic molecular formation. In conclusion, molecules have influenced the births and distributions of all stars and galaxies, often by serving as coolants but in other ways as well. Moreover astrochemistry produces species that sometimes have never been manufactured in detectable quantities in terrestrial laboratories. Chemistry of the first objects is not complete, as the chemistry of life in the Universe is a central and unsolved question. For the post-life chemistry as Coyne (2001) pointed out: *In some extraordinary chemical process the human brain came to be, the most complicated machine that we know...*

5. Acknowledgements

I would like to thank Professor Chela-Flores and Professor Joseph Seckbach for the organization of this conference and of these proceedings, and Flavia Puy for discussions on the origin of life. This work was supported by the Swiss National Science Foundation and the University of Geneva.

6. References

Barkana, R., and Loeb A. (2001) In the beginning: The first sources of light and the reionization of the Universe, *Phys. Report* **349**, 125–238.

Bennett, C., Halpern, M., Hinshaw, G., Jarosik, N., Kogut, A., Limon, M., Meyer, S., Page, Spergel, D., Tucker, G., Wollack E., Wright, E., Barnes, C., Greason, M., Hill; R., Komatsu, E., Nolta, M., Odegard, N., Peirs, H. and Verde L. (2003) First year Wilkinson microwave anisotropy probe (WMAP) observations: Preliminary Maps and Basic Results *Astrophysical J. Supp.* **148**, 1–27.

Chakrabarti, S., and Chakrabarti, S. (2000) Can DNA bases be produced during molecular cloud collapse? *Astrophy and Astrophys. Letters* **354**, L6–L8.

Coyne, G. (2001) Origins and creation, in *First steps in the origin of life in the Universe* Eds J. Chela-Flores, T. Owen and F. Raulin, Kluwer Academic Publishers, pp. 359–364.

Cyburt, R., Fields, B., and Olive K. (2003) Primordial nucleosynthesis in light of WMAP, *Phys. Letters B* **567**, 227–234.

Galli, D., and Palla, F. (1998) The chemistry of the early Universe, *Astrophysics and Astrophysics* **335**, 403–420.

Harwitt, M., and Spaans, M. (2003) Chemical composition of the early Universe, *Astrophy. J.* **589**, 53–57.

Lepp, S. and Shull, M. (1984) Molecules in the early Universe, *Astroph. J.* **280**, 465–469.

Lepp, S., Stancil, P., and Dalgarno, A. (2002) Atomic and molecular processes in the early Universe, *J. of Phys. B: Atomic, Mo. and Optical Phys.* **35**, R57–R80..

Pfenniger, D., and Puy D. (2003) Possible flakes of molecular hydrogen in the early Universe, *Astrono. and Astrophys.* **398**, 447–454.

Puy, D., Le bourlot, J., Léorat, J., Pineau des Forets, G., and Alecian, G. (1993) Formation of primordial molecules and thermal balance in the early Universe, *Astrono. and Astrophys.* **267**, 337–346.

Puy, D., and Signore, M. (1996) Primordial molecules in an early cloud formation, *Astrono. and Astrophys.* **305**, 371–378.

Puy, D., and Pfenniger, D. (2003) Differential chemistry in the early Universe submitted to *Astronomy and Astrophys.*

SEARCHING FOR OLDEST LIFE ON EARTH: A PROGRESS REPORT

STEPHEN MOORBATH[1] and BALZ SAMUEL KAMBER[2]
[1]Department of Earth Sciences, Oxford University, Parks Road, Oxford OX13PR, UK and [2]ACQUIRE, The University of Queensland, St Lucia Qld 4072, Brisbane, Australia

1. Introduction

Valid claims for reliable recognition of Earth's earliest biological remnants or by-products must be based on the following minimum requirements: i) geologically correctly identified rocks; ii) correctly and precisely dated rocks; iii) accurately localised rocks of which a type specimen is kept in a collection accessible to international researchers; iv) chemical, isotopic or morphological data which can discriminate between a biological and non-biological origin; v) samples which have been totally decontaminated from extraneous, unrelated biomatter.

These requirements are just as essential for ancient terrestrial rocks as for meteoritic and planetary materials. Because of inadequate investigation of materials of potential biological interest by one or more of the above criteria, it is evident that most (if not all) recent claims for presence of biosignatures in Earth's most ancient rocks are either wrong or, at best, highly debatable. Here we summarise a few examples from recent literature.

2. Claims for Biogenicity Based on Morphology

Putative bacterial or cyanobacterial microfossils from the 3.47 Gyr Apex chert from Western Australia were long thought to provide the oldest morphological evidence for life on Earth (Schopf, 1993). These microstructures have recently been re-interpreted (Brasier *et al.* 2002) as secondary, non-biological artefacts formed from amorphous graphite within multiple generations of metalliferous hydrothermal vein chert and volcanic glass, thus offering no support for primary biological morphology. Regardless of the significance of reduced carbon in these rocks, this particular example illustrates the importance of correctly identifying the host sample, which Schopf (1999) described as graded sediments that were deposited near a river mouth. However, field mapping has shown them to be hydrothermal veins that crosscut a stack of volcanic pillow basalts (Brasier *et al.* 2002).

Repeated claims (Pflug and Jaeschke-Boyer, 1979; Pflug, 2001) for spherical micro-fossils, named *Isuasphaera isua* (Pflug) in chert from the 3.7–3.8 Gyr Isua greenstone belt (IGB) from southern West Greenland, have recently been re-investigated (Appel *et al.* 2003). The site was revisited in 2001 and it was shown that extreme stretching deformation of the metamorphosed chert could not possibly have preserved spherical objects from the time

J. Seckbach et al. (eds.), Life in the Universe, 63–66.

of chert deposition, and are therefore not primary features. These spherical objects were clearly formed by post-tectonic processes, probably related to pre-Quaternary weathering. This is in broad agreement with high-resolution SEM data (Westall and Folk, 2003) on samples of nearby Isua cherts and banded iron formation (BIF). The latter contain abundant remains of fossilised cyanobacteria, fungi and spores, as well as varied carbonaceous particles, between grains and on fracture surfaces. These are regarded as less than 8000 years old, because the area was until recently covered by ice.

3. Claims for Biogenicity Based on Carbon Isotopes

Strong claims for biogenicity of ^{13}C-depleted graphite microparticles in the mineral apatite from >3.8 Gyr-old metamorphosed chemical sediments on the island of Akilia of southern West Greenland have long been defended (Mojzsis *et al.* 1996; Nutman *et al.* 1997). The situation regarding the age of the carbon-bearing rocks, as well as the nature of their geological relationship to the dated rocks, was summarised in the previous Trieste Conference Volume (Moorbath, 2001), and is not repeated here. Since then, it has been demonstrated (Fedo and Whitehouse, 2002) that the crucial carbon-bearing rocks on Akilia Island, misidentified (Mojzsis *et al.* 1996; Nutman *et al.* 1997) as chemical sediments, are actually banded quartz-pyroxene rocks of mixed magmatic and metasomatic (fluid-penetrated) parentage with no intrinsic biological relevance whatever.

The overall debate about interpretation of C isotope ratios has taken a major step forward both from study of ancient reduced carbon and modern seafloor hydrothermal systems. In genuine chemical sediments (including BIF) of the 3.7–3.8 Gyr IGB, some 150 km northeast of Akilia Island, ^{13}C-depleted graphite particles have been claimed as evidence for biological activity (Schidlowski, 1988; Mojzsis *et al.* 1996). However, it has been shown (Lepland *et al.* 2002; van Zuilen *et al.* 2003) that graphite in IGB rocks occurs abundantly only in secondary carbonate veins formed at depth in the crust by injection of hot fluids reacting with older crustal rocks (metasomatism). During these reactions, graphite is formed by disproportionation of $FeCO_3$ at high temperatures ($\approx 450°C$) by the reaction $6FeCO_3 \rightarrow 2Fe_3O_4 + 5CO_2 + C$. Van Zuilen and co-workers call for a reassessment of earlier interpretations of life at 3.8 Gyr made on these carbonate-rich rocks, and conclude that the isotopic composition of graphite, in general, does not serve as a reliable biomarker in strongly metamorphosed rocks.

Discovery of highly reduced gases (H_2 and methane) in vent fluids of the presently active Mid-Atlantic Rainbow hydrothermal system (Charlou *et al.* 1998) has opened the possibility of studying an active site of abiogenic hydrocarbon production. Holm and Charlou (2001) reported abiogenic (de novo) synthesis (of the Fischer-Tropsch type) of linear saturated hydrocarbons with chainlengths between 16 and 29 C atoms. Horita and Berndt (1999) measured C-isotope compositions of dissolved methane from Rainbow vent fluids and found a similar extent of isotopic fractionation as in microbial methane. Hence, isotopically light C in graphite on its own as preserved in ancient rocks is not sufficient evidence to imply biogenicity.

There is one occurrence of abundant reduced carbon in a 3.7–3.8 Gyr IGB metamorphosed sediment (not hydrothermally altered mafic rock) which was claimed to contain biogenic ^{13}C-depleted graphite, putatively derived from the detritus of planktonic organisms

(Rosing, 1999). This is so far the most plausible prospect for early life in IGB rocks, but these rocks are highly deformed and strongly metamorphosed (at garnet-grade). Furthermore, because these rocks have anomalous tungsten isotope ratios the possibility needs to be considered that the reduced carbon could be meteoritic in origin (Schoenberg *et al.* 2002).

In short, several much publicised claims for biosignatures in early Archaean rocks of West Greenland are highly questionable, whilst some are certainly wrong. In particular, we emphasise that morphological studies as well as stable isotope analyses of minute amounts of retrieved 'organic' material need to exclude the effects of contamination by modern organisms.

4. Discussion

There is still no undisputed evidence for when and where life started on Earth. Methods for investigating the oldest rocks for biosignatures require far greater geological and geochemical sophistication than evident in most published efforts. A major limitation is the uncertainty surrounding the true nature of host rocks that are claimed to contain biosignatures. However, progress is being made in establishing morphological, mineralogical, chemical and isotopic fingerprints of microbial carbonates (rocks that form due to the decay of microbial organic matter). Presently, the oldest known such deposits have been documented from the 3.4 Gyr Warrawoona Formation in Western Australia (Van Kranendonk *et al.* 2003). Microbial carbonates allow for detailed direct chemical comparisons between ancient sedimentary rocks with biopotential and modern examples.

The cumulative effects of the putative, massive Late Heavy Bombardment of the Earth at around 3.9–4.0 Gyr on life (if present) were arguably devastating. But we know for sure from evidence of surviving rocks (e.g. the IGB sedimentary and volcanic rocks) that by 3.7–3.8 Gyr geological processes at and near the surface of the Earth were becoming recognisably uniformitarian, perhaps allowing life to get started in leisurely fashion under relatively quiescent conditions on mineral surfaces in metaphorical Darwinian "warm little ponds".

5. References

Appel, P.W.U., Moorbath, S. and Myers, J.S. (2003) *Isuasphaera isua* (Pflug) revisited, *Precambr. Res.* **126** pp. 309–312.

Brasier, M.D., Green, O.R., Jephcoat, A.P., Kleppe, A.K., van Kranendonk, M.J., Lindsay, J.F., Steele, A. and Grassineau, N.V. (2002) Questioning the evidence for Earth's oldest fossils, *Nature* **416**, pp. 76–81.

Charlou, J.L., Fouquet, Y., Bougalt, H., Donval, J.P., Etableau, J., Jean-Baptiste, P., Dapoigny, A., Appriou, P. and Rona, P.S. (1998) Intense CH_4 plumes generated by serpentinisation of ultramafic rocks at the intersection of the 15°20'N fracture zone and the Mid-Atlantic Ridge, *Geochim. et Cosmochim. Acta* **62**, pp. 2323–2333.

Fedo, C.M. and Whitehouse, M.J. (2002) Metasomatic origin of quartz-pyroxene rock, Akilia, Greenland, and implications for Earth's earliest life, *Science* **296**, pp. 1448–1452.

Holm, N. and Charlou, J.O. (2001) Initial indications of abiotic formation of hydrocarbons in the Rainbow ultramafic hydrothermal system, Mid-Atlantic Ridge, *Earth Planet. Sci. Lett.* **191**, pp. 1–8.

Horita, J. and Berndt, M.E. (1999) Abiogenic methane formation and isotopic fractionation under hydrothermal conditions, *Science* **285**, pp. 1055–1057.

Lepland, A., Arrhenius, G. and Cornell, D. (2002) Apatite in early Archaean Isua supracrustal rocks, southern West Greenland: its origin, association with graphite and potential as a biomarker, *Precambr. Res.* **118**, pp. 221–241.

Mojzsis, S.J., Arrhenius, G., McKeegan, K.D., Harrison, T.M., Nutman, A.P. and Friend, C.R.L. (1996) Evidence for life on Earth before 3,800 million years ago, *Nature* **384**, pp. 55–59.

Moorbath, S. (2001) Geological and geochronological constraints for the age of the oldest putative biomarkers in the early Archaean rocks of West Greenland, In: J. Chela-Flores, T. Owen and F. Raulin (eds.) *First Steps in the Origin of Life in the University*, Dordrecht: Kluwer Academic Publishers, pp. 217–222.

Nutman, A.P., Mojzsis, S.J. and Friend, C.R.L. (1997) Recognition of $\geq$3850 Ma water-lain sediments in West Greenland and their significance for the early Archaean Earth, *Geochim. et Cosmochim. Acta* **61**, pp. 2475–2484.

Pflug, H.D. (2001) Earliest organic evolution. Essay to the memory of Bartholomew Nagy, *Precambr. Res.* **106**, pp. 79–91.

Pflug, H.D. and Jaeschke-Boyer, H. (1979) Combined structural and chemical analysis of 3,800 Myr-old microfossils, *Nature* **280**, pp. 483–486.

Rosing, M.T. (1999) ^{13}C-depleted carbon microparticles in >3,700-Ma sea-floor sedimentary rocks from West Greenland, *Science* **283**, pp. 674–676.

Schidlowski, M. (1988) A 3,800-million-year isotopic record of life from carbon in sedimentary rocks, *Nature* **333**, pp. 313–318.

Schoenberg, R., Kamber, B.S., Collerson, K.D. and Moorbath, S. (2002) Tungsten isotope evidence from ~3.8 Gyr metamorphosed sediments for early meteorite bombardment of the Earth, *Nature* **418**, pp. 403–405.

Schopf, J.W. (1993) Microfossils of the early Archaean Apex Chert: New evidence of the antiquity of life, *Science* **260**, pp. 640–646.

Schopf, J.W. (1999) *The Cradle of Life*, Princeton University Press, New York.

Van Kranendonk, M.J., Webb, G.E. and Kamber, B.S. (2003) New geological and trace element evidence from 3.45 Ga stromatolitic carbonates in the Pilbara craton: support of a marine, biogenic origin and for a reducing Archaean ocean, *Geobiology* **1**, pp. 91–108.

Van Zuilen, M.A., Lepland, A., Terranes, J., Finarelli, J., Wahlen, M. and Arrhenius, G. (2003) Graphite and carbonates in the 3.8 Ga-old Isua supracrustal belt, southern West Greenland, *Precambr. Res.* **126**, pp. 331–348.

Westall, F. and Folk, R.L. (2003) Exogenous carbonaceous microstructures in early Archaean cherts and BIFs from the Isua greenstone belt: implications for the search for life in ancient rocks, *Precambr. Res.* **126**, pp. 313–330.

THE EUROPEAN EXO/ASTROBIOLOGY NETWORK ASSOCIATION

A. BRACK
Centre de Biophysique Moléculaire, CNRS,
Rue Charles Sadron, 45071 Orléans cedex 2, France

1. Introduction

The question of the chemical origins of life is engraved in the European scientific patrimony as it can be traced back to the pioneer ideas of Charles Darwin, Louis Pasteur, and more recently to Alexander Oparin. During the last decades, the European community of origin of life scientists has organized seven out of the twelve International Conferences on the Origins of Life held since 1957. This community enlarged the field of research to life in extreme environments, including the early Earth, and to the search for extraterrestrial life, i.e. exobiology in its classical definition or astrobiology if one uses a more NASA-inspired terminology. The present contribution aims to describe the European networking activities in this field of research.

2. The Science Background

The science of exo/astrobiology, although very broad, yet forms a coherent whole crossing many disciplines. For example, laboratory and theoretical work on the origin of life is based on our understanding of the geochemical conditions of the early Earth, which in turn depends on conditions in the early solar system, again dependent on the chemistry of the gas and dust clouds between the stars. We neither know how life started on earth nor if it exists outside the Earth. Answering both these questions will be scientifically very exciting and also have the deepest meanings for philosophy and the place of humanity in the Universe. Our understanding of the early Earth is making real progress; we are finding that life can exist in extreme environments; there are novel ideas as to how life originates; there are real searches to find life outside the Earth, both on planets and moons in our own Solar System and on planets in other planetary systems. In the search for life outside the Earth, the discovery of water on Mars and of a liquid ocean on Europa, a moon of Jupiter, give us definite solar system targets, and the discovery in 1995 of giant planets orbiting nearby stars has revitalised the prospects for finding Earth-like planets which can be studied for the signature of life (Brack, 2001).

Collaborative researches are necessarily developped in the different fields covered by exo/astrobiology (Westall et al., 2000; Brack et al., 2001):

- Terrestrial life as a reference (origins of life, geological and climatic context, ingredients for primitive life, life in a test tube, diversity of bacterial life, panspermia).

J. Seckbach et al. (eds.), Life in the Universe, 67–69.

- Exploring the Solar System (Mars, Europa, Titan, Comets).
- Search for life beyond the Solar System (exoplanets, Corot and Darwin missions).

3. Exo/Astrobiology Networking Activities in Europe

Because of the vastly interdisciplinary nature of the research (astronomical instrumentation, cosmochemistry, galactic astronomy, planetary science, solar science, atmospheric physics, geology, paleontology, chemistry, environmental biology, biology, information theory, philosophy, to name but some) that must be undertaken to answer the overall question of the origin and distribution of life, progress necessitate widespread collaborations. In the United States, NASA is funding the NASA Astrobiology Institute, a network of 16 lead teams, both to increase the funding for astrobiology and to promote interdisciplinary work.

In Europe, several nations, e.g., Finland, France, Germany, Russia, Spain, Sweden, Switzerland and United Kingdom, have already established national networks. Collaborative links are already producing real results in Europe, such as the ROSE, Response of Organisms to Space Environment, consortium of ESA-funded experiments for the International Space Station, and the ESA Topical Team ROME on Responses of Organisms to Martian conditions. Within the framework of the European Commission COST Actions to foster cooperation in a specific research area, COST D27 "*Origin of life and early evolution*" has been approved in June 2001 for a period of 5 years. The main objective of this action is to develop the chemistry of the origins and early evolution of life, with special attention to cosmochemistry, prebiotic chemistry of small molecules, directed evolution and origin of the genetic code. Hosted by the International Space Science Institute in Bern, the International Space Science Team "Prebiotic matter in space" is a consortium of 13 scientists, each representing a specific research field crucial to revealing the origin of life as a consequence of the evolving Universe (Ehrenfreund et al., 2002).

4. The European Exo/Astrobiology Network Association (EANA)

The European Exo/Astrobiology Network Association, EANA, was created in 2001 to coordinate the different European centres of excellence in exo/astrobiology or related fields. The specific objectives of EANA are:

- To bring together European researchers interested in exo/astrobiology programmes and to foster their cooperation.
- To attract young scientists to this quickly evolving, interdisciplinary field of research.
- To create a website establishing a database of expertise in different aspects of exo/astrobiology.
- To interface the Network with European bodies such as ESA, ESF, the European Commission and with non European institutions active in the field.
- To popularise exo/astrobiology to the public and to students.

The value of sharing resources on an international scale was highlighted at the Inaugural Meeting of the European Exo/Astrobiology Steering Group at the British National Space Centre, London in October 1999 and at the Strategy-oriented Meeting at CNES, Paris in October 2000. At these two meetings, senior representatives of the European Science Foundation and the European Space Agency acknowledged that exobiology in Europe should be strengthened, formalized as a Network and supported. The First European Exo/Astrobiology Workshop, held in Frascati, Italy, May 2001, was attended by 200 scientists. The Second European Workshop was organised in Graz, Austria, in September 2002. The workshop, attended by 320 participants, was oriented particularly to the planetology aspects of astrobiology, in acknowledgement of the expertise of the local organisers. The Third European Workshop was hosted in November 2003 by the Centro de Astrobiologia in Madrid and dedicated to the search for traces of life on Mars. It was attended by 260 participants.

EANA is run by an Executive Council consisting of national members presently representing 17 European nations active in the field, e.g. Austria, Belgium, Denmark, Finland, France, Germany, Hungary, Italy, Poland, Portugal, Romania, Russia, Spain, Sweden, Switzerland, The Netherlands, United Kingdom, on the basis of one representative per nation, and elected members in a number equal to the number of active nations.

EANA is affiliated to the NASA Astrobiology Institute. The formal affiliation was signed in 2002 at the Graz Workshop by Rosalind Grymes, Deputy Director of NAI, during a reception hosted by the Governor of Styria in the historical Eggenberg Castle. EANA is member of the Federation of Astrobiology Organizations, FAO, including the Australian Centre for Astrobiology (ACA), the Astrobiology Society of Britain (ASB), the Spanish Centro de Astrobiologia (CAB), the French Groupement de Recherche en Exobiologie (GDR Exobio), the American NASA Astrobiology Institute (NAI) and the Swedish Astrobiology Center (SWAN). The FAO has been created to facilitate international exchange between students and to harmonise the planning of joint astrobiology meetings.

The EANA Web Page, *http://www.spaceflight.esa.int/exobio*, is hosted as part of the ESA Virtual Institute at ESA/ESTEC in Noordwijk, The Netherlands.

5. References

Brack, A. (2001) Life: origins and possible distribution in the Universe, In: P. Murdin (ed.) Encyclopedia Astronomy & Astrophysics. IOP, Bristol, pp. 1411–1421.

Brack, A., Horneck, G., and Wynn-Williams, D. (2001) Exo/Astrobiology in Europe. Origins Life Evol. Biosphere 31, 459–480.

Ehrenfreund, P., Irvine, W., Becker, L., Blank, J., Brucato, J.R., Colangeli, L., Derenne, S., Despois, D., Dutrey, A., Fraaije, H., Lazcano, A., Owen, T. and Robert, F. (2002) Astrophysical and astrochemical insights into the origin of life. Rep. Prog. Phys. 656, 1427–1487.

Westall, F., Brack, A., Hofmann, B., Horneck, G., Kurat, G., Maxwell, J., Ori, G.G., Pillinger, C., Raulin, F., Thomas, N., Fitton, B., Clancy, P., Prieur, D. and Vassaux, D. (2000) An ESA study for the search for life on Mars. Planet. Space Sci. 48, 181–202.

III. Physical Constraints on the Origin of Life

THE ORIGIN OF BIOMOLECULAR CHIRALITY
Search for Efficient Chiroselective Autocatalytic Reactions

J. RIVERA ISLAS[1], J. C. MICHEAU[2] and T. BUHSE[1]
[1]Centro de Investigaciones Químicas, Universidad Autónoma del Estado de Morelos, Av. Universidad No 1001, Col. Chamilpa, 62210 Cuernavaca, Morelos, México and [2]Laboratoire des IMRCP, UMR au CNRS No 5623, Université Paul Sabatier, 118, route de Narbonne, F-31062 Toulouse Cedex, France.

1. Introduction

Life is characterized by broken mirror symmetry (Pályi *et al.*, 1999). On the molecular level, proteins are composed almost exclusively of L-amino acids while nucleic acids only contain D-sugars. Without this chiral asymmetry, prebiotic molecular complexity leading to the formation of biologically active polymers could probably not have evolved (Joyce *et al.*, 1984; Avetisov and Goldanskii, 1991). Nevertheless, more than 1½ century after Pasteur's discovery, the origin of biomolecular chiral asymmetry is still a mystery. Meanwhile, it is accepted that homochirality has already appeared early during chemical evolution and that a homochiral molecular environment was rather a pre-condition than a consequence of life (Keszthelyi, 1995; Avalos *et al.*, 2000). Parity violation (MacDermott, 1993) and other chiral factors such as circularly polarized light are omnipresent and can lead under favorable conditions to enantio-meric enrichment. However, this enhancement usually remains tiny and can be annihilated by long-term racemization processes.

Hence amplifying mechanisms have to be considered that were strong and selective enough to recognize small enantiomeric imbalances and to overcome noise and racemization as concerned for chemical evolution scenarios. Respective theoretical model reactions (Frank, 1953; Decker, 1979; Kondepudi and Nelson, 1985) can give rise to so-called chiral symmetry breaking, a bifurcation-like process in which the racemic state becomes unstable when some critical external constraints have been reached (Fig. 1). These approaches contribute to the perception of the nonlinear dynamics involved in chiral amplification but, on the other hand, remain elusive for the explicit design and feasibility of new laboratory experiments that can help to test the validity of the proposed hypothesis and to develop new concepts. Such studies are highly needed even if they are not of entire prebiotic relevance, since only experimental systems can reveal the richness by which nature has constructed nonlinear scenarios exhibiting the required properties.

J. Seckbach et al. (eds.), Life in the Universe, 73–77.

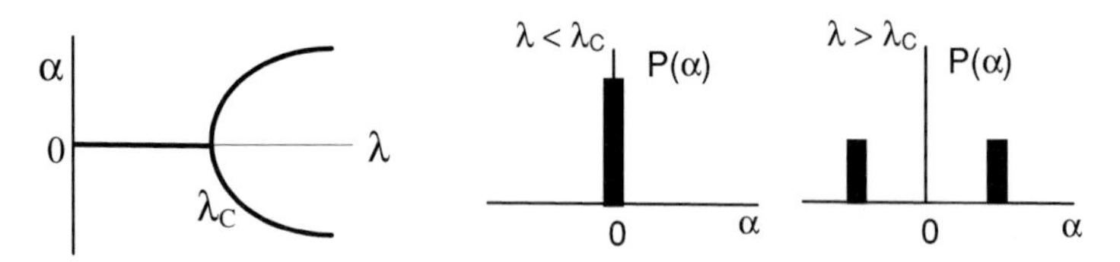

Figure 1. Sketch of a possible bifurcation scenario in a chirally autocatalytic reaction system (after Kondepudi and Askura, 2001). After the external constraint λ has reached a critical value (λ_C), the racemic state ($\alpha = 0$) of the system becomes unstable and it evolves inevitably into an optically active state ($\alpha \neq 0$). Accordingly, repeated experimental runs will show the transition from a monomodal (racemic) to a bimodal (optically active) probability distribution for α if $\lambda > \lambda_C$.

2. Experimental Systems Showing the Presence of Possible Chiral Symmetry Breaking

2.1. STIRRED CRYSTALLIZATION

A number of compounds exhibit spontaneous resolution upon crystallization like binaphhyl, 4,4′-dimethylchalcone, tri-o-thymotide, ethylmethylanilinium iodide, benzodiazepine, α-amino δ-caprolactam Ni(II) chloro complexes and sodium chlorate or bromate (Jacques *et al.*, 1981). All these systems are characterized by the presence of both a "labile configuration" in solution phase and of conglomerate crystals that can serve as preferential seeds for later crystallization.

A major breakthrough was reached when the role of stirring was emphasized by Kondepudi *et al.* (1990). When stirred, secondary nucleation occurs which gives rise to a strong chiroselective autocatalytic effect leading in each experiment to a virtually homochiral population of sodium chlorate or bromate crystals although starting from entirely achiral conditions. During the stirred crystallization of sodium chlorate, it was shown (Kondepudi *et al.*, 1995) that the bimodal probability distribution (Fig. 1) of the crystal enantiomeric excess sensitively depends on the stirring rate. It is suggested that chiral symmetry breaking can be expected in stirred crystallization of any achiral or rapidly inter-converting compound that crystallizes in enantiomeric forms. However, for extended chiral propagation, it would be required that chirality will be established at a molecular state – for instance in asymmetric C-atoms.

2.2. NONLINEAR EFFECTS IN ASYMMETRIC SYNTHESIS

Chirally autocatalytic effects constitute the next generation of asymmetric synthesis (De Min *et al.*, 1988; Feringa and van Delden, 1999). As a property of life, such processes could be of fundamental importance in the genesis of chiral asymmetry in nature. The first clearly successful reaction (1) involving a chiral autocatalyst has been reported by Soai *et al.* (1995, 2000). The autocatalytic addition of diisopropylzinc (i-Pr_2-Zn) to a prochiral pyrimidyl aldehyde (CHO) yielding a chiral pyrimidyl-alcohol after hydrolysis of the direct reaction product (COZn-*i*-Pr) features several important aspects.

$$\text{CHO} + i\text{-Pr}_2\text{-Zn} \longrightarrow \text{COZn-}i\text{-Pr} \qquad (1)$$

The system exhibits strong amplification of enantiomeric excess shown by the initial addition of chiral alcohol or minute amounts of other chiral species like mandelic acid, [2,2]-paracyclophanes, helicene, octahedral cobalt complexes, deuterated chiral molecules, circularly polarized light, photolyzed DL-leucine, chiral sodium chlorate and quartz crystals (Soai and Sato, 2002). Repeated experiments without chiral initiator lead to a bimodal-like probability distribution of e.e. (without any racemic realization) (Soai *et al.*, 2003). The basic autocatalytic mechanism underlying these unprecedented properties is still subject of investigations (Blackmond *et al.*, 2001; Singleton and Vo, 2002; Buono and Blackmond, 2003).

From kinetic data, a chemically reasonable tentative mechanism has been designed (Fig. 2) that can reproduce experimentally observed chiral amplification (Buhse, 2003). The model describes the underlying dynamics of the system in terms of a template-directed self-replication in which a dimer species, Zn-$(COZn)_2$, acts as the active autocatalytic species. Inverse kinetic data treatment allowed us to extract the main reaction parameters and to predict dynamic and structural principles of this autocatalytic chiral amplification reaction. The dimeric chiral catalyst permits a chiroselective 3-point attachment of the prochiral substrate. This feature is not only important for the chiroselective action but has also kinetic consequences as it provides a cubic autocatalytic dynamic behavior that is the basis for the occurrence of bifurcation phenomena.

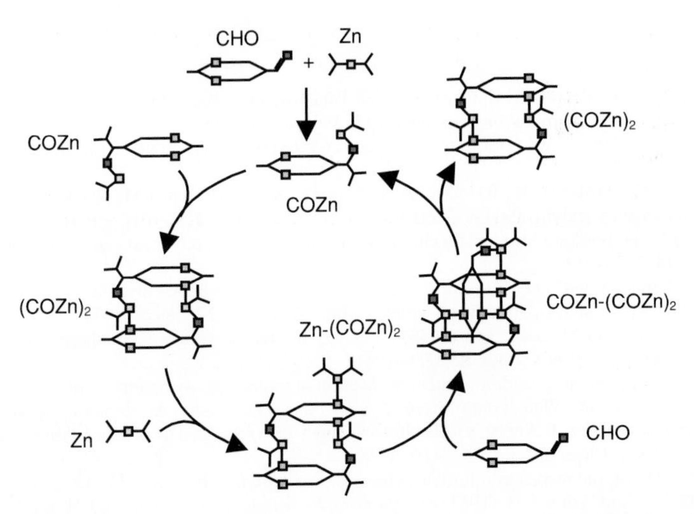

Figure 2. Schematized reaction network proposed for the *Soai*-type autocatalytic alkylzinc addition. Strong cubic-type autocatalysis of the net type $A + 2\,B \rightarrow 3\,B$ emerges by the formation of the catalytic dimer species that consists of two product molecules.

3. Conclusion

The Soai-type reaction appears as a powerful machine to produce highly enriched chiral molecules in terms of a self replication-like mechanism. From this analysis, some prebiotically relevant concepts can be proposed that we have designated as 'smartness': 1) The generation of template species from two sub-units of chiral product molecules provides the possibility of asymmetric synthesis via 3-point attachment of a prochiral precursor. 2) The dimerization is a fundamental process for chiral amplification because it is related to mutual inhibition between enantiomers. 3) Ternary organometallic complexes are able to provide both kinetic nonlinearity and chiral recognition from strongly asymmetrical interactions between the ligands. 4) The amplification of the more kinetically cooperative catalysts is likely to have occurred within primordial reacting systems: catalysts perform an autoselection through self-replication.

4. Summary

Chiroselective autocatalytic reaction systems provide experimental examples in which chiral symmetry breaking could occur. Among the few that have been already discovered, the autocatalytic addition of diisopropylzinc to a prochiral pyrimidyl-aldehyde yielding the chiral pyrimidyl-alcohol (the *Soai*-type reaction) appears to be the most promising. Although the reaction is driven far from prebiotically relevant conditions, its kinetic analysis permits to propose a number of concepts that can be useful for further investigations in the disputed domain of the origin of biomolecular homochirality.

5. References

Avalos, M., Babiano, R., Cintas, P., Jiménez, J.L. and Palacios, J.C. (2000) From parity to chirality: Chemical implications revisited, *Tetrahedron: Asymmetry* **11**, 2845–2874.

Avetisov, V.A. and Goldanskii, V.I. (1991) Homochirality and stereospecific activity: Evolutionary aspects, *BioSystems* **25**, 141–149.

Blackmond, D.G., Mc Millan, C.R., Ramdeehul, S., Schorm, A. and Brown, J.M. (2001) Origins of asymetric amplification in autocatalytic alkylzinc additions, *J. Am. Chem. Soc.* **123,** 10103–10104.

Buhse, T. (2003) A tentative kinetic model for chiral amplification in autocatalytic alkylzinc additions, *Tetrahedron: Asymmetry* **14**, 1055–1061.

Buono, F.G. and Blackmond, D.G. (2003) Kinetic evidence for a tetrameric transition state in the asymetric autocatalytic alkylation of pyrimidyl aldehydes, *J. Am. Chem. Soc.* **125**, 8978–8979.

De Min, M., Levy, G. and Micheau, J.C. (1988) Chiral resolutions, asymmetric synthesis and amplification of enantiomeric excess, *J. Chim. Phys.* **85,** 603–619.

Decker, P. (1979) Spontaneous generation and amplification of molecular asymmetry through kinetical bistability in open systems, In: D.C. Walker (ed) *Origin of optical activity in nature*, Amsterdam, Elsevier, pp. 109–124.

Feringa, B.L. and van Delden, R.A. (1999) Absolute asymmetric synthesis: The origin, control, and amplification of chirality, *Angew. Chem. Int. Ed.* **38**, 3419–3438.

Frank, F.C. (1953) On spontaneous asymmetric synthesis, *Biochim. Biophys. Acta* **11**, 459–463.

Jacques, J., Collet, A. and Wilen, S.H. (1981) *Enantiomers, Racemates and Resolution*, J. Wiley, New York. Joyce, G.F., Visser, G.M., van Boeckel, C.A.A., van Boom, J.H., Orgel, L. and van Westrenen, J. (1984) Chiral selection in poly(C)-directed synthesis of oligo(G), *Nature* **310**, 602–604.

Keszthelyi, L. (1995) Origin of the homochirality of biomolecules, *Quart. Rev. Biophys.* **28**, 473–507.

Kondepudi, D.K. and Nelson, G.W. (1985) Weak neutral currents and the origin of biomolecular chirality, *Nature* **314**, 438–441.

Kondepudi, D.K., Kaufman, R. and Singh, N. (1990) Chiral symmetry breaking in sodium chlorate crystallization, *Science* **250**, 975–977.

Kondepudi, D.K., Bullock, K.L., Digits, J.A. and Yarborough, P.D. (1995) Stirring rate as a critical parameter in chiral symmetry breaking crystallization, *J. Am. Chem. Soc.* **117**, 401–404.

Kondepudi, D.K. and Asakura, K. (2001) Chiral autocatalysis, spontaneous symmetry breaking, and stochastic behavior, *Acc. Chem. Res.* **34**, 946–954.

MacDermott, A.J. (1993) The weak force and the origin of life, In: C. Ponnamperuma and J. Chela-Flores (eds.) *Chemical evolution: Origin of life*, Hampton, Deepack Publishing, pp. 85–99.

Pályi, G., Zucchi, C. and Caglioti, L. (eds.) (1999) *Advances in BioChirality*, Elsevier, Amsterdam.

Singleton, D.A. and Vo, L.K. (2002) Enantioselective synthesis without discrete optically active additives, *J. Am. Chem. Soc.* **124,** 10010–10011.

Soai, K., Shibata, T, Morioka, H., Choji, K (1995) Asymmetric autocatalysis and amplification of enantiomeric excess of a chiral molecule, *Nature* **378**, 767–768.

Soai, K., Shibata, T. and Sato, I. (2000) Enantioselective automultiplication of chiral molecules by asymmetric autocatalysis, *Acc. Chem. Res.* **33**, 382–390.

Soai, K. and Sato, I. (2002) Asymmetric autocatalysis and its application to chiral discrimination, *Chirality* **14**, 548–554.

Soai, K., Sato, I., Shibata, T., Komiya, S., Hayashi, M., Matsueda, Y., Imamura, H., Hayase, T., Morioka, H., Tabira, H., Yamamoto and Kowata, Y. (2003) Asymmetric synthesis of pyrimidyl alkanol without adding chiral substances by the addition of diisopropylzinc to pyrimidine-5-carbaldehyde in conjunction with asymmetric autocatalysis, *Tetrahedron: Asymmetry* **14**, 185–188.

SALAM HYPOTHESIS AND THE ROLE OF PHASE TRANSITION IN AMINO ACIDS

WENQING WANG, NAN YAO, YU CHEN and PENG LAI
Department of Applied Chemistry, Department of Physics, Peking University, Beijing 100871, China

Abstract. Salam hypothesis was elaborated in the First Trieste Conference on Chemical Evolution. We spent twelve years on a number of experiments (i.e. temperature-dependence of X-ray diffraction, neutron diffraction, atomic force microscopy of the surface structure, specific heat measurement, DC and AC magnetic susceptibilities, longitudinal ultrasonic attenuation and velocity, ^{1}H CRAMPS and ^{13}C CP/MAS solid state NMR studies, Raman spectra & natural optical rotation) to verify the second-order phase transition and the role in amino acids. The configuration transformation from D- to L-type was refuted by gas chromatographic analysis of Chirasil-Val capillary column of DL, D- and L-valine. This paper focuses attention on the role of phase transition in amino acid, which might play a bifurcation-type mechanism in a chirally-pure state instead of a direct configuration change from D- to L-amino acid below T_c.

1. Introduction

The homochirality of natural amino acids and sugars remains a puzzle for theories of the chemical origin of life (Bada, 1995, Meiiring, 1987, Mason, 1985). The cause for the homochirality of biologically relevant molecules is assumed to be the intrinsic chirality present at the elementary particle level (Kondepudi 1985, Quack 1989). The idea started with the possibility of defining a science of life based on atomic physics. Salam (1991,1992) proposed the subtle energy difference of chiral molecules induced by Z^0 interactions in terms of quantum mechanical cooperative and condensation phenomena, which could give rise to second-order phase transitions (including D to L transformations) below a critical temperature T_c. The value of T_c is around 250K deduced from Ginzberg-Landau equation.

Bonner (2000) and L. Keszthelyi (2001) have discussed the extraterrestrial origin of the homochirality of biomolecules and the amplification of tiny enantiomeric excess. Attention is called upon the mechanism for production of extraterrestrial handedness based on Salam's condensation hypothesis. Figureau *et al.* (1995) reported that their experiments have failed to validate Salam's predicted phase transitions. A recent report by Compton *et al.* (2003) presented arguments against the Salam hypothesis in terms of their observing small feature in the experiments.

J. Seckbach et al. (eds.), Life in the Universe, 79–82.

2. Experimental

2.1. SAMPLE PREPARATION/CHARACTERIZATION

Elemental analysis of C, H and N of D-/L-alanine single crystals showed that D-/L-alanine were pure single crystals without crystal water, and the crystals of D-/L- alanine were determined by X-ray diffraction crystallography at 300K, 293K, 270K, 223K, 250K and D-valine was measured at 293K, 270K, 223K and 173K (Wang, 2003).

2.2. SECOND-ORDER PHASE TRANSITION AND THE BIFURCATION MECHANISM

In previous work, a number of experiments were designed to search for the parity violation of the electroweak force at the phase transition of single crystals of D- and L-alanine and valine. An obvious second–order phase transition was shown in specific heat measurements of D- and L-valine by the differential scanning calorimeter. The phase transitions were reversible and reproducible. The differential peak was shown to be on the order 6.5×10^{-6} eV/ molecule • K = d ΔC_p/dT = d ΔE_{PV}/dT which was reflected in the slope difference of two Δ C_p vs T curves of D- and L-valine at T_c (Wang, 2000).

Magnetisation measurements on a SQUID showed a difference in the magnetic susceptibility (χ_ρ) as a function of temperature between the D- and L-alanine. DC and AC magnetic susceptibility measurements proved that the electron spin of α-H atom of D-/L-alanine is in a certain direction only in DC magnetic field. We studied the temperature dependence of AC mass susceptibility which showed that the curves χ_ρ vs T of both D-/L-alanine were horizontal without discontinuity (Wang et al. 2002). Based on the above experiments, we conjectured that the electron spin of α-H in D-alanine (D-valine) causes a flip at T_c by decreasing the hydrogen atom's total spin from one to zero then released the energy 6×10^{-6} eV. It coincides with the measured parity-violating energy difference 6.2×10^{-5} eV at T_c of valine (Rith *et al.* 1999).

The temperature-dependent DC-magnetic susceptibilities of D-/L-alanine show a contrary parity-violating role and bifurcation mechanism from 250 to 200K (Fig. 1).

^{1}H CRAMPS ssNMR was performed to study the temperature-dependent proton nuclei dynamics of D-/L-alanine and found that α-H nucleus is the active center in the parity-violating phase transition (Wang et al., 2003).

3. Results and Discussion

Our experimental results support Salam prediction of second-order phase transition involving dynamic symmetry breaking as an amplification mechanism. Salam proposed that α-H atoms give up their loose electrons and act as metallic hydrogen under T_c which was verified in ^{1}H CRAMPS ssNMR measurements of alanine and valine. Zanasi *et al.* (1998) have calculated the parity-violating potential E_{pv} of valine enantiomers (−16.525 Hartree) which was larger than that of alanine enantiomers (−12.318 Hartree). The higher PVED values result in an increase of T_c in Salam model (Buschmann *et al.*, 2000; Zanasi *et al.*, 1999). The

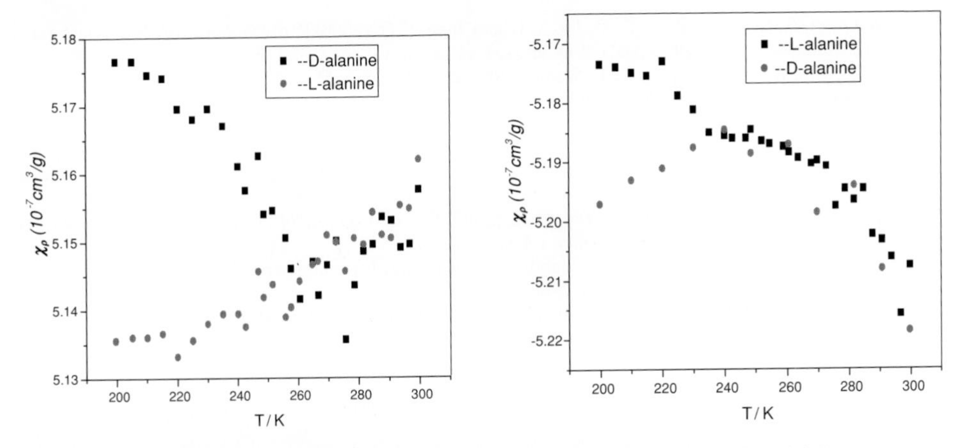

Figure 1. Temperature dependence of magnetic susceptibility at a field of (a) 1.0 T (b) -1.0 T.

transition temperature shift is observed in ^{1}H CRAMPS ssNMR. The T_c of valine is around 270K and that of alanine around 250K. A bifurcation-type mechanism in a chiraly-pure state is preferred to a direct configuration change from D- to L-amino acid at the critical temperature.

4. Acknowledgments

This research was supported by the grant of NSFC (20310202026).

5. References

Avalos, M., Babiano, R., Cintas, P., Jimenez, J. L. and Palacios, J. C. (2000) From Parity to chirality: chemical implications revisited, *Tetrahedron: Asymmetry.* **11**, 2845–2874.

Bada, J. L. (1995) Origins of homochirality, *Nature* **374**, 594–595.

Bonner, W.A. (2000) Parity violation and the evolution of biomolecular homochirality, *Chirality* **12**, 114–126.

Buschmann, H., Thede, R. and Heller, D. (2000) New developments in the origins of homochirality of biological relevant molecules, *Angew. Chem. Int. Ed.* **39**, 4033–4036.

Sullivan, R., Pyda, M., Pak, J., Wunderlich, B., Thompson, J.R., Pagni, R., Pan, H., Barnes, C., Schwerdtfeger, P. and Compton, R. (2003) Search for electroweak interactions in amino acids crystals. II. The Salam Hypothesis, *J. Phys. Chem. A* **107**, 6674–6680.

Figureau, A., Duval, E. and Boukenter, A. (1995) Can biological homochirality result from a phase transition? *Orig Life Evol. Biosphere* **25**, 211–217.

Kondepudi, D. K. and Nelson, G. W. (1985) Weak neutral currents and the origin of biomolecular chirality, *Nature* **314**, 438–441.

Keszthelyi, L. (2001) Homochirality of biomolecules: Counter-arguments against critical notes, *Origins of Life and Evolution of the Biosphere* **31**, 249–256.

Mason, S. (1985) Origin of biomolecular chirality, *Nature* **314**, 400–401.

Meiring W. J. (1987) Nuclear β-decay and the origin of biomolecular chirality, *Nature* **329**, 712–714.

Quack, M. (1989) Structure and dynamics of chiral molecules, *Angew. Chem. Int. Ed. Engl* **288**, 571–589.

Rith, K. and Schäfer, A. (1999) The mystery of nuclear spin, *Scientific American,* 58–63.

Salam, A. (1991) The role of chirality in the origin of life, *J. Mol. Evol.* **33**, 105–113.

Salam, A. (1992) Chirality, phase transitions and their induction in amino acids, *Phys. Lett. B.* **288**, 153–160.

Wang, W. Q., Yi, F., Ni, Y. M., Zhao, Z. X., Jin, X. L. and Tang, Y. Q. (2000) Parity violation of electroweak force in phase transitions of single crystals of D- and L-alanine and valine, *J. Biol. Phys.* **26**, 51–65.

Wang, W. Q., Min, W., Bai, F., Sun, L., Yi, F., Wang, Z. M., Yan, C. H., Ni, Y. M. and Zhao, Z. X. (2002) Temperature-dependent magnetic susceptibilities study on parity-violating phase transition of D- and L-alanine crystals, *Tetrahedron: Asymmetry* **13**, 2427–2432.

Wang, W. Q., Min, W., Liang, Z., Wang, L.Y., Chen, L. and Deng, F. (2003) NMR and parity violation: low-temperature dependence in ^{1}H CRAMPS and ^{13}C CP/MAS ssNMR spectra of alanine enantiomer, *Biophysical Chemistry* **103**, 289–298.

Wang W. Q., Gong, Y., Wang, Z. M. and Yan, C. H. (2003) Crystal structure of D-alanine at the temperature of 293, 270, 223 and 173K, *Chinese J. Struct. Chem.* **22(5)**, 539–543.

Zanasi, R. and Lazzeretti, P. (1998) On the stabilization of natural L-enantiomers of α-amino acids via parity-violating effects, *Chem. Phys. Lett.* **286**, 240–242.

Zanasi, R., Lazzeretti, P., Ligabue, A. and Soncini, A. (1999) Theoretical results which strengthen the hypothesis of electroweak bioenantioselection, *Phys. Rev. E* **59**, 3382–3385.

A MECHANISM FOR THE PREBIOTIC EMERGENCE OF PROTEINS
The Role of Proton Gradient and High Temperature in the Polymerization of Amino Acids Embedded in Bilayers

H.P. DE VLADAR[1], R. CIPRIANI[2], B. SCHARIFKER[3] and J. BUBIS[4]
[1]Centro de Biotecnología. Fundación Instituto de Estudios Avanzados, AP 17606, Parque Central. Caracas 1015-A, Venezuela. [2]Departamento de Estudios Ambientales. Universidad Simón Bolívar, Caracas-Venezuela. [3]Departamento de Química. Universidad Simón Bolívar, Caracas-Venezuela. [4]Departamento de Biología Celular. Universidad Simón Bolívar, Caracas-Venezuela.

1. Introduction

The first living organisms were not necessarily the result of the assembly of fully structured biochemical mechanisms involving macromolecules, but at least some life-related processes, as we know them today, probably appeared alongside the structural integration and early evolution of these proto-organisms. Along these ideas, we consider that spatial compartmentalization and protein functionality are tightly related in their origin. A possible scenario of this relationship is the early polymerization of amino acids (AA) embedded in amphiphilic membranes. The resulting membrane-embedded proto-proteins could have played an important role modulating the transport of elements or ions between the internal compartment and the environment. This scenario is congruent with selectivity arguments of AA (Hitz & De Luisi, 2000) (i.e., 20 out of nearly 70 originally available (Croning & Chang, 1993; Engel & Nagy, 1982)) and their homochirality (Hitz et al., 2001). Other mechanisms of peptide bond formation, such as alumina-catalyzed reactions (Bujdak & Rode, 2002), polymerization on clay surfaces (Bujdak & Rode, 1996) and polymerization mediated by thioesters (De Duve, 1996), can also lead to this scenario.

In this paper we present theoretical evidence that supports the possibility of polymerization of AA embedded in amphiphilic membranes by means of non-equilibrium, ion flux mechanism, totally independent of any DNA-mediated process.

2. Non-Equilibrium Fluxes Between Compartments

Our model is a system formed by two aqueous environments, namely the "inside" of the closed amphiphilic membrane, and the outside or "environment", with proton concentrations C_H^{in} and C_H^{env} respectively. A bilayered membrane of thickness Δx separates these environments. AA at a concentration C_{AA} are embedded in this membrane. In order to allow the amino acid polymerization to take place, the system is displaced from equilibrium,

J. Seckbach et al. (eds.), Life in the Universe, 83–87.

as occurs with the dynamics state of other living processes (e.g. Nicolis & Prigogine 1977).

We want to know if a simple entropic coupling is enough to displace the polymerization reaction affinity, A, toward the polymerization of AA. To test our hypothesis, we consider the near-equilibrium equations (De Groot, 1968) for the fluxes in our system:

$$\vec{J}_w = \frac{L_w}{T}\frac{\partial C_w}{\partial x} + \frac{L_i}{T}\frac{\partial C_H}{\partial x} \tag{1}$$

$$\vec{J}_H = \frac{L_i}{T}\frac{\partial C_w}{\partial x} + \frac{L_H}{T}\frac{\partial C_H}{\partial x} \tag{2}$$

$$J_{CH} = \frac{L_{CH}}{T}A \tag{3}$$

In these equations, $\vec{J}_q$ denotes the proton ($q = H$) and water ($q = w$) fluxes, both perpendiculars to the membrane's surface. J_{CH} denotes the chemical flux. The quantities L_i are the phenomenological coefficients (De Groot, 1968), and T is the absolute temperature. We constrain the system to the concentration difference between inside and environment through boundary conditions, and we let free $\vec{J}_w$ and J_{CH} to relax under the constraints. Following the law of minimum local entropy production (De Groot, 1968), our system attains a stationary state when

$$L_i\frac{\partial C_H}{\partial x} = -L_w\frac{\partial C_w}{\partial x} \tag{4}$$

$$A = 0 \tag{5}$$

The proton concentration profile is obtained by considering equations (5) and (2), and the boundary conditions $C_H(0) = C_H^{in}$ and $C_H(\Delta x) = C_H^{env}$:

$$C_H(x) = C_H^{in} + \frac{\Delta C_H}{\Delta x}x \tag{6}$$

with $\Delta C_H = C_H^{env} - C_C^{in}$. Replacing equation (4) into (1), leads to a constant distribution of water molecules, whose value is determined by the physical properties of the membrane. For a completely hydrophobic membrane, we have $C_w(x) = 0$.

3. Equilibrium Displacement of the Condensation Reaction

The polymerization reaction of the AA is of the form

$$H^+ + P_n + P_m \underset{K_f}{\overset{K_C}{\Leftrightarrow}} P_{n+m} + H_2O \tag{7}$$

in which P_k denotes an oligopeptide consisting of k AA residues. The quantities K_C and K_f are the condensation and fragmentation constants, independent of polymer size k. Following the Blatz-Tobolsky (1945) model for reversible polymerization, we can determine the net concentrations of peptidized residues by measuring the quantity of peptide bonds: $P_B = \sum_k (k-1)C_{P_k}$, and reactive groups (i.e., either N-term or C-term): $C_R = \sum_k C_{P_k}$:

$$\dot{C}_R = -\dot{C}_H = -\frac{1}{2}K_C C_H C_R^2 + \frac{1}{2}K_f C_w C_B, \tag{8}$$

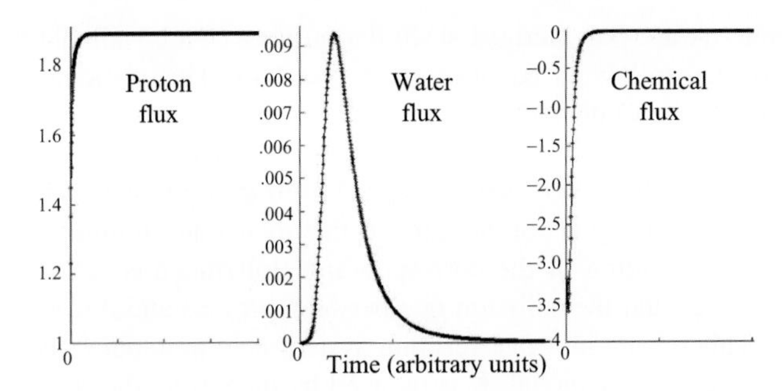

Figure 1. Proton flux (left) across the membrane rapidly atttains a stationary state determined by the pH difference across the membrane. Water (center) is initially produced by the polymerization reaction, but later it is eluted by diffusion. The chemical flux (right) attains a stationary state without chemical production, resulting from the limited number of AA able to polymerize.

Assuming that C_{AA} is constant, the equilibrium point of the process in equation (8) is

$$C_B = C_{AA} + \frac{\mathrm{K}\,\bar{C}_w(x_0)}{2\,\bar{C}_H(x_0)} - \left[\frac{\mathrm{K}\,\bar{C}_w(x_0)}{2\,\bar{C}_H(x_0)}\left(\frac{\mathrm{K}\,\bar{C}_w(x_0)}{2\,\bar{C}_H(x_0)} + 2C_{AA}\right)\right]^{1/2} \tag{9}$$

in which the concentrations $\bar{C}_w(x_0)$ and $\bar{C}_H(x_0)$ are those of the stationary state.

4. Discussion

In this model, the proton flux between the inside and the environment rapidly attains a stationary state and the proton concentration increases linearly from the inside to the environment (fig. 1). During the first stages of the process, the flux of water increases as a direct consequence of the synthesis of water molecules from the condensation reaction. Later, the molecules of water elute to the environment. At the stationary state of the system the flux of water is null. The results of a simulation (fig. 2) demonstrate that the stationary state of the chemical equilibrium corresponds to a null chemical flux. Since the membrane is totally hydrophobic, and the stationary concentration of water is zero, the displacement of the chemical affinity is maximum. Hence, the displaced reaction always reaches a stationary

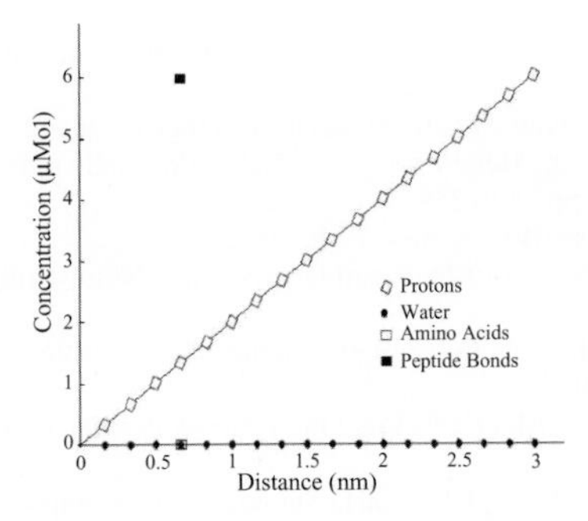

Figure 2. Stationary concentration profiles (in arbitrary units) of the four chemical species. AA and peptides are embedded at $x_0 = 2/3$ nm from the inner membrane's interface.

state in which all AA are polymerized since the number of all available reactive groups decreases to zero, i.e. following equation (9) $\bar{C}_B = C_{AA}$. This stationary distribution is attained regardless of the value of K_f.

The chemical rate theory (Hanggi et al., 1990) states that the velocity of reaction is $K = \kappa \exp(\Delta G^*/RT)$, where ΔG^* is the energy of the transition state. In a non-equilibrium process, this transition energy is not the same as that in an equilibrium process, thus ΔG^* is determined by the solution of the corresponding Boltzmann equation (Hanggi et al., 1990). The proton flux and the collision rate between protons and AA are high, for what ΔG^* decreases. This results in a fast-reacting system able to attain fully condensed AA in relative short times. In our model, K is reduced by increasing the collision rate and by increasing temperature. For this reason, our model is even more plausible in environments with high temperatures (Kasting, 1993; Rohlfing, 1976) or moderately high temperatures (i.e., below 100 °C) (Miller & Lazcano, 1995), such as those abundant scenarios in which the origin of life might have occurred.

5. Conclusions

We have presented theoretical evidence that supports the possibility of polymerization of AA embedded in membranes by means of a mechanism based on non-equilibrium fluxes: a maintained proton flux and a constant eflux of water, coupled with the vanishing chemical affinity, displaces the chemical equilibrium toward AA condensation. The process leads to the polymerization of all the AA embedded in the membrane, *regardless of the value of the fragmentation constant* K_f. Hence, the reaction *results in a large number of oligopeptides.* Systems with these properties allow linking chemical and biological evolution, since vesicles mixtures with diverse polymers on their surface become a rich scenario on which natural selection can act.

6. References

Blatz, P.L. and Tobolsky, T.B. (1945) Note on the Kinetics of Systems Manifesting Simultaneous Polymerization Phenomena, J. Phys. Chem. **49**, 77–80.

Bujdak, J. and Rode, B.M. (2002) Preferential amino acid sequences in alumina-catalyzed peptide bond formation, J. Inorg. Biochem. **90**, 1–7.

Bujdak, J. and Rode, B.M. (1996) The Effect of Smectite Composition on the Catalysis of Peptide Bond Formation, J. Mol. Evol. **43**, 326–333.

Cronin, J.R. and Chang, S. (1993) Organic matter in meteorites: molecular and isotopic analyses of the Murchison meteorite, In: M. Greenberg, C. X. Mendoza-Gomez, and V. Pirronello (eds.) *The Chemistry of Life's Origins*, Kluwer Academic Publishers, pp. 209–258.

De Duve, C. (1996) *Vital Dust*, Basic Books, New York, USA.

De Groot, S.R. (1968) *Thermodynamics of Irreversible Processes*, North-Holland Publishing Co., Amsterdam, Holland.

Engel, M.H. and Nagy, B. (1982) Distribution and enantiomeric composition of amino acids in the Murchinson meteorite, Nature **296**, 837–840.

Hanggi, P., Talkner, P. and Borkovec, M. (1990) Reaction-rate theory: fifty years after Kramers, Rev. Mod. Phys. **62**, 251–342.

Hitz, T., Blocher, M., Walde, P., and Luisi, P.L. (2001) Stereoselectivity Aspects in the Condensation of Racemic NCA-Amino Acids in the Presence and Absence of Liposomes, Macromolecules **34**, 2443–2449.

Hitz, T. and Luisi, P.L. (2000) Liposome-Assisted Selective Polycondensation of α-Amino Acids and Peptides, Biopolymers **55**, 381–390.

Kasting, J.F. (1993) Earth's Early Atmosphere, Science **259**, 920–926.

Miller, S.L. and Lazcano, A. (1995) The origin of life—did it occur at high temperatures? J. Mol. Evol. **41**, 689–692.

Nicolis, G. and Prigogine, I. (1977) *Self-Organization in Nonequilibrium Systems: From Dissipative Structures to Order through Fluctuations*, Wiley-Interscience, New York, USA.

Rohlfing, D.L., (1976) Thermal Polyamino Acids: Synthesis at Less Than 100 °C, Science **193**, 68–70.

FUNCTIONAL, SELF-REFERENTIAL GENETIC CODING

ROMEU CARDOSO GUIMARÃES[1] and CARLOS HENRIQUE COSTA MOREIRA[2]
[1]Dept. Biologia Geral, Inst. Ciências Biológicas, Univ. Federal Minas Gerais, Belo Horizonte MG 31270-901 Brasil, [2]Dept. Matemática, Inst. Ciências Exatas, Univ. Federal Minas Gerais, Belo Horizonte MG 30123-970 Brasil

Abstract. A model for the genetic code structure and organization is presented. It is self-referential due to starting without an mRNA to be translated. Recruitment of tRNAs occurs in pairs, one fishing the other, their anticodons being simultaneously codons for each other. Genes – mRNAs – arise in the process of formation of the code. It is also functional, depicting various consistent correlations with protein properties. First attributions – Gly, Pro, Ser – are the outliers from the hydropathy correlation, protein-stabilizing and RNA-binding amino acids. These properties allow formation of a stable RNP system, the source-product relationship being established. The succession of entries also obeys the following criteria: (a) synthetases class II to class I; (b) protein conformations from aperiodic to helices and then strands; (c) ordering of protein sequences with heads and tails, respectively, stable and unstable, called a non-specific punctuation system, in the second stage; (d) DNA-binding amino acids in the third stage; (e) late development of the specific punctuation system, that of initiation defining the stop signs, and (f) of the hexacodonic expansions of Leu and Arg.

1. Introduction

A model for the structure of the genetic code is presented, describing how the matrix was formed and organized.

This work had a first presentation in Guimarães (1996) and a description of the first stages in Guimarães and Moreira (2002). A concise account of the full model (Guimarães and Moreira, 2004) is shown here.

2. Self-reference and Function

The model is based on successive recruitment of pairs of anticodons of the palindromic type – with the same bases at the extremities (Table I).

It is indicated that the triplets AGA : UCU fished each other while Ser was octacodonic, before concession of YCU to Arg, and that the coherence of synthetases class I for Gln and Val was established after concession of YUG to Gln.

J. Seckbach et al. (eds.), Life in the Universe, 89–91.

TABLE 1. Anticodon pairs and synthetase class couples. pDiN underlined, stages numbered with letter indices, classes in Roman numerals.

AAA Phe UUU Lys 2 b II : II (both atypic)	AGA Ser UCU Ser 1 b II : II	ACA Cys UGU Thr 3 c I : II	AUA Tyr UAU Ile 4 a I : I
GAG Leu CUC Glu 2 a I : I	GGG Pro CCC Gly 1 a II : II	GCG Arg CGC Ala 3 a I : II	GUG His CAC Val 3 b I : I (Gln)

In Stage 1 the system is entirely self-referential – anticodons in a pair are simultaneously codons for each other – and there is no need for an mRNA to be translated.

In Stage 2, mRNAs are formed and structured, with heads and tails (non-specific punctuation; Guimarães, 2001). So, genes (mRNAs) are formed and defined concomitantly with formation of the coding system, when the proteins (now 'products') feed back upon the RNAs, through binding to them and stabilizing the nucleoproteins. Thereafter, the system developed auto-catalytic features.

The model is also functional, due to the consistency with various other protein properties. Stage 3 contains the amino acids most characteristic of DNA-binding motifs. The specific punctuation system and the hexacodonic expansions of Leu and Arg are placed in Stage 4.

Besides the characters specified, note the trajectories: from the GC-rich core to the periphery of the matrix; from synthetases class II to I, and the consistency with amino acid biosynthesis derivations. Sectors of pDiN are self-contained, the only cross-sector attribution being the hexacodonic Arg (Figure 1).

+AA Phe (2 b) Leu (4 c)	**+GA** Ser (1 b)	+CA Cys (3 c) Trp X (4 b)	+UA Tyr (4 a) X (4 b)
+AG Leu (2 a)	**+GG** Pro (1 a) (Gly)	+CG Arg (3 a) (Ala)	+UG His (3 b) Gln
+AC Val (3 b)	+GC Ala (3 a)	**+CC** Gly (1 a)	**+UC** As (2 a) Glu
+AU Ile (4 a) Met fMet (4 b)	+GU Thr (3 c)	**+CU** Ser (1 b) Arg (4 c)	**+UU** Asn (2 b) Lys

Figure 1. Formation of the genetic anticode. Bases in hydrophilicity A-G-C-U order. Sectors of principal dinucleotide (pDiN) types: mixed (Mx; quadrants +RY : +YR) and homogeneous (Ho, bold; +RR : +YY); +, the 5′ bases. Stages numbered with letter indices, in parenthesis. The amino acids in parenthesis are indicated to be previous occupiers of the respective boxes. Further explanations in the text.

3. Stages

Stage 1: (1 a) Gly $^{+CC:+GG}$ (Gly) Pro, (1 b) Ser $^{+GA:+CU}$ Ser.

The sector of Ho pDiN (quadrants: +RR : +YY) starts being filled. Amino acids are protein-stabilizers, RNA-binders and characteristic of coils and turns of proteins. Attributions are the outliers of the hydropathy correlation. All synthetases are class II. Gly may have preceded Pro in the +GG box, since Pro is derived biosynthetically from Glu.

Stage 2: (2 a) Asp then Glu $^{+UC:+AG}$ Leu, (2 b) Asn then Lys $^{+UU:+AA}$ Phe.

The hydropathy correlation is established. Synthetases make one couple of class I and the atypical couple: Lys class II, but class I in some organisms; Phe class II, but acylating in the class I mode.

Non-specific punctuation and mRNA structure are formed: amino acids of Stage 1 become heads of proteins, those of Stage 2 added as tails. Enter 3/8 of the amino acids of the set of protein helix-forming, 1/7 of the strand-forming.

Stage 3: (3 a) Ala $^{+GC:+CG}$ (Ala) Arg, (3 b) His then Gln $^{+UG:+AC}$ Val, (3 c) Thr $^{+GU:+CA}$ Cys then Trp.

Start the sector of mixed pDiN (quadrants: +RY : +YR). Amino acids are characteristic of DNA-binding motifs. Synthetases make one class I couple (Gln / Val) and the two couples of different classes – among the 8 pairs of boxes. Ala may have preceded Arg in the +CG box, which is a simpler alternative to other proposals for the predecessor to Arg.

Stage 4:

(4 a) Ile then Met $^{+AU:+UA}$ Tyr. All synthetases are class I.

(4 b) The specific punctuation system is formed: fMet with slipped pDiN C$\underline{\text{AU}}$ and codons +$\underline{\text{UG}}$; X anticodons U$\underline{\text{CA}}$ and Y$\underline{\text{UA}}$, pairing with the initiation codon, thereby competing with fMet, are deleted.

(4 c) The hexacodonic expansions of Leu and Arg synthetases class I, respectively, into Y$\underline{\text{AA}}$ and Y$\underline{\text{CU}}$, are formed, driven by codon usage.

4. Acknowledgments

Support from CNPq and FAPEMIG to Romeu Cardoso Guimarães.

5. References

Guimarães, R.C. (1996) Anticomplementary order in the genetic coding system, *Abstracts of the International Conference on the Origin of Life*, ISSOL, Orléans, 11: 100.

Guimarães, R.C. (2001) Two punctuation systems in the genetic code, In: J. Chela-Flores, T. Owen and F. Raulin (eds.) *First Steps in the Origin of Life in the Universe*, Kluwer, Dordrecht, pp. 91–94.

Guimarães, R.C. and Moreira, C.H.C. (2002) Genetic code structure and evolution: aminoacyl-tRNA synthetases and principal dinucleotides, In: G. Pályi, C. Zucchi and L. Caglioti (eds.) *Fundamentals of Life*, Elsevier, Paris, pp. 249–276.

Guimarães, R.C. and Moreira, C.H.C. (2004) The functional and self-referential genetic code, In: G. Pályi, C. Zucchi and L. Caglioti (eds.) *Progress in Biological Chirality*, Elsevier, Oxford, pp. 83–118.

IMPORTANCE OF BIASED SYNTHESIS IN CHEMICAL EVOLUTION STUDIES

A. NEGRON-MENDOZA[1], S. RAMOS-BERNAL[1], and F. G. MOSQUEIRA[2]
[1]Instituto de Ciencias Nucleares, UNAM, A.P. 70-453, Mexico, D.F. and
[2]D. Gral. Divulgación de la Ciencia, UNAM. A.P. 70-487, México D.F.

Abstract. The emergence of life needed a physical and chemical preamble. There is a large variety of experimental data to support the hypothesis for the abiotic formation of organic compounds. Although much knowledge has been obtained, many questions remain. One important factor in chemical evolution is related to the importance of random chemical synthesis *versus* more selective pathways forming compounds of biological relevance. Biased synthesis mechanisms would induce a much smaller space sequence in comparison to a space sequence derived from a purely random synthesis. Such condition in turn would render more feasible the emergence of a chemical system compatible with life.

In this paper we exemplify a biased synthesis that could have been relevant in primitive scenarios. We perform experiments simulating an organic compound adsorbed in soil and exposed to an energy source. Carboxylic acid is adsorbed in a clay mineral and is exposed to radiation. The chemical reaction induced in this system follows a preferential pathway of decomposition over others; both solid surfaces and radiation play an important role.

1. Introduction

In the study of the origin of life it is important to consider three facts. (1) There is life on Earth. (2) There is only one kind of life on the Earth: The central machinery in all organisms is built out of the same set of molecular components. Therefore, all known living things are rooted to a common ancestor. (3) Even the simplest living systems are highly organized and very complex (Cairn-smith, 1985). This may suggest that somehow the chemical components that formed a primitive chemical system compatible with life were fairly abundant. On the other hand, how could it be so if plausible prebiotic synthesis shows a variety of chemical products? For instance, if we analyzed a chemical reaction that goes randomness, it may form (a) by-products, (b) structural isomers and, (c) optical isomers. They decrease the yield of a specific compound. It is important then to explore chemical mechanisms able to narrow the variety of prebiotic chemical compounds liable to be used in the construction of primitive living systems.

In this work, we emphasize the importance of biased synthesis in prebiotic chemistry for the formation of biological molecules to reproduce a narrow set of monomers. We study a simple chemical system: a carboxylic acid adsorbed in a clay mineral and exposed to radiation.

J. Seckbach et al. (eds.), Life in the Universe, 93–96.

Our interest is twofold: (1) to show experimentally that biased synthesis might have occurred in geologically relevant conditions on the primitive Earth. (2) To evaluate the role that a solid surface plays to induce biased reaction.

1.1. IMPORTANCE OF A SOLID SURFACE IN PREBIOTIC SYNTHESIS

Despite the importance of gas and solution phase reactions in prebiotic monomer synthesis, it is difficult to envision the synthesis of complex macromolecules and their assembly into proto-cells without a multiphase system. Solid phase contribution is of great relevance to the experimental simulation of the prebiotic Earth (Ramos-Bernal and Negron-Mendoza, 1998). The most geologically relevant solid surfaces to promote chemical reactions on the primitive Earth were clays. They have several roles: (1) Adsorption of monomer that causes highly concentrate systems. (2) They facilitate condensation and polymerization reactions. (3) They have a surface that may serve as template for specific adsorption and replication of organic molecules.

To induce chemical reactions in these surfaces it is important to consider the energy source. In multiphase systems this source needs to be available, abundant and efficient to promote chemical reactions. Natural energy sources such as radioactive decay, cavitation and triboelectric energy (i.e., energy of mechanical stress) may have a great importance in those multiphase scenarios. This importance is due that they are penetrating energy sources (Negron-Mendoza et al., 1996).

2. Experimental Procedures

The experimental part is divided in two stages: (1) radiolysis of aqueous solutions of the carboxylic acid, and (2) study of the radiolysis of the water-carboxylic acid-clay system.

The carboxylic acids studied were malonic, succinic, acetic and aconitic acids, all related to metabolic pathways. The solid surface was sodium montmorillonite of Wyoming bentonite. The aqueous solutions of carboxylic acids were 0.1 M, at natural pH. The oxygen was removed by passing argon through the solutions. The irradiations were carried out in a gamma source. The radiation doses were from 46 to 300 kGy.

Samples with clay. One hundred milligrams of clay were mixed with 3 ml of the acid 0.1 M; this amount is below the cation exchange capacity of the clay. The samples were adjusted to pH 2, and they were shaken for 60 minutes. After this time, the samples were irradiated as described above. After irradiation, the sample was centrifuged and the supernatant was removed and analyzed by GC. A measured amount of the supernatant was evaporated until dryness. Methyl esters were then prepared according to Negrón-Mendoza *et al.*, 1983. The analysis was done by gas chromatography and GC-mass spectrometry.

3. Results and Discussion

3.1. SAMPLES WITHOUT CLAY

The irradiation of aqueous carboxylic acid produced many compounds. The central feature of these series of experiments was the production of the acid dimer, as the main way of

TABLE 1. Products Formed from the Irradiation of Aqueous Carboxylic Acids.

Acid	Principal reaction without clay	Main products	Principal reaction with clay	Main products
Acetic	Dimerization	Succinic acid	Decarboxylation	Methane, CO_2
Aconitic	Addition	Tricarballylic and citric acids	Decarboxylation	Itaconic acid, CO_2
Malonic	Dimerization	Succinic and carboxysuccinic acids	Decarboxylation	Acetic acid, CO_2
Succinic	Dimerization	Succinic dimer	Decarboxylation	Butanoic acid, CO_2

decomposition of the targeted compound. For example, in the radiolysis of acetic acid, the main product was succinic acid, the dimeric product. The decomposition of the target compound increased as a function of the dose.

3.2. SAMPLES WITH CLAY

In presence of clay the number of products identified decreased considerably. The main pathway of decomposition was a decarboxylation reaction; this is the loss of a carbon atom as CO_2. The production of the dimer decreased as a function of the dose.

Gaseous products were also detected and identified in all the systems under study: CO_2 and H_2. The production of CO_2 was greater in samples with clay. Its formation increased as function of the dose. The source of H_2 was the radiolysis of water and from the abstraction reactions produced during the radiolysis. Table 1 summarized the main results.

The results showed that the radiolysis of the clay-acid system follows a defined path rather various. The main lane of decomposition was the decarboxylation of the target compound rather than condensation/dimerization reactions.

4. Discussion and General Remarks

In an aqueous system the radiation interacts with the water molecules. Very reactive species are formed by this interaction (H, OH, e_{aq}, H_2 and H_2O_2). These attack in a secondary way to the carboxylic acid molecules, yielding the observed products.

In the systems without clay, the main reactions induced by radiation take place via free radicals. The main reaction was the dimerization. In contrast, in presence of clay, the results showed that there are changes in the mechanism. First, the number of products diminishes and the generation of CO_2 increases lineally with the radiation dose. Thus, the main reaction is the decarboxylation and products obtained are CO_2 and the corresponding acid with one carbon atom less than the targeted compound.

The present study represents a further attempt to gain more insight into the role-played by radiation-induced reactions in solid surfaces in chemical evolution studies. As in previous studies (Negrón-Mendoza and Ramos-Bernal, 1998), the results obtained suggest that the clay alter the reaction mechanism in a preferential way for some reactions, acting as moderator in energy transfer process.

We have showed experimentally that radiation-induced reactions in carboxylic acids are an example of preferential synthesis, while adsorbed in a solid surface. The radiolysis of the clay-acid system goes along a definitive path (oxidation) rather than following several modes of simultaneous decomposition. This behavior is important for prebiotic synthesis because solid surfaces can drive the reaction in a preferable path. These experiments prove that non-random products are produced under plausible prebiotic conditions, and in the presence of clay unequal or biased probability of reactions appear. This work was partially supported by a grant IN115501-3.

5. References

Cairns-Smith, A.G. (1985) *Seven Clues to the Origin of Life* Cambridge University Press, London pp. 4–6.

Negrón-Mendoza, A., Albarrán, G. and Ramos-Bernal, S. (1996) Clays as natural catalyst in Prebiotic Processes, In: J. Chela-Flores, and F. Raulin (eds.) *Chemical Evolution: Physics of the Origin and Evolution of Life*, Kluwer Academic Publishers, Dordrecht, pp. 97–106.

Negrón-Mendoza, A. and Ramos-Bernal, S. (1998) Radiolysis of Carboxylic Acids Adsorbed in Clay Minerals. *Radiation Physics and Chemistry*. **52**, 395–397.

Ramos-Bernal S. and. Negrón-Mendoza, A. (1998) Surface Chemical Reactions During the Irradiation of Solids: Prebiotic Relevance. *Viva Origino* **26**, 169–176.

WHEN DID INFORMATION FIRST APPEAR IN THE UNIVERSE?[1]

JUAN G. ROEDERER
Geophysical Institute, University of Alaska-Fairbanks
Fairbanks, AK 99775, USA

1. Introduction

Most scientists would assume that information has been playing a role right from the beginning—the Big Bang. As the Universe evolved, after the gradual condensation of atoms and molecules and the formation of planetary systems, "islands" of increasing complexity and organization appeared, containing discrete aggregates of condensed matter with well-defined boundaries and increasingly complex interactions with each other and their environment. Viewed this way, it indeed seems that the process of cosmic evolution itself is continuously generating information [Chaisson, 2001].

On second thought, however, aren't we talking here of information *for us the observers* or *thinkers*? Did information as such really play an active role in the fundamental physical processes that shaped the Universe? Was information and information-processing involved at all in the evolution of the Universe *before* living organisms started roaming around and interacting with it, and intelligent beings began studying it? When and where did information begin to play an active role, actually controlling processes in the Universe?

It is obvious that to address and answer these questions objectively we must first discuss in depth the concept of information and its meaning. We must find a definition of this ubiquitous and seemingly trivial concept that is truly objective and independent of human actions and human-generated devices, and which relates to its most fundamental property: that the mere presence of a *pattern* can trigger a macroscopic change in a system, and that this can happen repeatedly in a self-consistent manner. Traditional information theory does not help. It works mainly with communications and control systems and is not so much interested in an independent formal definition of information and its meaning as it is in a precise mathematical expression for the information content of a given message and its degradation during transmission, processing and storage. Shannon's theory is not apt to express the information content in many non-technical human situations, nor is it adequate to measure information in biochemical systems. Consider the case of a genome: each one of the innumerable combinations of nucleotides has an equal a priori probability of appearance in random chemical synthesis, and all sequences of the same length have the same Shannon information content. Yet the corresponding molecules would have drastically

[1] Abridged version of Roederer J. G., When and where did information first appear in the Universe? In *Bioinformatics*, J. Seckbach (ed.), Kluwer Acad. Publ., Dordrecht, The Netherlands, 2004.

J. Seckbach et al. (eds.), Life in the Universe, 97–100.

different functional significance; only a minute fraction would be biologically meaningful. In other words, what counts is *how information becomes operational*. For this purpose, the concept of *pragmatic information* was introduced, linking the pattern in a "sender" with the pattern-specific change triggered in a "recipient" (Küppers [1990]). It should be clear that information as a stand-alone concept has no absolute meaning: what counts is what information *does*, not what it is made of.

I shall try to accomplish the task of finding a comprehensive and objective definition of information by choosing the process of *interaction* as the underlying basic, primordial concept. I shall identify two fundamentally different classes of interactions between the bodies that make up the universe as we know it, with the concepts of information and information processing appearing as the key discriminators between the two.

2. Physical and Biological Interactions

It is our experience from daily life (and from precise observations in the laboratory) that the presence of one object may alter the state of other objects in some well-defined ways. We call this process an *interaction*—without attempting any formal definition (i.e., taking it as a "metaphysical primitive"). We just note that in an interaction a *correspondence* is established between certain properties of one object (its position, speed, form, etc.) and specific changes of the other. The interactions observed in the *natural* environment (i.e., leaving out all human-made artifacts) can be divided into two broad classes.

The first class comprises the physical interactions between two bodies, in which we observe that the presence of one modifies the properties of the other (its motion, structure, temperature) in a definite way that depends on the relative configuration of both components of the system (their relative positions, velocities, etc.). Primary physical interactions between elementary particles are two-way (true *inter*-actions) and reversible. A most fundamental characteristic is the fact that during the interaction, there is a direct transfer of energy from one body to the other, or to and from the interaction mechanism itself. In other words, the changes that occur in the two interacting bodies are coupled energy-wise. I shall call the physical interactions between inanimate objects *force-field driven interactions*.

For the second class of interactions, consider a dog walking around an obstacle. The dog responds to the visual perception of the obstacle, a complex process that involves information-processing and decision-making at the "receiver's" end with a definite *purpose*. At the obstacle's (the sender's) side, we have scattering and/or reflection of incident light waves; no information and no purpose are involved—only physical processes are at work here. There is no energy coupling between sender and receiver; what counts is *not* the energy of the electromagnetic waves but the *pattern* of their spatial distribution (determined by the obstacle's shape).

We call this class of interactions *information-based* or information-driven [Roederer, 2000]. Of course, the responsible mechanisms always consist of physical and/or chemical processes; the key aspect, however, is the *control* by information and information-processing operations. There is no direct energy coupling between the interacting bodies, although energy must be supplied locally for the intervening processes. In the first example the electromagnetic waves (light) themselves do not drive the interaction—it is the *information* in the patterns of the wave trains, not their energy, which plays the controlling role; the energy needed for change must be provided locally (by the organism). Quite generally, in all natural information-based interaction mechanisms the information is the *trigger* of physical and chemical processes, but has no relationship with the energy or energy flows needed by the latter to unfold. Physical and chemical processes provide a medium for information, but do not represent the information *per se*. The mechanisms responsible for this class of natural interactions must *evolve*; they do not arise spontaneously (in fact, Darwinian evolution itself embodies a gradual, species-specific information extraction from the environment). This is why natural information-driven interactions are all *biological* interactions.

3. Information and Life

Let us now formalize our description of information-based interactions, and point out both the analogies and differences with the case of physical interactions between inanimate bodies. First of all, we note that information-based interactions occur only between bodies or, rather, between systems the complexity of which exceeds a certain, as yet undefined degree. We say that system A is in information-based interaction with system B if the configuration of A, or, more precisely, the presence of a certain spatial or temporal *pattern* in system A (called the sender or source) causes a specific alteration in the structure or the dynamics of system B (the recipient), whose final state depends *only* on whether that particular pattern was present in A. The interaction mechanism responsible for the intervening dynamic physical processes may be integral part of B, and/or a part of A, or separate from either. Furthermore: (a) both A and B must be *decoupled energy-wise* (meaning that the energy needed to effect the changes in system B must come from sources other than energy reservoirs or flows in A); (b) *no lasting changes* must occur as a result of this interaction in system A (which thus plays a catalytic role in the interaction process); and (c) the interaction process must be able to occur *repeatedly* in consistent manner (one-time events do not qualify). In other words, in an information-based interaction a specific one-to-one *correspondence* is established between a spatial or temporal feature or pattern in system A and a specific change triggered in system B; this correspondence depends only on the presence of the pattern in question, and will occur every time the sender and recipient are allowed to interact (in this basic discussion I will not deal with stochastic effects).

We should emphasize that to keep man-made artifacts and technological systems (and also clones and artificially bred organisms) out of the picture, both A and B must be *natural* bodies or systems, i.e., not deliberately manufactured or planned by an intelligent being. We further note that primary information-based interactions are *unidirectional* (i.e., they are really "actions", despite of which I shall continue to call them *inter*actions), going from the source or sender to the recipient. While in basic physical interactions there is an energy flow between the interacting bodies, in information-based interactions any energy involved must be provided (or absorbed) by reservoirs *external* to the interaction process. Energy distribution and flow play a defining role in complex systems [Chaisson, 2001]; however, for information-based interactions, while necessary, they are only subservient, not determinant, of the process per se.

We must now provide a more formal definition: information is *the agent that mediates the above described correspondence*: it is what links the particular features of the pattern in the source system A with the specific changes caused in the structure of the recipient B. In other words, information represents and defines the uniqueness of this correspondence; as such, it is an irreducible entity. We say that "*B* has received information from *A*" in the interaction process. Note that in a natural system we cannot have "information alone", detached from any interaction process past, present or future: information is always there *for a purpose*—if there is no purpose, it isn't information. Given a complex system, structural order alone does not represent information—information appears only when structural order leads to specific change elsewhere in a consistent and reproducible manner, without involving any direct transfer or interchange of energy.

There are two and *only two* distinct types of natural (not made) information systems: biomolecular and neural. Bacteria, viruses, cells and the multicellular flora are governed by information-based interactions of the biomolecular type; the responses of individual cells to physical-chemical conditions of the environment are ultimately controlled by molecular machines and cellular organelles manufactured and operated according to blueprints that evolved during the long-term past. In plants they are integrated throughout the organism with a chemical communications network. For faster and more complex information processing in multicellular organisms with locomotion, and to enable memory storage of current events, a nervous system evolved which, together with sensory and motor systems, couples the organism to the outside world in real time.

Since in the natural world only biological systems can entertain information-based interactions, we can turn the picture around and offer the following definition [Roederer, 1978]: *a biological system is a natural (not-human-made) system exhibiting interactions that are controlled by information.* The proviso in parentheses is there to emphasize the exclusion of artifacts like computers and robots. By adopting the process of interaction as the primary algorithm, I am really going one step further and recognize information as the defining concept that *separates* life from any natural (i.e., not made) inanimate complex system. This answers the question asked in the title of this chapter: *information appears in the Universe only wherever and whenever life appears.*

4. References

Chaisson, E. J. (2001) *Cosmic Evolution: the Rise of Complexity in Nature*, Harvard University Press, Cambridge Mass.

Küppers B.-O. (1990) *Information and the Origin of Life*, The MIT Press, Cambridge Mass.

Roederer, J. G. (2000) Information, life and brains, in: J. Chela-Flores, G. Lemarchand and J. Oró (eds.), *Astrobiology*, Kluwer Acad, Publ., Dordrecht, The Netherlands, pp. 179–194.

IV. From the Miller Experiment to Chemical and Biological Evolution

PREBIOTIC ORGANIC SYNTHESIS AND THE EMERGENCE OF LIFE: FROM THE MILLER EXPERIMENT TO THE START OF BIOLOGICAL EVOLUTION

L. DELAYE, A. BECERRA, A. M. VELASCO, S. ISLAS and A. LAZCANO
Facultad de Ciencias, UNAM Apdo. Postal 70–407 Cd. Universitaria, 04510 México, D.F., México

1. Introduction

Laboratory experiments have shown how easy it is to produce a number of biochemical monomers under reducing conditions. Such empirical support began to accumulate in 1953, when Stanley L. Miller, then a graduate student working with Harold C. Urey at the University of Chicago, achieved the first successful synthesis of organic compounds under plausible primordial conditions The action of electric discharges acting for a week over a mixture of CH_4, NH_3, H_2, and H_2O; racemic mixtures of several proteinic amino acids were produced, as well as hydroxy acids, urea, and other organic molecules (Miller 1953). The easiness of formation in one-pot reactions of amino acids, purines, and pyrimidines strongly suggest these molecules and many others were components of the prebiotic broth (cf. Miller and Lazcano, 2002).

However, the leap from biochemical monomers and small oligomers to living cells is enormous. There is a major gap between the current descriptions of the primitive soup and the appearance of non-enzymatic replication. Solving this issue is essential to our understanding of the origin of the biosphere: regardless of the chemical complexity of the prebiotic environment, life could not have evolved in the absence of a genetic replicating mechanism insuring the maintenance, stability, and diversification of its basic components.

2. A Soup without Nucleic Acids?

The organic monomers of abiotic origin described in the previous sections would have accumulated in the primitive environment, providing the raw material for subsequent reactions. As shown by numerous experiments, clays, metal cations, organic compounds bearing like highly reactive derivatives of HCN (such as cyanamide, dicyanamide, and cyanogen) or imidazole derivatives may have catalyzed polymerization reactions (Wills and Bada, 2000). Selective absorption of molecules onto mineral surfaces has been shown to lead to a successful surface-bound template polymerization up to 53 nucleotides (Ferris et al., 1996), and other processes like evaporation of tidal lagoons (Wills and Bada, 2000) and eutectic

J. Seckbach et al. (eds.), Life in the Universe, 103–106.

freezing of dilute aqueous solutions (Kanavarioti et al., 2001) could have also assisted concentration of precursors.

Although the properties of RNA molecules make them an extremely attractive model for the origin of life, their existence in the prebiotic environment is unlikely. Thus, it is possible that the RNA world itself was the end product of ancient metabolic pathways that evolved in unknown pre-RNA worlds, in which informational macromolecules with different backbones may have been endowed with catalytic activity, i.e. with phenotype and genotype also residing in the same molecules, so that the synthesis of neither protein nor related catalysts is necessary (Orgel, 2003).

The chemical nature of the first genetic polymers and the catalytic agents that may have formed the pre-RNA worlds are completely unknown and can only be surmised (Orgel, 2003). Nonetheless, it is easy to imagine that as polymerized molecules became larger and more complex, some of then began to fold into configurations that could bind and interact with other molecules, expanding the list of primitive catalysts that could promote nonenzymatic reactions. Some of these catalytic reactions, specially those involving hydrogen-bond formation, may have assisted in making polymerization more efficient. As the variety of polymeric combinations increased, some compounds may could have developed the ability to catalyze their own imperfect self-replication and that of related molecules. This could have marked the first molecular entities capable of multiplication, heredity, and variation, and thus the origin of both life and evolution (Bada and Lazcano, 2002). This scheme is necessarily speculative, but its intrinsic heuristic value cannot be underscored. Experiments with ribozymes appear to support the possibility that random mixtures of catalytic and replicative macromolecules were available in the primitive Earth, and provide an excellent laboratory model for understanding the evolutionary transition from the non-living to living.

3. On the Early Evolution of Protein Biosynthesis

Protein biosynthesis is a complex, exquisitely tuned process involving a large number of different components, which must have evolved in a step-wise fashion through a series of simpler stages. However, no such intermediate stages or simplified versions of ribosome mediated protein synthesis have been discovered among extant organisms. However, the fact that RNA molecules are capable of perfoming by themselves all the reactions involved in peptide-bond formation suggests that protein biosynthesis evolved in an RNA world (Zhang and Cech, 1998), i.e., that the first ribosome lacked proteins and was formed only by RNA. This possibility is strongly supported by the crystallographic data that has shown that ribosome catalytic site where peptide bond formation takes place is composed solely of RNA (Nissen et al., 2000).

Clues to the genetic organization of primitive forms of translation are also provided by paralogous genes, which are sequences that diverge not through speciation but after a duplication event. For instance, the presence in all known cells of pairs of homologous genes encoding two elongation factors, which are GTP-dependent enzymes that assist in protein biosynthesis, provide evidence of the existence of a more primitive, less-regulated version of protein synthesis took place with only one elongation factor. The experimental evidence of *in vitro* translation systems with modified cationic concentrations lacking both elongation factors and other proteinic components (Spirin, 1986) strongly supports the possibility of an older ancestral protein synthesis apparatus prior to the emergence of elongation factors.

4. Pushing Back the Molecular Fossil Record?

How protein bisoynthesis actually evolved is still unknown. However, it is highly unlikely that the first proteins were complex enzymes with exquisitely, finely tuned catalytic activity. Although the first peptides that were synthesized biologically (i.e., via ribosome-mediated translation) could have been selected by two properties that are not mutually exclusive, namely chaperone-like properties or catalytic activity, it can be assumed that a very significant characteristic would have been the possibility of stabilizing RNA-catalytic conformations. This property would be the primitive equivalent to the stabilizing effect of the protein subunit of RNase P plays *in vivo*, and can be explored by a detailed analysis of extant RNA-binding sites.

Structural analysis of conserved motifs of some polymerases also appear to provide insights into evolutionary stages older than DNA itself. All DNA polymerases whose tertiary structure has been determined share a common overall architectural feature comparable to a right hand shape. Detailed analysis of the three-dimensional structures of the pol I, pol II, and reverse transcriptase families have shown that their palm subdomain, which catalyzes the formation of the phosphodiester bond, is homologous in all of them, while the fingers and thumb subdomains are different in all four of the families for which structures are known (Steitz, 1999). Homologous palm subdomains have also been identified in the viral T7 DNA- and RNA polymerases (Jeruzalmi and Steitz, 1998), indicating that it can catalyze the template-dependent polymerization of ribo- and of deoxyribonucleotides. As argued elsewhere (Delaye and Lazcano, 2000), the ample phylogenetic distribution of the catalytic palm subdomain and the flexible template- and substrate specificities of polymerases suggests that the conserved palm subdomain described above is one of the oldest recognizable components of an ancestral cellular polymerase, that may have acted both as a replicase and a transcriptase during the RNA/protein world stage.

5. Conclusions

Although there have been considerable advances in the understanding of chemical processes that may have taken place before the emergence of the first living systems, life's beginnings are still shrouded in mystery: how the transition from the non-living to the living took place is still unknown. However, analysis of RNA-binding motifs and conserved motifs of polymerases, among others, may provide insights into the RNA/protein world, providing insights into very early stages of biological evolution.

Furthermore, the high levels of genetic redundancy detected in all sequenced genomes imply not only that duplication has played a major role in the accretion of the complex genomes found in extant cells, but also that prior to the early duplication events revealed by the large protein families, simpler living systems existed which lacked the large sets of enzymes and the sophisticated regulatory abilities of contemporary organisms. These redundancies support the idea that primitive biosynthetic pathways were mediated by small, inefficient enzymes of broad substrate specificity (Jensen, 1976). Larger substrate ranges may had not been a disadvantage, since relatively unspecific enzymes may have helped ancestral cells with reduced genomes overcome their limited coding abilities.

6. Acknowledgments

Support from UNAM-DGAPA Proyecto PAPIIT IN 111003-3 to A. L. is gratefully acknowledged.

7. References

Bada, J. L. and Lazcano, A. (2002) Some like it hot, but not the first biomolecules. *Science* **296**, 1982–1983.

Becker, L., Blank, J., Brucato, J., Colangeli, L., Derenne, S., Despois, D., Dutrey, A., Ehrenfreund, P., Fraaije, H., Irvine, W., Lazcano, A., Owen, T., Robert, F. (2002) Astrophysical and astrochemical insights into the origin of life. *Rep. Prog. Physics* **65**, R1–R56.

Guerrier-Takada, C., Gardiner, K., Marsh, T., Pace, N., and Altman, S. (1983) The RNA moiety of ribonuclease P is the catalytic subunit of the enzyme. *Cell* **35**, 849–857.

Jensen, R.A. (1976) Enzyme recruitment in the evolution of new function, *Ann. Rev Microbiol.* **30**, 409–425.

Jeruzalmi, D. and Steitz, T. A. (1998) Structure of T7 RNA polymerase complexed to the transcriptional inhibitor T7 lysozyme. *EMBO Jour.* **17**, 4101–4113.

Kanavarioti, A., Monnard, P. A., and Deamer, D. W. (2001) Eutectic phases in ices facilitate non-enzymatic nucleic acid synthesis. *Astrobiology* **1**, 481.

Miller, S. L. (1953) A production of amino acids under possible primitive Earth conditions. *Science* **117**, 528.

Miller, S. L. and Lazcano, A. (2002) Formation of the building blocks of life. In: J. W. Schopf (ed), Life's Origin: The beginnings of biological evolution (California University Press, Berkeley), pp. 78–112.

Nissen, P., Hansen, J., Ban., N., Moore, P. B. and Steitz, T. A. (2000) The structural basis of ribosome activity in peptide bond synthesis. *Science* **289**, 920–930.

Orgel, L. E. (2003) Some consequences of the RNA world hypothesis. *Origins Life Evol. Biosph.* **33**, 211–218.

Spirin, A. S. (1986) Ribosome structure and protein synthesis (Benjamin/Cummings, Menlo Park), 414 pp.

Steitz, T. A. (1999) DNA polymerases: structural diversity and common mechanisms. *J. Biol.Chem.* **274**, 17395–17398.

Wächtershäuser, G. (1988) Before enzymes and templates: theory of surface metabolism. *Microbiological Reviews* **52**, 452–484.

Wills, C. and Bada, J. (2000) The Spark of Life: Darwin and the primeval soup (Cambridge), 291 pp.

Zhang, B. and Cech, T. R. (1998) Peptidyl-transferase ribozymes: trans reactions, structural characterization and ribosomal RNA-like features. *Chem. Biol.* **5**, 539–553.

ORIGIN AND EVOLUTION OF VERY EARLY SEQUENCE MOTIFS IN ENZYMES*

H. BALTSCHEFFSKY[1], B. PERSSON[2], A. SCHULTZ[1], J. R. PÉREZ-CASTIÑEIRA[3] and MARGARETA BALTSCHEFFSKY[1]

[1]Department of Biochemistry and Biophysics, Arrhenius Laboratories, Stockholm University, S-106 91 Stockholm, Sweden, [2]IFM Bioinformatics, Linköping University, S-581 83 Linköping and Centre for Genomics and Bioinformatics, Karolinska Institutet, S-171 77 Stockholm, Sweden, [3]Instituto de Bioquímica Vegetal y Fotosíntesis, CSIC-Universidad de Sevilla, Sevilla, Spain

1. Introduction

Conserved amino acid sequence motifs in enzymes often indicate involvement in the binding of metal ion(s) and/or in the binding and/or reactions of substrate(s). The four very early proteinaceous amino acids are glycine (G), alanine (A), aspartic acid (D) and valine (V) as was demonstrated with the clarification of the stepwise evolution of the genetic code (Eigen and Schuster, 1979), and in agreement with the quantities obtained in the earlier, classical work (Miller, 1953 and on) showing proteinaceous amino acid production under possible prebiotic conditions. Aspartic acid stands out as the unique very early amino acid containing an additional, highly reactive free charge, suitable *i.a.* for cation binding. Active site motifs with a very high content of any or all of these four amino acids may well be of early evolutionary significance.

2. Very Early Sequence Motifs

Some "early" proteins appear to harbor or reflect very early sequence motifs. For example, certain enzymes involved in inorganic pyrophosphate (PP_i) metabolism, as well as in that of ATP, seem to be of particular significance in this connection. Special attention has recently been given to the integrally membrane-bound, proton-pumping PP_i synthase, which in bacterial photophosphorylation is the first and still only known alternative to the ubiquitous ATP synthase in biological electron transport coupled phosphorylation (Baltscheffsky *et al.*, 1999). The putative active site of the PP_i synthase from the purple photosynthetic bacterium

* This paper is written in honor of Professor Stanley L. Miller.

J. Seckbach et al. (eds.), Life in the Universe, 107–110.

Rhodospirillum rubrum, in loop 5–6 (between transmembrane segments 5 and 6) has two nonapeptidyl sequences (DVGADLVGK and DNVGDNVGD), which are strongly conserved in the homologous enzyme family and which contain unusually many very early amino acids. Importantly, these sequences have charged amino acids regularly arranged in positions 1, 5 and 9, five of these six being aspartic acid. We have discussed the possible roles of these charged amino acids in the putative active site for the binding and reactions of the inorganic phosphates (Baltscheffsky *et al.*, 1999, 2001) and have described this with a preliminary sketch (Baltscheffsky *et al.*, 2001). Notably, other loops may also be involved in the active site for these reactions (Schultz and Baltscheffsky, 2003).

Recently, we counted the total number of occurrences of the sequence pattern **D**-A/G/V-A/G/V-A/G/V-**D**-A/G/V-A/G/V-A/G/V-**D** (as a possible predecessor of the nonapeptidyl sequence DVGADLVGK) in the Swissprot and TrEMBL databases (Boeckmann *et al.*, 2003) of October 2003. In total, we found 35 occurrences, which is four-fold more than could be expected by chance, indicating that this pattern is overrepresented, probably due to its functional or structural importance. We also checked the patterns

D–**G**–A/G/V–A/G/V–**D**–**G**–A/G/V–A/G/V–**D** and
D–A/G/V–A/G/V–**G**–**D**–A/G/V–A/G/V–**G**–**D**.

These patterns showed to be even more overrepresented to the extent of 15-fold and 27-fold, respectively. The extremely strong overrepresentation of the G-D pattern appears to be in agreement with our preliminary observation that this pattern occurs particularly frequently in certain enzymes involved in energy transfer. We also compared the somewhat more general patterns D/E-$(A/G/V)_n$-D/E-$(A/G/V)_n$-D/E, where n was set to 2, 3 or 4. For $n = 2$, we arrived at the expected number of sequences, while for $n = 3$ and $n = 4$, the occurrences were between four- and seven-fold, also indicating some evolutionary selection caused by functional or structural requirements.

A search for nonapeptidyl sequences in the Swissprot and TrEMBL databases, with D in positions 1, 5 and 9 and the other, very early A/G/V in the other positions, revealed that since 1996, several such sequences have been shown. DGGGDGGGD was an early found (Wen and Tseng, 1996) and dominating sequence, but also, from the year 2000 and on, sequences such as DAGGDAGGD, DAAGDAAGD, DVGGDAGGD and DVAGDVGGD have been reported. Several duplications at the tetrapeptide level and subsequent mutations would appear to have occurred. We interpret these findings as supporting our belief that the nonapeptide DVGADLVGK in PP_i synthase may have evolved from DVGADLVGD or, earlier, DVGADVGAD. The introduction of K should have been useful in connection with the binding of anionic phosphate oxygen.

Among the "very early" sequence motifs at active sites, that are found in some other PP_i metabolizing enzymes, and also, more or less similar, in some ATP metabolizing enzymes, we shall discuss those found in phosphofructokinases, which show the motif GGDD in those operating with PP_i and GGDG in those operating with ATP (Moore *et al.*, 2002). In the ^{174}GGDD^{177} sequence of the PP_i-dependent phosphofructokinase from *Borrelia burgdorferi*, ^{174}GGD are three conserved active site residues and D^{177} is one of four residues suggested to be directly involved in PP_i discrimination, by preventing ATP from binding.

3. Stepwise Motif Evolution

The considerations presented above indicate our belief that strongly conserved sequence motifs with a high content of the four "very early" proteinaceous amino acids may be most useful for extrapolating molecular evolution backwards, possibly all the way from present-day structures to those operating soon after the origin of life. In connection with the early, stepwise evolution of the genetic code, we shall briefly consider two questions.

The first concerns early polymers in molecular evolution. It has been suggested that high-molecular-weight polyphosphates may have preceded the nucleic acids and proteins (Kornberg, 1995). It is tempting to speculate that, in connection with the early operation of the first two triplets of the genetic code, the homopolymers polyglycine and polyalanine were produced. Could any of the three homopolymers discussed here, as intermediate stages between monomers and heteropolymers, have been useful in early evolution?

The second question is based on the apparently uninterrupted function of the triplet genetic code from the time of its early start to the present day and age. On the basis of the information now available about DNA, RNA and proteins it might be useful to emphasize both the continuity and what may be called the constructivity based on the functioning of the triplet code in biological evolution of life on the Earth.

4. A Principle of Constructive Continuity

In connection with early discussions of the origin of the genetic code (Crick, 1968) and the evolution of the genetic apparatus (Orgel, 1968) a Principle of Continuity was used as a plausible guide. This principle required that each stage in evolution develops "continuously" from the previous one, and it was argued that certain features of the contemporary genetic system emerged very early (Orgel, 1968).

All known life on Earth is based on DNA genomes, RNA, and protein enzymes. Today there are strong reasons to assume that these forms of life were preceded by a simpler life form, which in a so called "RNA world" was based primarily on RNA (Joyce, 2002). Whether a transition from such an RNA world first led to DNA genomes or to protein synthesis, the selection of proteinaceous amino acids in connection with the formation of the peptide bonds in protein synthesis may be assumed to have been steered by essentially the same triribonucleotide, or triplet, code which operates in the current forms of life.

The triplets coding for the two first of the four very early proteinaceous amino acids were probably GGC for glycine and GCC for alanine (Eigen and Schuster, 1978), and a transition mutation in the mid-letters gave GAC, coding for aspartic acid, and GUC, coding for valine, respectively. The continuity from the two first used triplets over the very early stepwise evolution to the 4-, 8-, 16-, 32-, and to the 64-triplets to their present-day utilization in biological evolution has led to the remarkable "tree of life", with which all known life is usually expressed (albeit with some slight variations). The increasing complexity clearly visible in over-all evolution and its constructivity may be briefly expressed in the following Principle of Constructive Continuity:

The triribonucleotide (triplet) code has from its origin and early evolution provided genetic information to all life on the Earth. This continuity of information transfer at the triplet level over time has withstood various continuities and discontinuities at other levels, divergence and convergence, determinism and chance, in gene and genome evolution. The function of the triplet code has persisted through both major destructive (catastrophic) and constructive (anastrophic) changes, while the biological evolution, in discrete and "plausible" steps, has resulted in increasingly complex ("higher") organisms, in a still ongoing constructive continuity at the fundamental level.

We consider both the Principle of Constructive Continuity and the anastrophe concept (Baltscheffsky, 1997) to be realistic, rather than, for example, "optimistic" or "philosophical", and hope they will be useful in connection with investigations of various evolutionary matters. Will the proposed Principle be applicable for other parts of the cosmic evolutionary process, on the one side physical and chemical evolution and on the other side social, human and individual evolution? This is an open question.

5. Acknowledgments

Support from Carl Tryggers Stiftelse för Vetenskaplig Forskning, Magnus Bergvalls Stiftelse, Stiftelsen Wenner-Grenska Samfundet, Linköping University and the Swedish Research Council is gratefully acknowledged.

6. References

Baltscheffsky, H. (1997) Major "anastrophes" in the origin and early evolution of biological energy conversion. J. theor. Biol. **187**, 495–501.

Baltscheffsky, M., Schultz, A. and Baltscheffsky, H. (1999) H^+-PPases: a tightly membrane-bound family. FEBS Lett. **457**, 527–533.

Baltscheffsky, H., Schultz, A., Persson, B. and Baltscheffsky, M. (2001) Tetra- and nonapeptidyl motifs in the origin and evolution of photosynthetic bioenergy conversion. Possible implications for the molecular origin of phosphate metabolism, In: J. Chela-Flores, T. Owen and F. Raulin (eds.) *First Steps in the Origin of Life in the Universe*, Kluwer Academic Publishers, Dordrecht, pp. 173–178.

Boeckmann, B., Bairoch, A., Apweiler, R., Blatter, M.C., Estreicher, A., Gasteiger, E., Martin, M.J., Michoud, K., O'Donovan, C., Phan, I., Pilbout, S. and Schneider, M. (2003) The SWISS-PROT protein knowledgebase and its supplement TrEMBL in 2003. Nucleic Acids Res. **31**, 365–370.

Crick, F.H.C. (1968) The origin of the genetic code. J. Mol. Biol. **38**, 367–379.

Eigen, M. and Schuster, P. (1978) The hypercycle. Naturwiss. **65**, 341–369.

Joyce, G.F. (2002) The antiquity of RNA-based evolution. Nature **418**, 214–221.

Kornberg, A. (1995) Inorganic polyphosphate: Toward making a forgotten polymer unforgettable. J. Bacteriol. **177**, 491–496.

Miller, S.L. (1953) A production of amino acids under possible primitive Earth conditions. Science **117**, 528–529.

Moore, S.A., Ronimus, R.S., Robertson, R.S. and Morgan, H.W. (2002) The structure of a pyrophosphate-dependent phosphofructokinase from the lyme disease spirochete *Borrelia burgdorferi*. Structure **10**, 659–671.

Orgel, L.E. (1968) Evolution of the genetic apparatus. J. Mol. Biol. **38**, 381–393.

Schultz, A. and Baltscheffsky, M. (2003) Properties of mutated *Rhodospirillum rubrum* H^+-pyrophosphatase expressed in *Escherichia coli*. Biochim. Biophys. Acta (in press).

Wen, F.S. and Tseng, Y.H. (1996) Nucleotide sequence of the gene presumably encoding the adsorption protein of filamentous phage phi Lf. Gene **172**, 161–162.

THE LIPID WORLD: FROM CATALYTIC AND INFORMATIONAL HEADGROUPS TO MICELLE REPLICATION AND EVOLUTION WITHOUT NUCLEIC ACIDS

ARREN BAR-EVEN, BARAK SHENHAV, RAN KAFRI and DORON LANCET
Department of Molecular Genetics and the Crown Human Genome Center, the Weizmann Institute of Science, Rehovot 76100, Israel

1. Lipid World And The GARD Model

A widespread notion is that life arose from a single molecular replicator, probably a self-copying polynucleotide, in an RNA World (Joyce, 2002). We have proposed an alternative Lipid World scenario as an early evolutionary step in the emergence of cellular life on Earth (Segre *et al.*, 2001). This concept combines the potential chemical activities of lipids and other amphiphiles, with their capacity to undergo spontaneous self-organization into supramolecular structures, such as micelles and bilayers. In quantitative, chemically-realistic computer simulations of our Graded Autocatalysis Replication Domain (GARD) model (Segre *et al.*, 1998), we have shown that prebiotic molecular networks, potentially existing within assemblies of lipid-like molecules, manifest a behavior similar to self reproduction or self-replication.

Because amphiphile assemblies may form readily and spontaneously under prebiotic conditions (Deamer, 1985), the Lipid World scenario may represent an intermediate "mesobiotic" phase, bridging an a-biotic random collection of organic molecules with a biotic protocell that contains long biopolymers, as well as more intricate information storage, catalysis and replication (Shenhav *et al.*, 2003). In our model, lipid-like amphiphiles may possess a very large variety of chemical structures, including head-groups that resemble amino-acids or nucleotides. Catalysis is proposed to be exerted by such diverse chemical moieties, enhancing amphiphile exchange rates as well the formation of more complex head-groups with similarity to peptides or oligonucleotides. In a more recent version of our model (Polymer GARD or P-GARD), a path is delineated for the gradually increased dependence on linear molecular sequences (Shenhav *et al.*, in press).

A most central notion in our P-GARD model is that life began with monomers only, and later proceeded to evolve biopolymers. This is in strict contrast to the widespread view, epitomized by the RNA World concept, whereby the abiotic formation of biopolymers is a prerequisite for life. In other words, RNA World or protein World proponents assert that a molecular entities devoid of biopolymers cannot possess the basic attributes of life. This credo derives from the premise that a living entity has to contain information and transmit it to progeny via replication, and that such a sequence of events cannot occur

J. Seckbach et al. (eds.), Life in the Universe, 111–114.

without long informational molecules that form similes by base pairing, i.e. without nucleic acids.

The unorthodoxy of the Lipid World notion thus goes beyond the question of what chemical entities formed the first replicators – nucleotides, amino acids or lipids. In fact, Lipid World in its most general sense does not prescribe a particular chemistry, and only suggests that the first events en route to life were mediated by molecules capable of readily forming non-covalent aggregates. This allows a much wider diversity of compounds to take part in the first events leading to life, hence generates a more plausible scenario for life's beginning in a highly heterogeneous primordial "soup".

In the basic model of monomer GARD, quasi-stationary assembly compositions (composomes) are shown to survive many splits, before giving rise to other quasi-stationary state or being scattered into "random drift" composition (Segre *et al.*, 2000). Two major ideas underlie the GARD model. First is the notion of compositional information, namely that information can be stored in the distribution of compounds within the entity, rather by a sequence embodied within each compound. In other words, information is not a characteristic of an individual molecule, but of the system as a whole. Second is the capacity of that information to be transmitted along generations with acceptable fidelity. The transition between one composomal state and another can be viewed as an evolution-like process, in which one organism gives rise to another due to stochastic progression.

2. Fitness Parameters and Populations of Assemblies

Previously, we have mainly dealt with individual GARD assemblies, and at each split, one was randomly selected for further scrutiny. However, evolution is mostly about populations of organisms, displaying "ecological" relationships such as competition. We therefore have initiated a study of the properties of GARD assembly populations. Because of computing power limitations it is imperative to generate a phenomenological description of every composome in the realm of a higher level of analytical hierarchy. This is done by describing every composome with three kind of emergent property parameters (Shenhav *et al.*, in press). First is growth time, T_i, the time elapsing between assembly divisions, while the assembly is in a specific composomal state i. Second is the survival perameter (S_i), namely the probability of an assembly to preserve its composomal state following a split. Last is the emergence likelihood (E_i), which is the probability of entering a composomal state from another one or from random drift (a state in which the assembly possesses no composomal state). In a given population of assemblies and for specific environment attributes, these fitness parameters will provide a good approximation of the composomes population dynamics. In fact, it is possible to display this dynamics in a set of linear differential equations, using an "ecological" matrix that can be constructed using the three fitness parameters.

Our simulations showed that in most cases there is a dominant composome, in term of the S_i and E_i, though other composomes might be much faster (small T_i) and therefore can be present in the population in a significant fraction (Fig. 1). Moreover, repeated simulations demonstrated that if the system is left to evolve without any constraint, the population distribution reaches a steady state, that could be interpreted as a stable "eco-system". In any given set of parameters this state is unique, regardless of the initial population seeded. Imposing population constraints, such as constant population, prevents the system from converging

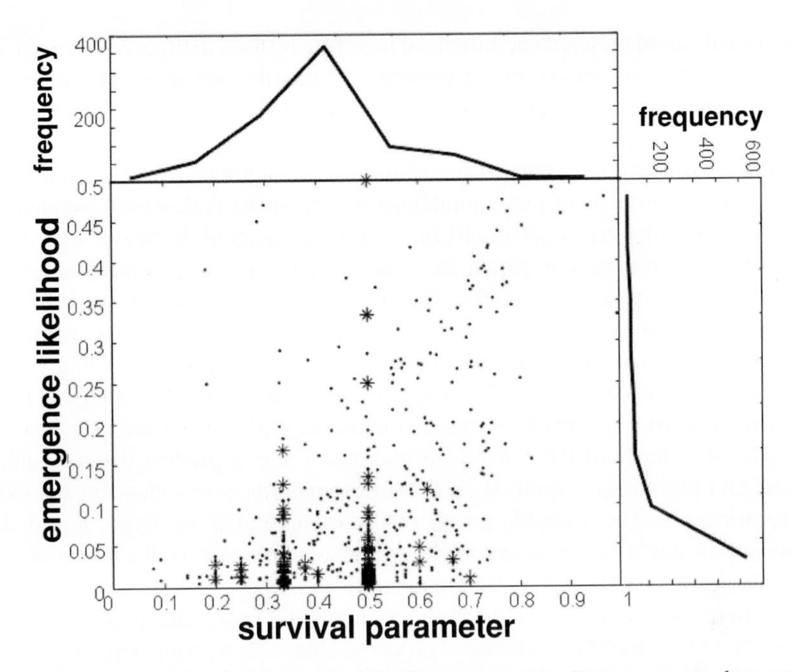

Figure 1. Figure 1. Correlation between two of the fitness parameters. Dots represents slow composome (large T_i) and asterisks denote the top 20% fastest composomes (low T_i). Overall, there are 786 composomes, obtained from 100 Dimer-GARD simulations with different catalytic enhancement matrices, all drawn from the same catalytic values distribution (Segre *et al.*, 2000). The survival parameter (S_i) values behave roughly as a normal distribution, while the emergence likelihood (E_i) values obey an exponential-like distribution. There is a rough linear correlation between those two fitness parameters. It is interesting to note that the dominant composomes (having both large Si and Ei) tend to be slow, allowing the fast growing ones, located mostly in the middle bottom of the graph, to successfully compete, resulting with their significant representation in the population.

on one state for long periods, though quasi-stationary population distribution can survive for substantial time before giving rise to an other. Interestingly, in the linear approximation of the "eco-system" dynamics that we have begun to pursue, the dominant population distribution corresponds to the all-positive eigen-vector of the "ecological matrix", in analogy to the dominant GARD composome being related to the all-positive eigen-vector of the catalytic-enhancements beta matrix (Segre *et al.*, 2000).

3. Towards Increasing Complexity—Synthesis of Polymers

Our Monomer-GARD simulations (Segre *et al.*, 2000) have a limited capability of creating metabolism-like systems with relatively high complexity, which is one of the basic features of evolution. In order to generate possible trajectories towards ever increasing complexity, we have introduced into our simulations a more elaborate form of metabolism, namely the capability of monomer to join each other to form oligomers. The basic interactions logic remains unchanged, and the only additional type of allowed interaction among molecules is the catalytic enhancement upon the oligomerization and bond-breaking rates.

First we have used monovalent dimer and therefore restricted oligomerization to dimers only (Shenhav *et al.*, in press). Fig. 1 presents the distribution of the fitness parameters calculated for the 786 composomes obtained in one hundred such Dimer-GARD simulations with different catalytic enhancement matrices.

Introducing divalent monomers, capable of generating polymer with unlimited length, highlights a central problem of such simulations, namely how to relate the dependence of the catalytic capacity of longer molecules to the catalytic capacity of the monomers composing it. At the one end are models in which the catalytic enhancement of a molecule is the sum of those of its monomers, and therefore the oligomerization contributes minimally to the overall catalysis in the system. An alternative scenario is that the catalytic characteristics of an oligomer are unrelated to those of its monomers, i.e. potencies are randomly assigned. The models of Dyson (Dyson, 1999) and Kaufmann (Kauffman, 1993) are good examples of these two options, respectively. At another extreme, oligomer-induced catalysis is non-linearly related to those of the constituent monomers, e.g. a product thereof, underlying synergism. Our preliminary results show that in a purely linear model, stable composomes do not form readily. The desirable model should compromise two opposing trends—the ability to sustain stable composomes while giving enough space for the evolution of more complex quasi-stationary states, in the course of time.

Our main research objective is to define a basic set of interactions rules so that primitive genetic memory will become an emergent property, alongside with an increased complexity of the system. We anticipate that this could be achieved by a carefully designed set of oligomerization rules. It is not impossible that without synergism the system will not develop significance memory and complexity. Implementation of extended oligomer-based models could help bridge the gap between simple and close-ended systems to more elaborate ones that are essentially open-ended and therefore can serve as a true platform for the origin of life.

4. References

Deamer, D.W. (1985) Boundary Structures are Formed by Organic Components of the Murchison Carbonaceous Chondrite. Nature **317**, 792–794.

Dyson, F. (1999) *Origins of life*, Cambridge University Press, Cambridge.

Joyce, G.F. (2002) The antiquity of RNA-based evolution, Nature **418**, 214–221.

Kauffman, S. (1993) *The origin of order*, Oxford University Press.

Segre, D., Lancet, D., Kedem, O., Pilpel, Y. (2001) Graded autocatalysis replication domain (GARD): Kinetic analysis of self-replication in mutually catalytic sets. Orig. Life Evo. Biosph. **28**, 501–514.

Segre, D., Ben-Eli, D. and Lancet, D. (2000) Compositional genomes: Prebiotic information transfer in mutually catalytic noncovalent assemblies. Proc. Natl. Acad. Sci. USA **97**, 4112–4117.

Segre, D., Ben-Eli, D., Deamer D.W. and Lancet, D. (2001) The lipid world. Orig Life Evol. Biosph. **31**, 119–145

Shenhav, B., Segre D. and Lancet D. (2003), Mesobiotic emergence: Molecular and ensemble complexity in early evolution. Adv. Complex Syst. **6**, 15–35.

Shenhav, B., Bar-Even, A., Kafri, R. and Lancet, D., Polymer GARD: simulation of covalent bond formation in reproducing molecular assemblies. Orig Life Evol. Biosph, in press.

COENZYMES IN EVOLUTION OF THE RNA WORLD

M. S. KRITSKY, T. A. TELEGINA, T. A. LYUDNIKOVA and Yu . L. ZEMSKOVA
A. N. Bach Institute of Biochemistry, Russian Academy of Sciences, Leninsky Prospekt 33 (Bldg.2), Moscow, 119071, Russian Federation

1. Introduction

One of the main problems of the origin of life studies is chemical interpretation of the emergence of organisms capable of both keeping genetic information and catalyzing chemical reactions. The discovery of catalytically active RNAs, ribozymes, has permitted to assume, that early life was based on polyribonucleotides which could have served as their own genes and also performed catalytic functions in the absence of genetically ordered proteins (Gilbert, 1986). Ribozymes isolated from organisms and selected *in vitro* from random polynucleotides demonstrate an impressive repertoire of catalytic activities, which could give rise to the development of replication and translation mechanisms (Bartel and Unrau, 1999). Some important reactions which protein enzymes do catalyze, in particular, redox processes that maintain energy metabolism in organisms, were not found in ribozymes. We believe that such catalytic inferiority of polynucleotides could be compensated by the attachment of coenzymes, the low molecular weight substances that combine in modern cells with apoproteins to form active enzymes. When energized by light, some coenzymes are efficient catalysts of electrons or chemical groups transfer in absence of apoproteins and thus could play a key role in RNA world metabolism (Kritsky *et al.*, 1998; Kritsky and Telegina, 2004).

2. Photochemical Activity of Coenzymes

The free molecules of coenzymes are significantly less reactive than when combined into a complex with their apoprotein, but excitation by light strongly enhances their reactivity. In addition to UV-radiation with a wavelength maximum at $250 \div 280$ nm absorbed by nucleic acid bases, flavins, pterins and reduced nicotinamides absorb UV-light of the area between 300 and 400 nm, and flavins are excited by visible light up to about 500 nm as well. Photoexcitation strongly enhances the reactivity of coenzymes, The absorption of photon by coenzyme heterocycle leads to the molecule transition to exited singlet states, which are efficiently converted into triplet states. From the standpoint of prebiotic chemistry, the latter are of more interest, because they can interact with other molecules in solution, whereas the excited singlet states are chemically quenched only in structurally ordered proteins.

J. Seckbach et al. (eds.), Life in the Universe, 115–118.

The midpoint redox potential value of flavins (E'_o) is -0.22 V, *i.e*. reduced flavins are rather strong reductans. After transition to triplet state the E'_o of the pertinent redox pair is shifted to $+1.85$ V because of energization of the oxidized form of the pair. Strongly electronophilic, it subtracts electrons from donors including those with highly positive E'_o values. Pterins demonstrate similar properties; their lowest triplet state gets an additional energy of about 2.50 eV, whereas reduced forms, especially tetrahydropterins are strong reductants. The photochemical reactions of excited flavins and pterins involve a free radical mechanism and lead to a formation of metabolic reductants, dihydroflavins and dihydro- and tetrahydropterins, as well as to adduce the donor molecule's radical (Heelis, 1982; Kritsky *et al*., 1997). The reduced nicotinamide coenzymes have been shown to sensitize the uphill electron transfer, too (Nikandrov *et al*., 1978).

In the presence of inorganic electron acceptors such as dioxygen (O_2) and nitrate or organic molecules, for example, Fe^{3+}-cytochrome *b*, excited flavins and pterins catalyze donor-to-acceptor electron transfer. Although O_2 sharply decreases the viability of reduced coenzymes and changes the mechanism of photoreactions due to the generation of singlet oxygen ($^1O_2$) and other reactive oxygen species, the reactions sensitized by excited flavins and pterins in oxygenic atmosphere lead to accumulation of free energy in products like they do in anaerobic conditions. Besides the reactions proceeding in aqueous solution, excited lipophilic flavin embedded in a lipid membrane was shown to sensitize translocation of redox equivalents across this membrane. Moreover, considering a strong difference of proton dissociation constants of the ground state and excited molecule, the excitation and relaxation cycle of flavins was predicted to act in light-dependent transmembrane translocation of proton (Schmidt, 1984).

3. The Attachment of Coenzymes to Polynucleotides

By using modern methods of *in vitro* randomization, selection and amplification of polynucleotides a number of RNA aptamer complexes, a type of synthetic oligonucleotides that non-covalently bind a particular target molecule, were synthesized and selected for their affinity to riboflavin, flavin mononucleotide (FMN), flavin adenine dinucleotide (FAD), nicotinamide mononucleotide (NMN), nicotinamide dinucleotide (NAD)and reduced forms of latter two molecules (Lauhon and Szostak, 1995; Fan *et al*., 1996; Roychowdhury-Saha *et al*., 2002). The nucleic acid strands with covalently attached coenzymes can arise from chemical and enzymatic reactions. For example, synthetic pteridine analogs of purine bases were incorporated into non-terminal positions of nucleic covalent linkage by automated DNA synthesis (Hawkins, 2001). In other experiments coenzymes were covalently attached at the ends of RNA molecules in reactions catalyzed by protein and RNA catalysts. NAD and FAD substituted adenosine triphosphate in the initiation of the template-dependent RNA synthesis catalyzed by *Escherichia coli* RNA-polymerase (Malygin and Shemyakin, 1979). A shortened version of a *Tetrahymena thermophila* ribozyme was shown to catalyze the self-incorporation of NAD (Breaker and Joyce, 1995). Catalytic RNA was shown to create 5′, 5′-pyrophosphate linkages with FMN and nicotinamide adenine dinucleotide phosphate (NADP) (Huang and Yarus, 1997). Terminal transferase from calf thymus accepted nicotinamide nucleoside 5′-triphosphate and efficiently added the NMN residue to the 3′-end of the oligodeoxyribonucleotide, and the T4 polynucleotide kinase accepted NMN as substrate.

The same study showed that reduced NADP was an excellent substrate for polynucleotide phosphorylase of *Micrococcus luteus* in polymerization and in primer extension reactions (Liu and Orgel, 2000).

At first glance coenzyme heterocycles show a worse spatial conformity to RNA structure than the common bases. However, the ketone groups of flavins and amido group of nicotinamide have been reported to participate in hydrogen base pairing with nucleic acid bases (Liu and Orgel, 2000; Fan *et al.*, 1996). Due to a closeness of pteridine structure to purine, the presence of this uncommon base was minimally disruptive for DNA double-stranded structure (Hawkins *et al.*, 2001).

4. On the Role of Photoreactive Coenzymes in Evolution

The simulation of processes on primitive Earth has shown that catalytically significant heterocycles of coenzymes were available from chemical evolution (Oró, 2001). Although the ability of RNA-attached coenzymes to drive primitive metabolism at the expense of captured energy of solar radiation can be regarded as a positive selection character, which favored the further development of the system, its negative consequence was an increased risk of photodegradation of the bases situated in vicinity of the coenzyme's heterocycle. It is noteworthy to recall that flavins and pterins, but not nucleic bases are active $^1O_2$ generators (Neverov *et al.*, 1996).

After having been rejected from polynucleotide carriers because of their relative structural unfitness as compared to canonic nucleotides, coenzymes have managed to survive in evolution by changing their host polymer. The development of genetically ordered polypeptides had to lead to the appearance of amino acid sequences displaying affinity to coenzymes. In its turn, this affinity could have become a strongly favorable characteristic encouraging the survival of those organisms, whose genotypes were capable to encode such polypeptides, since their presence provided a wider set of catalytic activities. Flavins and pterins have retained, however, their role of photochemically active redox reagents. In modern metabolism, besides their function in enzyme catalysis, these coenzymes act as photosensors governing the activity of some enzymes and protein transcription factors (De la Rosa *et al.*, 1989; Christie and Briggs, 2001; Iseki *et al.*, 2002; Cheng *et al.*, 2003). Considering the structural diversity of photochemically active coenzyme-binding proteins, it is reasonable to believe, that they have not arisen from a common ancestor, but are products of convergent evolution.

5. Summary

According to the RNA world hypothesis, RNAs in primitive organisms have served as their own genes and also performed catalytic functions in metabolism. However, some catalytic activities necessary for development of metabolism, *e.g.* oxidoreductases, were not found in ribozymes. It is suggested that catalytic inferiority of RNA world could be compensated by the activity of coenzyme molecules attached to polynucleotides and displaying photocatalytic functions.

6. Acknowledgement

Support from Russian Foundation for Basic Research Grant 01-04-48268a is acknowledged.

7. References

Bartel, D.P. and Unrau, P.J. (1999) Constructing an RNA world, Trends Biochem. Sci., **241**, 9–13.

Breaker, R.R. and Joyce, G.F. (1995) Self-replication of coenzymes by ribozymes, J. Mol. Evol., **42**, 551–558.

Cheng, P., He, Q., Yang., Y, Wang, L., and Liu, Y. (2003) Functional conservation of light, oxygen, or voltage domains in light sensing, Proc. Natl. Acad. Sci. USA, **100**, 5938–5943.

Christie, J.M. and Briggs, W.R. (2001) Blue light sensing in higher plants, J. Biol. Chem, **276**, 11457–11460.

De la Rosa, M.A., Roncel, M., and Navarro, J.A. (1989) Flavin-mediated photoregulation of nitrate reductase. A key point of control in inorganic nitrogen photosynthetic metabolism, Bioelectrochem. Bioenerg., **27**, 355–364.

Fan, P., Suri, A.K., Fiala, R., Live, D., and Patel, J.D. (1996) Flavin recognition by an aptamer targeted toward FAD, J. Mol. Biol., **258**, 480–500.

Gilbert, W. (1986) The RNA World, Nature, **319**, 618.

Hawkins, M.E. (2001) Fluorescent pteridine analogs—a window on DNA interaction, Cell. Biochem. Biophys., **34**, 257–281.

Iseki, M., Matsunaga, S., Murakami, A., Ohno, K., Shiga, K., Yoshida, K., Sugai, M., Takahashi, T., Hori, T., and Watanabe, M. (2002) A blue-light-activated adenylyl cyclase mediates photoavoidance of *Euglena gracilis*, Nature, **415**, 1047–1051.

Kritsky, M.S., Lyudnikova, T.A., Mironov, E.A., and Moskaleva, I.V. (1997) The UV radiation-driven reduction of pterins in aqueous solution, J. Photochem. Photobiol., **B**: Biol., **48**, 43–48.

Kritsky, M.S., Kolesnikov, M.P., Lyudnikova, T,A., Telegina, T.A., Otroshchenko, V,A., and Malygin, A.G. (2001) The nucleotide- and nucleotide-like coenzymes in primitive metabolism, photobiology and evolution. In: J. Chela-Flores, T. Owen., and F. Raulin (eds.). *First Steps in the Origin of Life in the Universe*. Kluwer Academic Publishers, Dordrecht, pp 237–240.

Kritsky, M.S. and Telegina, T.A. (2004) Role of nucleotide-like coenzymes in primitive evolution. In: J. Seckbach (ed.) *Origins: Genesis, Evolution and Diversity of Life,* Kluwer Academic Publishers, Dordrecht, pp 000–000 *(in press)*.

Lauhon, C.T. and Szostak, J.W. (1995) RNA aptamers that bind flavin and nicotinamide redox cofactors, J. Am. Chem. Soc., **117**, 1246–1257.

Liu, R. and Orgel, L.E. (2000) Enzymatic synthesis of polymers containing nicotinamide mononucleotide, Nucl. Acid. Res., **23**, 3742–3749.

Malygin, A.G. and Shemyakin, M.F. (1979) Adenosine, NAD and FAD can initiate template dependent RNA synthesis catalyzed by *Escherichia coli* RNA polymerase, FEBS Letts., **102**, 51–54.

Neverov, K.V., Mironov, E.A., Lyudnikova, T.A., Krasnovsky, A.A., Jr., and Kritsky, M.S. (1996) Phosphorescence analysis of the triplet states of pterins in connection with their photoreceptor functions in biological systems, Biochemistry (Moscow), **61**, 1149–1155.

Nikandrov, V.V., Brin, G.P., and Krasnovskii, A.A. (1978) Light-induced activation of NADH and NADPH, Biochemistry (Moscow) **43**, 636–645.

Roychowdhury-Saha, M., Lato, S.M., Shank, E.D., and Burke, D.H. (2002) Flavin recognition by an aptamer targeted toward FAD, Biochemistry, **41**, 2492–2499.

Oró, J. (2001) Cometary molecules and life's origin. In: J. Chela-Flores, T. Owen., and F. Raulin (eds.). *First Steps in the Origin of Life in the Universe*. Kluwer Academic Publishers, Dordrecht, pp 113–120.

Schmidt, W. (1984) The study of basic photochemical and photophysical properties of membrane-bound flavins: The indispensable prerequisite for the elucidation of primary physiological blue light action. In: H. Senger (ed.). *Blue Light Effects in Biological Systems*, Springer Verlag, Berlin, *etc*., pp. 81–94.

THE ROLE OF HEAT IN THE ORIGIN OF LIFE

PETER R. BAHN[1], A. PAPPELIS[2], and RANDAL GRUBBS[3]
[1]Bahn Biotechnology Co., RR2 Box 239A, Mt. Vernon, IL 62864 USA,
[2]Department of Plant Biology, Southern Illinois University, Carbondale, IL 62901 USA. [3]250½ Charles St., Carbondale, IL 62901 USA

We show here that a hot origin of life (Fox & Dose, 1977) is more consistent with experimental and observational evidence than is a cold origin of life (Bada & Lazcano, 2002; Miller & Lazcano, 1995).

In his classic 1953 origin-of-life experiment Stanley Miller (1953) generated amino acids by circulating a mixture of CH_4, NH_3, H_2, and H_2O vapor past an electric discharge with a temperature of hundreds to thousands of degrees C for a week. Lightning discharges have temperatures in the tens of thousands of degrees C (Uman, 2001). Miller's system also utilized a reservoir of boiling water at 100°C. It would be interesting to repeat Miller's experiment but with a very hot wire as the energy source instead of an electric discharge. We predict that this variation of his experiment would produce a similar array of amino acids. Harada & Fox (1964) did in fact subject CH_4, NH_3, and H_2O to passage through a glass tube heated to 1000°C and obtained a product amino acid distribution similar to Miller's famous experiment. That heat is important in the prebiotic synthesis of amino acids (Fox, 1995) was shown by Fox & Windsor (1970) who produced amino acids by heating NH_3 and H_2CO with a hot iron pipe at a temperature of 950°C. Oro (1960) synthesized the purine adenine by heating NH_4CN at 90°C for 24 hours. Fox & Harada (1961) synthesized the pyrimidine uracil by heating malic acid and urea in the presence of polyphosphoric acid at 140°C for 2 hours. Heat thus appears to have had an important role in the formation of prebiotic monomers.

On the hot Primitive Earth, the first biopolymers to be formed were probably made by the thermal dehydration condensation of prebiotic monomers into protobiopolmers. This result seems inevitable wherever there are heat sources of any sort and monomers. The heat sources could be thermal convection, meteoritic impact, lightning discharge, infrared radiation, shock wave, or others. For example, Fox & Harada (1958) prepared thermal proteins, or *proteinoids,* by heating mixtures of amino acids above 100°C for a few hours. Rohlfing (1976) prepared thermal proteins by heating mixtures of amino acids at 65°C for a few weeks. The thermal polymerization of sugars into polysaccharides has long been known as the process of caramelization. Fox & Bahn (1990) heated glucose and fructose in the presence of glutamic acid at 140°C for 12 hours to form polyglucofructose. Schwartz (1962) heated ribose at 150°C for 12 hours to form polyribose. Some progress has been made in the thermal polymerization of mononucleotides into oligonucleotides. Schwartz, Bradley, & Fox (1965) heated CMP with polyphosphoric acid at 85°C for 6 hours to make

J. Seckbach et al. (eds.), Life in the Universe, 119–120.

oligo-C. Matsuno and his colleagues (Ogasawara et al., 2000) heated mononucleotides under simulated hydrothermal vent conditions at 110°C for 1 hour to get oligonucleotides. With respect to lipids, Hargreaves, Mulvihill, & Deamer (1977) made phospholipids by heating fatty acids, glycerol, and phosphate at 65°C for one week. Thermal polymerization of prebiotic monomers into protobiopolymers is decidedly nonrandom (Bahn & Pappelis, 2001), with the various sequences that are formed being highly influenced by side-by-side interactions of the various monomers involved. *The first informational contents to be encoded in the first biopolymers on the Primitive Earth may have been for sequences that were most heat-stable out of all the possible sequences that could have occurred. The very first "genes" may have been for heat-stability.* The first RNA sequences in the RNA World may have been protogenes for maximal heat-stability. This possibility is supported by all the available evidence that the Last Common Ancestor of all life on Earth was a hyperthermophile Archean.

In conclusion, heat appears to have been a universal catalyst and powerful environmental shaper in the transformation of inorganic matter into living matter.

References

Bada, J.L. & Lazcano, A. (2002) Some Like it Hot, But Not the First Biomolecules, *Science* **296**, 1982–1983.

Bahn, P.R. & Pappelis, A., in Chela-Flores, J., Owen, T., & Raulin, F. (eds.), (2001) HPLC Evidence of Nonrandomness in Thermal Proteins, *First Steps in the Origin of Life in the Universe,* Kluwer Academic Publishers, Boston, pp 69–72.

Fox, S.W. (1995) Thermal Synthesis of Amino Acids and the Origin of Life, G*eochim. Cosmochim. Acta* **59** 1213–1214.

Fox, S.W. & Bahn, P.R. (1990) Process for Preparing Thermal Carbohydrates, *U.S. Patent Number 4,975,534.*

Fox, S.W. & Dose, K. (1977) *Molec. Evol. & the Orig. of Life,* Marcel Dekker Inc., New York.

Fox, S.W. & Harada, K. (1958) Thermal Copolymerization of Amino Acids to a Product Resembling Protein, *Science* **128**, 1214.

Fox, S.W. & Harada, K. (1961) Synthesis of Uracil Under Conditions of a Thermal Model of Prebiological Chemistry, *Science* **133**, 1923–1924.

Fox, S.W. & Windsor, C.R. (1970) Synthesis of Amino Acids by the Heating of Formaldehyde and Ammonia, *Science* **170**, 984–986.

Harada, K. & Fox, S.W. (1964) Thermal Synthesis of Natural Amino Acids from a Postulated Primitive Terrestrial Atmosphere, *Nature* **201,** 335–336.

Hargreaves, W.R., Mulvihill, S.J., & Deamer, D.W. (1977) Synthesis of Phospholipids and Membranes in Prebiotic Conditions, *Nature* **266**, 78–80.

Miller, S.L. (1953) A Production of Amino Acids Under Possible Primitive Earth Conditions, *Science* **117**, 528–529.

Miller, S.L. & Lazcano, A. (1995) The Origin of Life – Did it Occur at High Temperatures?, *J. Mol. Evol.* **41** , 689–692.

Ogasawara, H., Yoshida, A., Imai, E., Honda, H., Hatori, K., & Matsuno, K. (2000) Synthesizing Oligomers from Monomeric nucleotides in Simulated Hydrothermal Environments, *Orig. Life & Evol. Biosphere* **30**, 519–526.

Oro, J. (1960) Synthesis of Adenine from Ammonium Cyanide, *Bioch.Biophy.Res.Com* **2**, 407–412.

Rohlfing, D.L. (1976) Thermal Polyamino Acids: Synthesis at Less than 100°C, *Science* **193**, 68–70.

Schwartz, A.W. (1962) *The Thermal Polymerization of Ribose,* M.S. Thesis, Florida State Univ.

Schwartz, A.W., Bradley, E., & Fox, S.W., in Fox, S.W. (ed.), (1965) Thermal Condensation of Cytidylic Acid in the Presence of Polyphosphoric Acid, *The Origins of Prebiological Systems & of Their Molecular Matrices,* Academic Press, New York, pp 317–326.

Uman, M.A. (2001) *The Lightning Discharge,* Dover Publications, Inc., Mineola, NY.

A POSSIBLE PATHWAY FOR THE TRANSFER OF CHIRAL BIAS FROM EXTRATERRESTRIAL C^{α}-TETRASUBSTITUTED α-AMINO ACIDS TO PROTEINOGENIC AMINO ACIDS

MARCO CRISMA,[1] ALESSANDRO MORETTO,[1] FERNANDO FORMAGGIO,[1] BERNARD KAPTEIN,[2] QUIRINUS B. BROXTERMAN[2] AND CLAUDIO TONIOLO[1]
[1]Institute of Biomolecular Chemistry, CNR, Department of Organic Chemistry, University of Padova, 35131 Padova, Italy [2]DSM Research, Life Sciences, Advanced Synthesis and Catalysis, P.O. Box 18, 6160 MD Geleen, The Netherlands

1. Introduction

A growing evidence has recently been accumulated on the presence of chiral, C^{α}-tetrasubstituted α-amino acids with a significant L (S) enantiomeric excess (up to 15%) in carbonaceous chondritic meteorites. The amino acids analysed to date, extracted from several samples of the Murchison and Murray meteorites, include isovaline (Iva), C^{α}-methyl norvaline [(αMe)Nva], C^{α}-methyl valine [(αMe)Val], C^{α}-methyl isoleucine [(αMe)Ile], and C^{α}-methyl *allo*isoleucine [(αMe)*a*Ile] (Cronin and Pizzarello, 1997; Pizzarello and Cronin, 2000; Pizzarello *et al.*, 2003).

These amino acids are either extremely rare (Iva) or totally unknown (all of the others) in the terrestrial biosphere. Therefore, the reported enantiomeric excesses can be hardly ascribed to a terrestrial contamination of the samples. Moreover, C^{α}-tetrasubstitution protects amino acids against racemization.

These results have suggested that C^{α}-tetrasubstituted α-amino acids of extraterrestrial origin could have been the seeds for homochirality of life in our planet (Bada, 1997). However, this hypothesis implies that the enantiomeric excesses would have been somehow transferred from C^{α}-tetrasubstituted α-amino acids to protein amino acids.

Herewith we describe the results of a study aimed at determining if C-activated derivatives or peptides based on chiral, C^{α}-tetrasubstituted α-amino acids could have reacted with protein amino acids and favoured the incorporation of one of their enantiomers over the other.

2. Results and Discussion

We have investigated the stereochemical outcome of reactions involving N-acetylated, homo-chiral (L) homo-oligomers (from dimer through octamer) of C^{α}-tetrasubstituted α-amino acids [Iva, (αMe)Nva, (αMe)Leu and (αMe)Val] activated at the C-terminus as 5(4*H*)-oxazolones, and a large excess of the racemate of a proteinogenic amino acid

J. Seckbach et al. (eds.), Life in the Universe, 121–122.

methyl ester (*e.g.*, H-D,L-Val-OMe). The choice of the 5(4*H*)-oxazolone as the activated intermediate is rooted on the fact that oxazolones are formed to a significant extent from C^{α}-tetrasubstituted α-amino acid residues by nearly all activation methods used for peptide synthesis. In particular, oxazolones are generated very easily by intramolecular dehydration of C^{α}-tetrasubstituted peptides with a free C-terminus (Formaggio *et al.*, 2003).

Our results indicate that homo-peptide oxazolones based on chiral, C^{α}-tetrasubstituted α-amino acids are indeed able to exert chiral discrimination when reacting with the racemate of a protein amino acid. The diastereomeric excesses of the peptides formed in the various reactions range from 3 to 64%. However, the direction of the stereoselection is bimodal, in that peptide oxazolones with shorter main-chain length and less bulky side chains preferentially incorporate the proteinogenic amino acid of the same chirality as that of the C^{α}-tetrasubstituted residues, whereas longer main chains (which promote the onset of β-turns and 3_{10}-helices at the tetramer and pentamer levels, respectively) and bulky side chains in the peptide oxazolone direct the selection towards the proteinogenic amino acid of the opposite chirality. These findings can be explained on the basis of the reaction mechanism, in which the transition state is reactant-like (Crisma *et al.*, 1997), of the participation to the stereoselection by both the C-terminal and the penultimate residues of the peptide oxazolone (with opposite effects), and of the turn/helix handedness and stability.

Our results suggest that two extreme scenarios can be envisaged, depending on the relative abundance of C^{α}-tetrasubstituted α-amino acids of different bulkiness and on the extent to which relatively long peptide chains, containing mostly residues of this kind, may have formed on the early Earth. Either (i) short C^{α}-tetrasubstituted peptides, with simple side chains, by seeding the preferential incorporation of L proteinogenic amino acids of the same chirality, may have promoted the growth of longer and progressively more chirally uniform peptides through amplification processes, or (ii) longer C^{α}-tetrasubstituted peptides, with bulky side chains, may have depleted the racemic primordial soup of proteinogeic amino acids of their D isomers, thus generating an L excess from which the homochirality of life could have been developed.

3. References

Bada, J.L. (1997) Meteoritics—Extraterrestrial handedness?, *Science* **275**, pp. 942–943.

Crisma, M., Valle, G., Formaggio, F., Toniolo, C. and Bagno, A. (1997) Reactive intermediates in peptide synthesis: first crystal structures and *ab initio* calculations of 2-alkoxy-5(4*H*)-oxazolones from urethane-protected amino acids, *J. Am. Chem. Soc.* **119**, pp. 4136–4142.

Cronin, J.R. and Pizzarello, S. (1997) Enantiomeric excesses in meteoritic amino acids, *Science* **275**, pp. 951–955.

Formaggio, F., Broxterman, Q.B. and Toniolo, C. (2003) Synthesis of peptides based on C^{α}-tetrasubstituted α-amino acids, in *Houben-Weyl Methods of Organic Chemistry, Vol. E22c, Synthesis of Peptides and Peptidomimetics*. Goodman, M., Felix, A., Moroder, L. and Toniolo, C. (eds.), Stuttgart: Thieme, pp. 292–310.

Pizzarello, S. and Cronin, J.R., (2000) Non-racemic amino acids in the Murray and Murchison meteorites, *Geochim. Cosmochim. Acta* **64**, pp. 329–338.

Pizzarello, S., Zolensky, M. and Turk, K.A. (2003) Nonracemic isovaline in the Murchison meteorite: chiral distribution and mineral association, *Geochim. Cosmochim. Acta* **67**, pp. 1589–1595.

PREBIOTIC POLYMERIZATION OF AMINO ACIDS. A MARKOV CHAIN APPROACH

F.G. MOSQUEIRA[1], S. RAMOS-BERNAL[2], and A. NEGRON-MENDOZA[2]
[1]*Dirección General de Divulgación de la Ciencia, UNAM. Cd. Universitaria, A.P. 70-487, México, 04510 D.F., México. and* [2]*Instituto de Ciencias Nucleares, UNAM, A.P. 70-453, D.F. 04510, México*

Abstract. We develop a simple model using a Markov chain approach for the oligomerization of amino acids, based on electromagnetic differences among reacting amino acids. Such model can explain an experimental oligomerization process that shows a remarkable bias in its products. Such results may be of importance as it makes more accessible the replication of minimal chemical machinery compatible with life processes.

1. Introduction

The polymerization phenomena associated to the origin of life had to be a strongly biased process. Otherwise, the probability of nucleation of a 'minimum living chemical system' would be excluded (Mosqueira, 1988). We resort to the self-ordering principle (Fox and Dose, 1977) that establishes that the reactivity between different reacting amino acids is not even. Thus, we grouped them into four different sets (Dickerson and Geis, 1969).

We apply this model to the products formed in a dry heating mixture of the amino acids glycine, tyrosine and glutamic acid (Nakashima et al., 1977). Nakashima and collaborators reported only those trimers containing tyrosine and found only two trimers from 36 theoretically possible tripeptides, assuming complete randomness. Further, this reaction mechanism has been investigated by Hartmann et al. (1981). With such knowledge, our model is able to account for the production of the two trimers observed (Mosqueira et al., 2000).

2. Markov Chain Formalism Applied to the Thermal Polymerization of Amino Acids

We consider only nearest pair-wise interactions among reacting species. The main assumption is that the transition probability associated to the chemical incorporation of new specie is influenced only by the interaction between the incoming specie and the reactive end of the oligomers. It is not influenced by any other previous monomer already bonded in the n-oligomer. We use a Markov chain formalism to model the dry heating reaction of the

J. Seckbach et al. (eds.), Life in the Universe, 123–124.

specified amino acids. Let us define a finite Markov chain (Moran, 1984):

$$\mathrm{Prob}(X_{T+1} = m|X_0 = a, X_1 = b, \ldots, X_T k) = \mathrm{Prob}(X_{T+1} = m|X_T = k), \qquad (1)$$

where a, b, ... ,m are arbitrary designations of the states assumed by system X at time 0, 1, ..., T, T+1, ... Under such construction, it is possible to follow the state of the system as the oligomerization proceeds. For this we use the following equation: $P_{(n)}$=$AP_{(n-1)}$. Where $P_{(n)}$ and $P_{(n-1)}$ are the vectors of the n and n−1 states respectively, and A is the transition or reactivity matrix with elements a_{ij}. These are obtained from the four groups classification of amino acids (Dickerson and Geis, 1969): polar positive (p+), polar negative (p−), neutral (n), and non polar (np). So, we arrange the four possible electromagnetic interactions between amino acids into a 4 × 4 matrix as follows:

$$\begin{pmatrix} p^+p^+ & p^+p^- & p^+n & p^+np \\ p^-p^+ & p^-p^- & p^-n & p^-np \\ np^+ & np^- & nn & nnp \\ npp^+ & npp^- & npn & npnp \end{pmatrix} = A = (a_{ij}) \qquad (2)$$

The state of the system at any stage "k" is represented by row matrix with four elements: (p^+ p^- n np).

3. Remarks

To the present stage, we have applied this model to reacting mixtures containing only two groups of amino acids at a time. We have found that despite of the numerical values assigned to its elements, only three different situations arise that are discussed elsewhere (Mosqueira et al., 2002). Further, we applied our model through the three stages of the reaction mechanism (Hartmann et al. 1981), (i.e., initiation, elongation and termination) and successfully explain the production of only two trimers reported by Nakashima et al. (1977).

4. References

Dickerson, R.E. and Geis, I. (1969) *The Structure and Action of Proteins,* Harper & Row Publishers, New York.

Fox, S. and Dose, K. (1977) *Molecular evolution and the origin of life*, Marcel Dekker, Inc., New York.

Hartmann, J., Brand, M.C., and Dose, K (1981) Formation of specific amino acid sequences during thermal polymerization of amino acids, *BioSystems* **13**, 141–147.

Moran, P.A.P. (1984) *An Introduction to Probability Theory,* Clarendon Press, Oxford.

Mosqueira, F.G. (1988) On the Origin of Life Event, *Origins of Life and Evolution of Biosphere* **18**, 143–156.

Mosqueira, F.G., Ramos-Bernal, S., Negrón-Mendoza, A. (2000) A simple model of the thermal prebiotic oligomerization of amino acids, *BioSystems* **57**, 67–73.

Mosqueira, F.G., Ramos-Bernal, S., Negrón-Mendoza, A. (2002) Biased polymers in the origin of life, *BioSystems* **65**, 99–103.

THE ELECTROCHEMICAL REDUCTION OF CO_2 TO FORMATE IN HYDROTHERMAL SULFIDE ORE DEPOSIT AS A NOVEL SOURCE OF ORGANIC MATTER

M. G. VLADIMIROV[1], YU. F. RYZHKOV[2], V. A. ALEKSEEV[2], V. A. BOGDANOVSKAYA[3], V. A. OTROSHCHENKO[1], and M. S. KRITSKY[1]

[1]A.N. Bach Institute of Biochemistry, Russian Academy of Sciences, Moscow; [2]The Russian Federation State Research Center "Troitsk Institute for Innovation and Fusion Research", Troitsk, Moscow Region; [3]A.N. Frumkin Institute of Electrochemistry, Russian Academy of Sciences, Moscow, Russian Federation.

Earlier models for organic synthesis in the conditions of sulfide hydrotherms were based on the ability of sulfide minerals (chalcogenides) and hydrogen sulfide to act as electron donors in chemical reduction of water-dissolved CO_2 to organic molecules (for literature sources see Vladimirov *et al.*, 2004). The detailed analysis suggests a possibility of another mechanism: a direct electrochemical reduction of CO_2 at the surface of metal-sulfide minerals, for instance pyrite or galenite. When putting forward this hypothesis, we proceeded from the fact that electrode potentials of sulfide minerals can reach considerable magnitudes varying with composition and structure of mineral. Because chalcogenides are capable of conducting electrical current, the contact of mineral bodies of different structures must lead to arising of galvanic circuits with an electromotive force of about several volts. The functioning of such circuits has been registered under natural conditions.

The combination in sulfide minerals of semiconductor properties with a relatively low adsorption activity and the presence of traces of transient metals makes them a good cathode material. Our experiments demonstrated the increase of cathode current on rotating pyrite electrode in a range of potentials more negative than –800 mV in presence of CO_2 and revealed the dependence of this current from the CO_2 diffusion rate, temperature and ionic composition of aqueous phase. In these experiments, however, no formation of organic substances was observed.

According to mechanism of the CO_2 reduction established in the studies of cathode materials other than pyrite, the electron tunneling from cathode to a CO_2 molecule produces an active particle, the radical-ion $\cdot CO_2^-$. Further transformation of this ion-radical can lead to formation of carboxylic acids such as formate or a less reduced product, carbon monoxide. The rate of CO_2 reduction to organic molecules increases under high pressure, when aqueous media acquire a number of properties of non-aqueous solvents. Since the hydrotherms in earth crust exist under the pressure reaching hundreds of Bars, the further experiments were performed in the device we have designed to test the possibility of electrosynthesis of organic

J. Seckbach et al. (eds.), Life in the Universe, 125–126.

substance under high pressure. The device included an electrolyzer placed in an autoclave. The design of the autoclave enabled us to perform studies at pressures up to 3000 Bar and simultaneously permitted continuous passage of CO_2 through the cathode compartment of the electrolyzer. The housing of the electrolyzer was made from fluoroplastic and presented two separated electrode compartments of 130 cm^3 working volume each. The cathode was a monolith disc, 5 mm thick and 30 mm in diameter, cut out from a pyrite crystal. The catholyte and anolyte compartments were separated with an MK-40 cation exchange membrane. The counter electrode was a Pt plate of 30 mm in diameter and 0.2 mm thick and the reference electrode was a saturated AgCl electrode. In the course of 24 hour experiment, CO_2 was bubbled through the catholyte (0.1M $KHCO_3$). The pressure of CO_2 in the autoclave and in the electrolyzer was being maintained at a rate of 50 Bar at room temperature. The control of process and all measurements were made at a distance, outside the protective system of the device. The potential in the electrochemical cell was provided by means of a potentiostat and a programmer connected with a voltmeter and a register device (for details of experiment see Vladimirov *et al., in press*). On completion of the experiment, the content of organic products of CO_2 reduction was determined by a color reaction with chromotropic acid.

As the potential of cathode was increased up to –800 mV, substantial amounts of formate were revealed in the catholite. The current (Faradaic) efficiency of its accumulation grew up to –1000 mV where it was 0.12%. The yield of formate increased exponentially with potential. In none of the experiments formation of free formaldehyde was registered. The following experiments were performed as controls. (1) The reactor filled with 0.1M $KHCO_3$ solution was bubbled by CO_2 (50 Bar, 10 ml min^{-1}) during 24 hours without potential imposed on pyrite cathode. (2) The -1 Volt potential was imposed for 24 hours on the pyrite cathode placed in 0.1M $KHCO_3$ solution without CO_2 bubbling and under normal atmospheric pressure. (3) The -1 Volt potential was imposed for 24 hours on the pyrite cathode placed in 0.1M $KHCO_3$ solution and the reactor was bubbled by N_2 (50 Bar, 10 ml min^{-1}) during 24 hours. In none of these control experiments formation of formate or other organic products was registered.

Thus, the reduction of CO_2 on a pyrite cathode in aqueous solutions pH $\cong 7$ at potentials higher than –800 mV and a pressure of 50 Bar resulted in the formation of a simplest organic acid. Our data suggest that in neutral and weekly acidic solutions accompanying deep-lying deposits of sulfide ore (more than 500 m below crust surface) as well as in the regions of "black smokers" at the ocean floor there may occur formation of organic substances by direct electrochemical reduction of CO_2 in the cathode regions of mineral galvanic systems.

Support from Russian Foundation for Basic Research Grant 01-04-48268 is appreciated.

Reference

Vladimirov, M.G., Ryzhkov, Y.F., Alekseev V.A., Bogdanovskaya, V.A., Otroshchenko,V.A., and Kritsky, M.S. (2004) Electrochemical reduction of carbon dioxide on pyrite as a pathway for abiogenic formation of organic molecules, Origins of Life and Evolution of the Biosphere (in press).

TOWARDS A CHRONOLOGICAL ORDER OF THE AMINO ACIDS

Last Common Ancestor may have Arisen from Genome Fusion

W. J. M. F. COLLIS
Strada Sottopiazzo, 18, 14056 Boglietto (AT), ITALY

1. Introduction

The origin of the genetic code and early evolution of life on earth has fascinated researchers for many decades. The recent availability of complete genome sequences provides a molecular 'fossil record' which can be accessed by reconstructing amino acid sequences corresponding to ancient gene duplications.

The present study was stimulated by the idea that as evolution proceeds in small steps, individual amino acids should have been added to the genetic code, one at a time. Sampling conserved sequences may identify that subset of amino acids used at the time of gene duplication. Collection of many such samples can elucidate the chronological order in which amino acids were first coded into proteins. In this study we examine families common to most living organisms – so called Clusters of Othologous Groups of proteins (COGs) – those present in modern Bacteria, Eukarya, and Archaea.

2. Materials and Methods

Clustal 1.81 software was used to align over 3238 families of proteins downloaded from **www.ncbi.nlm.nih.gov/COG/**. Of these, 934 families were discarded because they contained less than 8 modern sequences. A partial sequence reconstruction was attempted on the remainder by the (rather crude) method of accepting an amino acid as being conserved if it is in the majority at a site in the alignment. A simple count of all such conserved amino acids is made for each family. Of these, a further 2158 families were discarded by accepting only those ancestral counts containing a significant number of amino acids – a high residue limit of 100 was used leaving just 146 partially reconstructed sequences.

3. Results

We expect all 20 amino-acids to be represented in about 89% of the sequences reconstructed, reflecting the assumption that the modern genetic code of 20 amino acids was inherited from LCA. Instead we find that in 32 cases both tryptophan and cysteine are missing, in 38 cases cysteine is missing but tryptophan is present, in 30 cases cysteine is present but not tryptophan. Only in 37 (25%) cases out of the total of 146 are all 20 amino acids present.

J. Seckbach et al. (eds.), Life in the Universe, 127–128.

4. Discussion

The absence of either tryptophan or cysteine is incompatible with the conjecture that amino acids were added one at a time to an earlier sub-set. Nor can systematic bias in the elaboration of the data could be responsible. For example if there were a bias towards aligning cysteine residues then we have a difficulty explaining why such residues are absent in the tryptophan rich ancestral sequences and vice versa.

One possibility may be that a speciation event occurred when neither tryptophan nor cysteine were part of the genetic code. In one species cysteine, and in the other tryptophan independently appropriated UGN codon(s) and the two organisms continued to evolve developing new proteins with distinct, possibly incompatible genetic codes. Unlike most other amino acids which share the same family box, cysteine and tryptophan have different properties, consistent with this hypothesis. Later genome fusion took place between these two species and diverging rapidly towards Bacteria, Eukarya, and Archaea. Perhaps the new fused genome provided significant advantages, sufficient to lead to the eventual extinction of all other forms of life.

Preliminary indications suggest that the two parents of the fusion specialized in somewhat different biochemistries. The typtophan encoding ancestor developed enzymes involved in membrane transport and ATPase. The cysteine encoding ancestor developed NAD/FAD and DNA repair enzymes.

The approach and conclusions of this study differ from those of Trifonov who has used amino acid frequencies to estimate chronology. Modern amino acid frequencies quite accurately reflect base composition and the codon assignments of the genetic code. The amino acid absence criterion used here, less dependent of any conjecture regarding either base composition or codon assignments is likely to be more reliable. Given the ubiquity of horizontal gene transfer, Trifonov's use of only 2 enzymes per family to reconstruct conserved motifs may also be inadequate.

5. Summary

A method to estimate the relative chronological order of the amino acids and ancient gene duplication events before the LCA is illustrated. The last two amino acids added to the genetic code, were tryptophan and cysteine but these two probably originated in distinct species whose genomes fused. Full details of the 146 alignments can be requested from the author (mr.collis@physics.org)

6. References

Higgins, D.G., Bleasby, A.J. and Fuchs, R. (1992) CLUSTAL V: improved software for multiple sequence alignment. Computer Applications in the Biosciences (CABIOS), **8(2)**:189–191.

Tatusov *et al.* (2000). The COG database: a tool for genome-scale analysis of protein functions and evolution. Nucleic Acids Res. **28**: 33–6.

Trifonov, E.N. (2000) Glycine clock: Eubacteria first, archaea next, protoctista, fungi, plants and animalia at last. Gene Therapy & Mol. Biol., **4**: 313–323.

Collis W J M F (2000) Evolution of Protein Synthesis. Origins of Life and Evolution of the Biosphere, **30**: 337.

ORIGIN AND EVOLUTION OF METABOLIC PATHWAYS

M. BRILLI and R. FANI
Department of Animal Biology and Genetics, Via Romana 17-19, I-50125 Florence, Italy

1. Introduction

Contemporary genomes are the result of 3.5–4 billions of years of evolution. Their history can be traced using the increasing number of available sequences and a panel of bioinformatic tools. As a result it is now possible to shed some light on the mechanisms involved in their evolution and responsible for the shaping of metabolic pathways. These analyses also enabled to shed some light on the mechanisms and forces that drove the evolution of the earliest genes and genomes. Whole-genome comparison demonstrates that a high proportion of the gene set of different prokaryotes results from ancient gene duplications. Thus, the duplication of DNA sequences appears to have played a very important role in the evolution of genes and genomes. It suggests that the earliest living organisms contained a small number of genes that gave rise to new genes by duplication followed by evolutionary divergence. In the 30s Haldane and Muller suggested that duplicated genes might acquire different mutations eventually arriving to code for products with different catalytic features.

Sequences evolved from a common ancestor are referred to as homologs. These, in turn, are called orthologs, if they originated after a speciation event or paralogs if they emerged after a gene duplication event. Serial duplications of homologous genes can give rise to a group of paralogous genes, a *paralogous gene family*. Duplication may involve DNA stretches whose length is variable, resulting in the duplication of fragments containing part of a gene, one or more genes, and complete genomes too.

1.1. GENE DUPLICATION AND THE EVOLUTION OF METABOLIC PATHWAYS

The so-called Oparin-Haldane hypothesis, proposed in the 20s, postulated that abiotically produced organic compounds accumulated on the Earth surface, and that life emerged in a nutrient-rich environment (the *primordial soup*). This view received an experimental support in 1953, when S.L. Miller, using a sterile apparatus simulating the *pre-biotic* environment where life originated, obtained aminoacids and other *biogenic* compounds (Miller, 1953). According to this hypothesis, the first living entities were heterotrophic. However, the progressive exhaustion of the prebiotic supply of nutrients resulted in a selective pressure that enabled the survival only of those microorganisms that have become able to produce autonomously those nutrients whose concentration was decreasing in the environment. Thus, the building up of new metabolic pathways represented a crucial event in molecular

J. Seckbach et al. (eds.), Life in the Universe, 129–132.

and cellular evolution, since the ancestral organisms became progressively less- dependent on the organic compounds prebiotically synthesized.

Several hypotheses on the origin and evolution of metabolic pathways have been proposed, but the most attractive one is the so-called *patchwork* hypothesis (Jensen, 1976), according to which metabolic pathways may have been assembled through the recruitment of primitive enzymes that could react with a wide range of chemically related substrates. Such relatively slow, non-specific enzymes may have enabled primitive cells containing small genomes to overcome their limited coding capabilities.

2. Results

2.1. NITROGEN FIXATION: A CASCADE OF GENE AND OPERON DUPLICATION

Nitrogen fixation is widespread in bacteria and archaea. In the free-living diazotroph *Klebsiella pneumoniae* the *nif* genes are clustered in a single chromosomal region and are organized into several operons. The enzyme responsible for nitrogen fixation is called nitrogenase. All known Mo-nitrogenase consist of two components: the *dinitrogenase*, a $\alpha_2\beta_2$ tetramer encoded by *nifD* and *nifK* genes, and the *dinitrogenase reductase*, a homodimer coded for by the *nifH* gene. Nitrogenase contains two metal clusters, one of which is the iron-molybdenum cofactor (FeMo-co), the site of dinitrogen reduction and whose synthesis requires the activity of an another tetrameric enzymatic complex (Nif N_2E_2) whose subunits are encoded by *nifE* and *nifN*.

The detailed analysis of the *nifDK* and *nifEN* pairs of genes showed that their products share common features (Fani et al., 2000): they code for tetrameric complexes, and the products of *nifE* and *nifN* are structurally related to the *nifD* and *nifK* products, respectively. Finally, those diazotrophs in which *nifDK* and/or *nifEN* have been characterized share the same gene organisation. The four genes are clustered in operons where the two genes of each pair are contiguous and arranged in the same order (*nifDK* and *nifEN*). Moreover, the four genes shared a high degree of sequence similarity suggesting that they belong to a paralogous gene family. Fani et al (2000) proposed a two-steps evolutionary model leading to these genes. The model proposes the existence of a single ancestral gene that underwent an in-tandem gene duplication event, which gave rise to a bicistronic operon. Then, this ancestral operon duplicated leading to the ancestors of the present-day *nifDK* and *nifEN* operons.

If the ability to fix nitrogen was a primordial property, then the duplication events leading to the two operons predated the appearance of the last common ancestor of Archaea and Bacteria. Thus the function(s) performed by the primordial enzyme would have evolved because of the composition of the atmosphere. Theories vary from strongly reducing to neutral; but it is generally accepted that O_2 was absent, an essential prerequisite for the evolution of an ancestral nitrogenase, as free oxygen inactivates it. The first living organisms were probably heterotrophic anaerobes and dependent on abiotically produced organic matter for their metabolism. Depending on the composition of the early atmosphere (neutral or reducing), the ancestor gene coded for an enzyme with a nitrogenase or a detoxyase activity, respectively. The first duplication event, leading to the ancestral bicistronic operon, followed by divergence, refined the specificity of the primitive nitrogenase/detoxyase, which might have also been involved in the biosynthesis of a Fe-Mo cofactor. Successively the ancestral operon duplicated and the following divergence lead to the appearance of the ancestors of the present day *nifDK* and *nifEN* operons which encoded different proteins involved in reducing substrates and biosynthesize FeMo cofactor, respectively. Thus, the ability to fix nitrogen appears to be an ancient property and might have arisen during an early period of cellular evolution. The corresponding genetic information could have been lost in many strains, possibly in the course of adaptation to changing environmental conditions and the associated selective pressures upon microorganisms. Nevertheless, the high degree of conservation of nitrogenase genes within archaeal and bacterial diazotrophs suggests that lateral transfer of *nif* genes might have occurred frequently.

2.2. HISTIDINE BIOSYNTHESIS: A PARADIGM FOR THE STUDY OF THE ORIGIN AND EVOLUTION OF METABOLIC PATHWAYS

Histidine biosynthesis is one of the best-characterized anabolic pathways. There is a large body of genetic and biochemical information, including operon structure, gene expression, and an increasingly larger number of sequences available for this route. This pathway has been extensively studied, mainly in *Escherichia coli* and *Salmonella typhimurium*. In all histidine-synthesizing organisms the pathway is unbranched and consists of nine intermediates, all of which have been described, and of eight distinct proteins that are encoded by eight genes, *hisGDCBHAF(IE)* that, in *E. coli*, are arranged in a compact operon. As previously reported, there are several independent indications of the antiquity of this pathway suggesting that the entire route was assembled long before the appearance of the Last Universal Common Ancestor (LUCA) of the three extant cell domains. Histidine biosynthesis plays an important role in cellular metabolism, since it is interconnected to both the *de novo* synthesis of purines and to nitrogen metabolism. How the *his* pathway originated remains an open question, but the detailed analysis of the structure and organization of the *his* genes in (micro)organisms belonging to different phylogenetic archaeal, bacterial and eucaryal lineages revealed that at least three molecular mechanisms played an important role in shaping the pathway, that is gene elongation, paralogous gene duplication(s), and gene fusion (Fani et al., 1994, Brilli and Fani, 2003). Moreover, genetic and sequence data show that once the entire pathway was assembled it underwent major rearrangements during evolution. A wide variety of different clustering strategies of *his* genes have been documented suggesting that many possible histidine gene arrays exist. According to the patchwork hypothesis on the origin and evolution of metabolic pathways, several histidine biosynthetic genes appears to have been assembled by recruitment of preexisting broad-specificity enzymes following gene duplications. Two of these genes, *hisA* and *hisF*, are particularly interesting from an evolutionary point of view. Their products catalyze sequential reactions in histidine biosynthesis, they are paralogous and share a similar internal structures (Fani et al. 1994). The detailed comparative analysis of the HisA and HisF proteins revealed that they might be subdivided into two paralogous modules half-the size of the entire genes. This finding led to the suggestion that the two genes are the outcome of two ancient paralogous duplication events. According to the model proposed, the first duplication involved an ancestral module (half the size of the present-day *hisA* gene) and led by a gene elongation event to the ancestral *hisA* gene, which in turn underwent a duplication, that gave rise to the *hisF* gene. Since the overall structure of the *hisA* and *hisF* genes are the same in all known (micro) organisms, it is likely that they were part of the genome of the last common ancestor and that the two duplication events occurred long before the separation of Archaea from Bacteria. The biological significance of the *hisA-hisF* structure relies on the structure of the encoded enzymes. The two proteins have a TIM-barrel structure, so the ancestral gene probably coded for a half-barrel enzyme. The elongation event leading to the ancestor of *hisA/hisF* genes enabled the covalent fusion of two half-barrel coding sequences, permitting the production of a complete barrel whose activity was then refined by natural selection acting on mutated alleles. The same happened for the whole-barrel gene after its duplication, resulting in modern-day HisA and HisF. The reminiscence of this last duplication event may lie in the physical closeness between the two genes in present-day genomes.

3. Summary

In the course of evolution different molecular mechanisms acted amplifying the coding and metabolic abilities of organisms. It is clear that the DNA duplication is a major force in that sense. The evidence for gene elongation, gene duplication and operon duplication events suggests in fact that the ancestral forms of life might have expanded their coding abilities and their genomes by "simply" duplicating a small number of mini-genes via a cascade of duplication events, involving DNA sequences of different size.

4. References

Alifano, P. Fani, R. Liò, P. Lazcano, A. Bazzicalupo, M. Carlomagno, MS. Bruni, CB. (1996) Histidine biosynthetic pathway and genes: structure, regulation, and evolution. Microbiological Review 60, 44–69.

Brilli, M. and Fani, R. (2003) Molecular evolution of *hisB* genes. Journal of Molecular Evolution **(in press)**.

Fani, R. Gallo, R. and Liò, P. (2000) Molecular evolution of nitrogen fixation: the evolutionary history of *nifD*, *nifK*, *nifE*, and *nifN* genes. Journal of Molecular Evolution 51, 1–11.

Fani, R. Liò, P. Chiarelli, I. and Bazzicalupo, M. (1994) The evolution of the histidine biosynthetic genes in prokaryotes: a common ancestor for the *hisA* and *hisF* genes. Journal of Molecular Evolution 38, 489–95.

Jensen, R.A. (1976) Enzyme recruitment in evolution of new function. Annual Review of Microbiology 30, 409–25.

Miller, S.L. (1953) A production of amino acids under possible primitive earth conditions. Science 117, 528–9.

CONSERVED OLIGOPEPTIDES IN THE RUBISCO LARGE CHAINS
An Evolutionary Perspective

P.B. VIDYASAGAR[1], PRATIP SHIL[2] and SARAH THOMAS[3]
[1]Professor(Biophysics), Department of physics, University of Pune, Pune, India-411007. [2]Research Fellow (Biophysics), Department of Physics, University of Pune, Pune, India-411007. [3]Lecturer, Bioinformatics Center, University of Pune, Pune, India-411007.

1. Introduction

The evolution of intelligent behavior observed in metazoan species requires a steady and sufficient supply of energy. Photosynthesis assumes the important role in the evolutionary mechanism in this respect. Since the basic mechanism of photosynthesis has remained unchanged in the evolutionary pathway, the study of the molecular evolution of the photosynthetic proteins has gained importance. Ribulose-1,5-bisphosphate carboxylase/oxygenase (RuBisCO) catalyses the first step in the CO_2 assimilation in Photosynthesis. RuBisCO exists as dimmer in case of prokaryotic autotrophs as revealed by the crystallographic structure in case of bacteria *Rhodospirilium rubrum* (Schinder et al 1986). It exists as a multimeric protein in case of the higher vascular plants viz Spinach (Chapman et al 1988) and Tobacco (Chapman MS et al 1987). In the present study amino acid sequence analysis of the RuBisCO large chains have been carried out using bioinformatics protocols.

2. Material and Methods

RuBisCO large chain complete sequences representative of different categories of plants have been retrieved from the databases. Multiple Sequences Alignment (MSA) was done using the CLUSTALW package. Phylogenetic tree has been constructed by using the treeing algorithm of the PHYLIP package (Felsenstein, 1985) from the profile output of the CLUSTALW. Pairwise alignment of the sequences has been carried out using the FASTA (version 3.0) program (Pearson, 1988) and the percentage amino acid identities have been determined. For the detailed analysis of the conserved regions the secondary structure prediction has been carried out using the web version of the Predict Protein package (Rost, 1993).

J. Seckbach et al. (eds.), Life in the Universe, 133–134.

3. Results

The MSA obtained from the CLUSTALW program reveals the occurrence of the conserved amino acid stretches i.e. oligopeptides, in all the species investigated in the study (Table 1)*. The FASTA Pairwise comparison of the sequences shows that the amino acid identity varies from 85%–96% among higher plants, 52–88% among algae and 37–82% among bacteria (Table 2). This indicated that the higher plants sequences are highly homologous while the bacterial sequences are the least. The Phylogenetic tree shows the clustering of the sequences broadly into three groups: A (archaeal sequences), B (bacterial and Algal sequences) and C (higher vascular plants) (Fig 1). Mismatch among the homologous regions are suggestive of the occurrence of point mutations in the evolutionary pathway. The pentapeptide "SGGIH" is conserved except in Archea. Other point mutations of relevance is the emergence of the mostly hydrophilic peptide "SDDGH" in higher plants from the hydrophobic counter parts "LGVDQ" in Archea and "IGVDQ" in bacteria (Table 3).

4. Conclusions

The analysis done here shows that the RuBisCO large chains are most diverse among the lower autotrophs viz. Archea, followed by the Bacteria and are highly conserved in case of the higher plants. However the efficiency of this enzyme has remained low along the evolutionary pathway. This low efficiency, as it is known, puts burden on the plants for producing the enzyme in large quantities. Since food production is the priority in case of autotrophs, energy is diverted for maintenance of the production-related machinery. This may be the reason that being autotrophs the plants did not develop intelligent behavior as animals. So, in the search of life elsewhere this point needs to be considered. One has to search for both autotrophs and heterotrophs if Darwinian evolution is assumed to be a universal phenomenon.

5. References

Chapman, S. and Won, S. (1987) Sliding layer conformational change limited by the quaternary structure of plant RuBisCO. Nature **329**, 354–356.
Chapman, S. and Won, S. (1988) Tertiary structure of Rubisco: Domains and their contacts. Science **241**, 71–74.
Schinder, G. and Lindquist, Y. (1986) Three-dimensional structure of ribulose-1, 5-bisphosphate carboxylase/ oxygenase from *Rhodospirillum rubrum* at 2.9 Å resolution. EMBO J. **5** no. 13, 3409–3415.
Rost, B. (1993) Prediction of protein structure at better than 70% accuracy. J Mol. Biol **232**, 584–599.
Felsenstein, J. (1998) Phylogeny interference package. Cladistics **5**, 164–166.
Pearson, W. and Lipman, D. (1988) Improved tools for biological sequence analysis. PNAS **85**, 2444–2448.
Pearson, W. R. (1998). Empirical statistical estimates for sequence similarity searches. Int. J. Mol. Biol. **276**, 71–84.

* [*For all Tables and Figures please visit* http://physics.unipune.ernet.in/~pratip].

ON THE QUESTION OF CONVERGENT EVOLUTION IN BIOCHEMISTRY

A. A. AKINDAHUNSI[1,2] and J. CHELA-FLORES[3]
[1]Department of Biochemistry, Biophysics and Macromolecular Chemistry, University of Trieste, Via Giorgieri, 1, I-34127, Trieste, Italy, [2]Permanent Institute: Department of Biochemistry, Federal University of Technology, Akure, Nigeria, [3]The Abdus Salam ICTP, I-34014, Trieste, Italy and Instituto de Estudios Avanzados, Caracas, Venezuela.

Abstract. Convergent evolution is an ubiquitous phenomenon, since it occurs at the levels of morphology, physiology, behavior (Eisthen and Nishikawa, 2002), and at the molecular level (Pace, 2001). Evolutionary convergence is significant for the central problem of astrobiology. Since all forms of life known to us are terrestrial, it is relevant to question whether the science of biology is of universal validity (Dawkins, 1983), and whether the molecular events that were precursors of the origin of life are bound to occur elsewhere in the universe wherever conditions are similar to the terrestrial ones. We discuss evolutionary convergence and its classification into functional, mechanistic, structural and sequence convergence (Doolittle, 1994), which should help us in the context of astrobiology defining bioindicators in the exploration of the solar system.

1. Introduction

The question "What would be conserved if the tape of evolution were played twice?" which is relevant to astrobiology has been raised repeatedly in the past (Fontana and Buss, 1994); it underlies one of the basic questions in astrobiology. Since all forms of life known to us are terrestrial organisms, it is relevant to the question of whether the science of biology is of universal validity (Dawkins, 1983; Chela-Flores, 2003).

The sharp distinction between chance (contingency) and necessity (natural selection as the main driving force in evolution) is relevant for astrobiology. Independent of historical contingency, natural selection is powerful enough for organisms living in similar environments to be shaped to similar ends. Our examples will favor the assumption that, to a certain extent and in certain conditions, natural selection may be stronger than chance (Conway-Morris, 1998; 2002). We raise the question of the possible universality of biochemistry, one of the sciences supporting chemical evolution.

2. Evolutionary Divergence and Convergence

The universal nature of biochemistry has been discussed from the point of view of the basic building blocks (Pace, 2001). One of the main points made in that paper is that it seems

J. Seckbach et al. (eds.), Life in the Universe, 135–138.

likely that the building blocks of life anywhere will be similar to our own. Amino acids are formed readily from simple organic compounds and occur in extraterrestrial bodies such as meteorites. Themes that are suggested to be common to life elsewhere in the cosmos are the capture of adequate energy from physical and chemical processes to conduct the chemical transformations that are necessary for life: lithotropy, photosynthesis and chemosynthesis. Other factors in favor of the universality of biochemistry are physical constraints (temperature, pressure and volume), as well as genetic constraints.

Divergence and convergence are two evolutionary processes by which organisms become adapted to their environments. Evolutionary convergence has been defined as the acquisition of morphologically similar traits between distinctly unrelated organisms (Austin, 1998). While there are many examples of molecular divergence, the same is not true of molecular convergence. Convergent evolution is said to occur when a particular trait evolves independently in two or more lineages from different ancestors. This is distinct from parallelism, which refers to independent evolution of a trait from the same ancestor. Although many of the best-known examples of convergence are morphological, convergence occurs at every level of biological organization. However, this paper focuses on molecular convergent evolution. We would like to proceed with examples of molecular convergence along the classification by Doolittle (1994) into Functional, Mechanistic, Structural and Sequence convergence.

2.1. FUNCTIONAL CONVERGENCE

This refers to molecules that serve the same function but have no sequence or structural similarity and carry out their function by entirely different mechanisms. Despite the fact that alcohol dehydrogenases in vertebrates and *Drosophila* bear no sequence similarity, and their tertiary structures are entirely different, they catalyze alcohol into acetaldehyde by different chemical reactions; they both remove hydrogen from alcohol (Doolittle, 1994).

2.2. MECHANISTIC CONVERGENCE

Mechanistic convergence occurs when the sequence and structure of molecules are very different but the mechanisms by which they act are similar. Serine proteases have evolved independently in bacteria (e.g. subtilisin) and vertebrates (e.g. trypsin). Despite their very different sequences and three-dimensional structures, they are such that the same set of three amino acids forms the active site. The catalytic triads are His 57, Asp 102, and Ser 195 (trypsin) and Asp 32, His 64 and Ser 221 (subtilisin), thus giving a consensus catalytic triads of the sort [Asp/Glu] His [Ser/Thr] (Doolittle, 1994, Tramontano, 2002).

2.3. STRUCTURAL CONVERGENCE

This refers to molecules with very different amino acid sequences that can assume similar structural motifs, which may carry out similar functions. For example, a helices and b sheets can be formed from a number of different amino acid sequences and are found in many proteins.

One example given by Doolittle (1994) is of the remarkable similarity in fibronectin type III and immunoglobulin domains (cf., Fig 1). They are composed of series of three

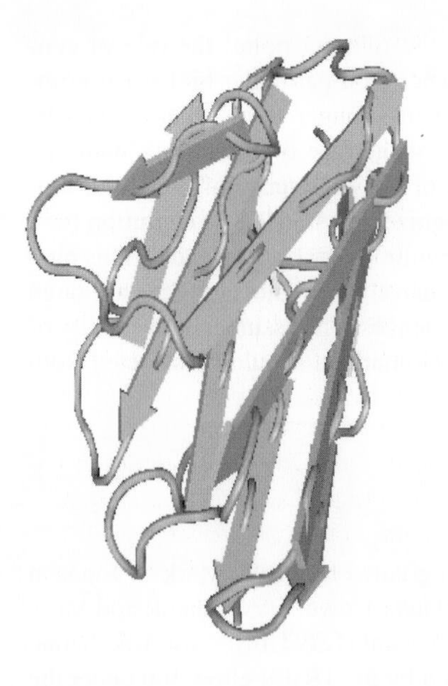

Figure 1a. Solution structure of the fibronectin type III domain from *Bacillus circulans* WL-12 chitinase A1 (Jee *et al.*, 2002).

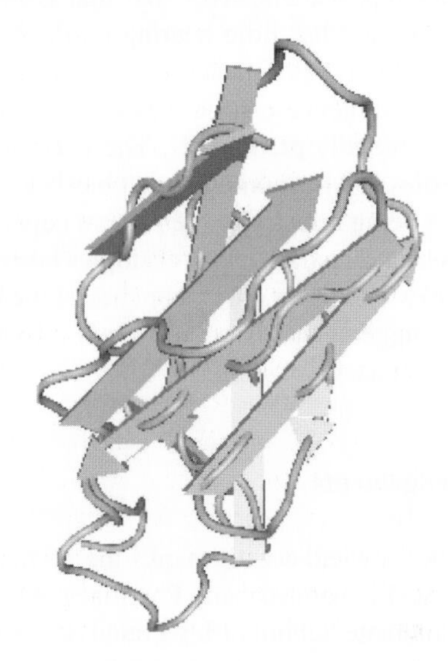

Figure 1b. Variable Light Chain Dimer of Ferritin Antibody (Nymalm *et al.*, 2002).

and four stranded b sheets that are virtually identical in structure despite a lack of sequence similarity between these two molecules.

2.4. SEQUENCE CONVERGENCE

In protein evolution, sequence divergence, rather than sequence convergence is the rule. In sequence convergence, one or more critical amino acids or an amino acid sequence of two proteins come to resemble each other due to natural selection. If the putative ancestral amino acids at a particular site were different in the ancestors of two proteins that now share an identical residue at that location, then convergent evolution may have occurred. The most frequently cited case of convergence and parallelism at the sequence level is the digestive enzyme lysozyme in a number of unrelated animals – the langur (a primate), the cow (an artiodactyls), and the hoatzin (a bird) – that have independently evolved the ability to utilize bacteria to digest cellulose (Kornegay, 1996; Zhang and Kumar, 1997). Specific residues have convergently evolved to allow digestion of cellulose-eating bacteria.

3. Discussion and Conclusions

We have assumed that natural selection seems to be powerful enough to shape terrestrial organisms to similar ends, independent of historical contingency. In an extraterrestrial environment, it could be argued that the evolutionary steps that led to human beings would probably never repeat themselves; but that is hardly the relevant point: the role of contingency in evolution has little bearing on the emergence of a particular biological property (Conway-Morris, 1998). Besides, it can be said in stronger terms that essentially, evolutionary convergence can be viewed as a 're-run of the tape of evolution', with end results that are broadly predictable. The inevitability of the emergence of particular biological properties is a phenomenon that has been recognized by students of evolution for a long time. It is being referred in the present paper as 'evolutionary convergence'. This phenomenon has been illustrated with examples taken exclusively from biochemistry, although its occurrence extends over other branches of the life sciences. The assumed universality of biochemistry suggests that in solar system missions, biomarkers should be selected from standard biochemistry.

4. Acknowledgements

The authors acknowledge with thanks the permission granted by Profs. Mark S. Johnson (Department of Biochemistry and Pharmacy, Abo Akademi University, Finland) and Masa Shirakawa (Graduate School of Integrated Science, Yokohama City University, Yokohama, Japan) for the use of the structures. AAA was supported by the TRIL Fellowship under the ICTP Programme for Training and Research in Italian Laboratories, Trieste, Italy.

5. References

Austin, D. F. (1998) Parallel and convergent evolution in the Convolvulaceae, In: P. Mathews and M. Sivadasan (eds.) *Diversity and Taxonomy of Tropical Flowering Plants*, Mentor Books, Calicut, India, pp. 201–234.

Chela-Flores, J. (2003) Testing Evolutionary Convergence on Europa. International Journal of Astrobiology (Cambridge University Press), in press.

Conway Morris, S. (1998) *The Crucible of Creation. The Burgess Shale and the Rise of Animals,* London, Oxford University Press, p. 202.

Conway Morris, S. (2002) First Steps towards defining galactic niches. Paper presented at the IAU Symposium 213 Bioastronomy 2002 *Life Among the Stars*. Summary of Proceedings, Hamilton Island, Great Barrier Reef, Australia July 8–12, Australian Centre for Astrobiology, Sydney, p. 12.

Dawkins, R. 1983, Universal Darwinism, In: D.S. Bendall (ed.) *Evolution from molecules to men*,., London, Cambridge University Press, pp. 403–425.

Doolittle, R. F. (1994) Convergent evolution: the need to be explicit. Trends Biochem. Sci., **19**, 15–18.

Eisthen, H.L. and Nishikawa K.C. (2002) Convergence: Obstacle or Opportunity? Brain Behav. Evol. **59** (5–6), 235–239.

Fontana, W. and Buss, L.W. (1994) What would be conserved if "the tape were played twice"? Proc. Natl. Acad. Sci. USA., **91**, 757–761.

Jee, J., Ikegami, T., Hashimoto, M., Kawabata, T., Ikeguchi, M., Watanabe, T and Shirakawa, M. (2002). Solution Structure of the Fibronectin Type III Domain from *Bacillus circulans* WL-12 Chitinase A1. J. Biol. Chem., **277**, 1388–1397.

Kornegay, J. (1996) Molecular genetics and evolution of stomach and nonstomachlysozymes in the hoatzin. J. Mol. Evol., **42**, 676–684.

Nymalm, Y., Kravchuk, Z., Salminen, T., Chumanevich, A.A., Dubnovitsky, A.P., Kankare, J., Pentikainen, O., Lehtonen, J., Arosio, P., Martsev, S. and Johnson, M.S (2002). Antiferritin VL homodimer binds human spleen ferritin with high specificity. J. Struct. Biol. **138**(3), 171–186.

Pace, N. R. (2001) The universal nature of biochemistry. Proc. Natl. Acad. Sci. USA **98**, 805–808.

Tramontano, A. (2002) Private communication.

Zhang, J. and S. Kumar (1997) Detection of convergent and parallel evolution at the amino acid sequence level. Mol. Biol. Evol., **14**, 527–536.

DIVERSITY OF MICROBIAL LIFE ON EARTH AND BEYOND
A Mini Review

JOSEPH SECKBACH
The Hebrew University of Jerusalem, ***Home:*** *P.O.B. 1132, Efrat 90435, Israel*

1. Introduction

Microorganisms occupy almost every habitable niche on Earth. They are abundant not only in "normal" environments but also thrive in very harsh habitats. These organisms, existing at the limits of life, have been designated "extremophiles." They may be found thriving in very severe growth conditions (from our anthropocentric point of view). Among the extremophiles are representatives of all three domains of life (*Bacteria*, *Archaea*, and *Eukarya*).

Life has existed on Earth at least for 3.5 to 3.8 billion years. Assumedly, the atmospheric conditions during the early period on Earth were different and more severe from those existing today. Life may have originated just below the Earth's surface and/or within deep dark oceanic hydrothermal vents.

Extremophilic microorganisms occupy almost every environmental niche, even those that are often considered to be totally inhabitable. These extremophiles may also serve as models for microbes that could perhaps live under the harsh conditions that exist in extraterrestrial environments. Celestial bodies that contain liquid water (as the main requirement for life) may harbor life. Among these places are Mars and the Jovian satellite, Europa. We should especially search for life on extrasolar Earth-like planets, where conditions may be suitable for life (as we know it). A review by Nisbet and Sleep (2001) covers some of the above aspects of early life.

2. The Extreme Environments

The study of the extremophiles and of organisms that live in habitats previously thought to be uninhabitable has become a major interest for those involved in the study of the origin of life. Although most extremophiles are *Bacteria* and *Archaea*, many species of *Eukarya* live at the edge of life's normal environment (Roberts, 1999). Also, oxygenic photosynthetic microorganisms (such as algae, in addition to the cyanobacteria) thrive in extreme environments. In this short paper I will briefly survey a few major groups of extremophiles and the importance of the understanding of their biology for searching for life elsewhere in the Universe (see also the Astrobiology web site that covers most of the extremophiles: http://www.astrobiology.com/extreme.html).

J. Seckbach et al. (eds.), Life in the Universe, 139–142.

3. Some Extremophiles and their Habitats as Models for Astrobiology

Microorganisms have been observed to grow in all ranges of temperature, from −20°C (Junge et al., 2004; Thomas and Dieckmann, 2002) up to 113°C (Blochl et al., 1997). The psychrophiles live in cold and freezing areas, while the thermophiles and hyperthermophiles live in warm and hot environments, respectively. Most phylogenetic models predict that the first microorganisms may have been hyperthermophiles (Rossi et al., 2003; Seckbach, 1994b). Following the Hadean era, Earth's subsurface remained hot due to the meteorite bombardments; the oceans might have been heated up to 100°C (Rossi et al., 2003).Hyperthermophiles could have been the first living pioneers within such hot environments, or the only survivors following such sterilizing hot events. It has been proposed that such hyperthermophiles may also serve as candidates for microorganisms in celestial bodies that may have similar physical conditions as occur in extremely hot environments on Earth. Hyperthermophiles may serve not only for scientific research but also for industrial application (Rothschild and Mancinelli, 2001).

Organisms living in cold habitats are called psychrophiles. They are distributed in places such as the Arctic and the Antarctic, where sea-ice organisms thrive in the ice (Thomas and Dieckmann, 2002; Junge et al., 2004). Other psychrophiles live in the permafrost of Siberia. Some microbes can resist a frozen period and may survive subzero temperatures, and then germinate under warmer conditions. Furthermore, bacterial spores are almost immortal and can retain their viability for millions of years under harsh conditions and finally be revived. Bacteria have been recovered and revived following a long stay on the moon and in space. They have tolerated severe cold, lack of an atmosphere, and high UV radiation and still have been revived upon returning to Earth. In Antarctic ice there are many planktonic organisms including bacteria, algae (most conspicuous are the pinnate diatoms) protists, flatworms and small crustaceans (Thomas and Dieckmann, 2002). Bacterial activity has been documented at −20°C—making the limits of life on Earth wider.

Thermophiles, or the heat loving microorganisms, are ubiquitously found in hot springs and hot locations, such as the Solfatara volcanic area near Naples, Italy. The red unicellular thermoacidophilic *Cyanidium caldarium* is an enigmatic alga (Seckbach, 1992. 1994a). This rhodophytan grows in very low pH areas, at temperatures up to 56–57°C. It thrives in media very well bubbled with pure CO_2 (Seckbach et al., 1970). Its cohorts in the family Cyanidiacean cohabit the same elevated temperature and low pH environments (Seckbach, 1992). Thermophilic cyanobacteria inhabit an even higher temperature range, living in neutral or alkaline hot springs at temperatures of up to ~70°C. *Bacteria* and *Archaea* are abundant in high temperatures and inhabit most hot environments.

Among the hyperthermophiles are the prokaryotes, *Bacteria* and *Archaea* that have been observed distributed from 80°C to temperatures higher than 100°C. These microbes occur in deep-sea hydrothermal vents, including in black smoker chimney structures (Takai et al., 2001). Stetter and his coworkers have determined the uppermost temperature of life in *Pyrolobus fumarii* at 113°C (Blochl et al., 1997), and this organism can survive incubation in an autoclave at 121°C for over 1 hour. Such hyperthermophiles may have been the initial microbes that evolved in the depth of the hot oceans, or the only survivors in the primeval environment (Rossi et al., 2003). In the depth of the subsurface and deep in the oceans these microbes could also have been protected from harmful UV radiation during the primeval era of Earth, as well as shielded from the meteorites' impact during the early states of the Earth.

The **acidophiles** thrive in lower ranges of pH (Seckbach, 2000) such as in sulfur hot springs, and in volcanic Solfatara soil (near Naples, Italy). On the other side of the pH scale are the **alkaliphiles** that live in alkaline environments, such as in the African soda lakes.

The **halophiles** are organisms living in very salty environments, such as in salt lakes, salt mines or saline ponds sometimes containing saturated salt solutions. In the seas and oceans the concentration of the salts is 34 g per liter, while the Dead Sea may reach 10 times as much dissolved salts. Some halophiles (e.g., the green alga *Dunaliella salina*) accumulate organic compounds, such as glycogen or β carotene. Its internal content of glycerol balances the external high osmotic pressure. High brine salinities may cause major dehydration stress for ice-trapped organisms. Square and triangular halophilic *Archaea* have been reported in salt media (Oren, 1999). A recent comprehensive volume on the halophilic world has been published by Oren (2002).

Those organisms that are able to tolerate dry conditions are the **xerophiles**. Bacteria (or their spores) have known to last in desiccating conditions for many years. *Bacillus sphaericus* has been isolated from extinct bees trapped in 25–40 million year (Myr) old amber. Vreeland et al. (2000) claimed to have isolated a 250-Myr-old halotorelant bacterium from a salt crystal.

Other microorganisms have been shown to resist **UV** radiation (e.g., *Deinococcus radiodurans).* The **barophiles** and **piezophiles** live under high pressures in the depths of the oceans (see hyperthermophiles, above) and deep underground. There are also microbes that consume oil spills in the oceans, grow in the presence of high concentrations of metal ions and even tolerate toxic compounds.

On Earth no living organism can exist without liquid water. The search for extraterrestrial life should be "follow the water": in any celestial body where there is liquid water there might be life. Meteorites from Mars that have landed on Earth have been reported to show some signatures of life, but the evidence has not been widely accepted. The NASA images of Mars suggest that in the past this planet had plenty of running water (e.g., rivers, lakes, canyons, oceans, etc.). The nature of the presumed nanofossils observed in ALH84001 is highly controversial, due to their petite size and laboratory-reconstructed artificial minerals that have the same appearance (Reitner, 2004). No consensus has been reached on whether life exists or even existed on Mars. Early Mars may have been an eminently habitable place. Had life existed on Mars, the ejecta during the impact era could have carried one or more cells within a rock or meteorite to infect the planet Earth. Another promising heavenly body is Europa, the Jovian moon. Deep under its surface there is a lake of liquid water. If in the Vostok station in Antarctica the subsurface lake contains microorganisms, similar microbes may also exist under the surface of Europa. Subsurface water on Mars has been assumed to exist as subsurface brine aquifers, while in early time the water on Mars was free on the surface.

Life may have existed on Venus within the early cytherean oceans; the possibility that life may have existed on the planet should therefore not be totally rejected (Seckbach and Libby, 1970). Terrestrial microorganisms can live in similar conditions as primordial Venusians under high CO_2, elevated temperature, and even tolerate sulfuric acid. One such microorganism is *Cyanidium caldarium* (see above) (Seckbach, 1992; 1994a; Seckbach and Libby, 1970).

4. Conclusions

The extremophiles grow in very severe environments that have been considered until recently totally inhabitable. They occur in very harsh habitats, thus enlarge our knowledge about the limits of life on Earth. If some celestial bodies have liquid water and other conditions required for life, such extremophiles may exist there.

5. References

Blochl, E., Rachel, R., Burggraf, S., Hafenbradl, D., Jannasch, H.W. and Stetter, K. O. (1997) *Pyrolobus fumarii*, Gen. And Sp. Nov., represents a novel group of archaea, extending the upper temperature limit for life to 113 degrees C. Extremophiles, **1**: 14–21.

Junge, K., Eicken, H. and Deming, J.W. (2004) Bacterial activity at −2 to −20°C in arctic wintertime sea ice. Appl. Environ. Microbiol., **70**: 550–557.

Nisbet, N. G. and Sleep, N. H. (2001) The habitat and nature of early life. Nature, **409**: 1083–1091.

Oren, A. (1999) The enigma of square and triangular halophilic Archaea. In: J. Seckbach (ed.) *Enigmatic Microorganisms and Life in Extreme Environments.* Kluwer Academic Publishers, Dordrecht, The Netherlands. pp. 338–355.

Oren, A. (2002) *Halophilic Microorganisms and their Environments.* vol. 5 of *Cellular Origins, Life in Extreme Habitats and Astrobiology (COLE)* Series editor J. Seckbach. Kluwer Academic Publishers, Dordrecht, The Netherlands.

Reitner, J. (2004) Organomineralisation—an Assumption to understand meteorite-related bacteria-shaped carbonates. In: J. Seckbach (ed.) *Origins: Genesis, Evolution and Diversity of Life.* Kluwer Academic Publishers, Dordrecht, The Netherlands, in press.

Roberts, D. M[c]L. (1999) Eukaryotic cells under extreme conditions. In: J. Seckbach (ed.) *Enigmatic Microorganisms and Life in Extreme Environments.* Kluwer Academic Publishers, Dordrecht, The Netherlands. pp. 163–173.

Rossi, M., Ciaramella, M., Cannio, R. Pisani, F. M., Moracci, M. and Bartolucci, S. (2003) Extremophiles 2002. J. Bacter. **183**: 3683–3689.

Rothschild, L.J. and Mancinelli, R.L. (2001) Life in extreme environments. Nature, **409**: 1092–1101.

Seckbach, J. (1992) The Cyanidiophyceae and the "anomalous symbiosis" of *Cyanidium caldarium.* In: W. Reisser (ed.) *Algae and Symbioses: Plants, Animals, Fungi, Viruses, Interactions Explored.* Biopress Ltd. Bristol, UK. pp. 339–426.

Seckbach, J. (ed.) (1994a) *Evolutionary Pathways and Enigmatic Algae*, Kluwer Academic Publishers, Dordrecht, The Netherlands.

Seckbach (2000) Acidophilic microorganisms, In: J. Seckbach (ed.) *Journey to Diverse Microbial Worlds.* Kluwer Academic Publishers, Dordrecht, The Netherlands. pp. 107–116.

Seckbach, J. (1994b) The first eukaryotic cells—acid hot-spring algae: Evolutionary paths from prokaryotes to unicellular red algae via *Cyanidium caldarium* (PreRhodophyta) succession. J. Biol. Phys. **20**: 335–345.

Seckbach, J. and Libby, F. W. (1970). Vegetative Life on Venus? Or investigations with algae which grow under pure CO_2 in hot media at elevated pressures. Space Life Sci., **2**: 121–143.

Seckbach, J., Baker, F.A. and Shugarman, P.M. (1970) Algae thrive under pure CO_2. Nature, **227**: 744–745.

Takai, K., Komatsu, T., Inagaki, F. and Horikoshi, K. (2001) Distribution of Archaea in a black smoker chimney structure, Appl. Environ. Microbiol. **67**: 3618–3629.

Thomas, D.N. and Dieckmann, G. S. (2002) Antarctic sea ice-a habitat for extremophiles. Science, **295**: 641–644.

Vreeland, R.H., Rosenzweig, W.D. and Powers, D.W. (2000) Isolation of a 250 million-year-old halotolerant bacterium from a primary salt crystal. Nature, **407**: 897–900.

Note: I thank Professors **Julian Chela-Flores, Aharon Oren** and **Tobias Owen** for their proofreading of this manuscript and providing practical revisions.

V. Alternative Scenarios for the Origin and Evolution of Life

MINERAL SURFACES AS A CRADLE OF PRIMORDIAL GENETIC MATERIAL

ENZO GALLORI, ELISA BIONDI AND MARCO FRANCHI
Department of Animal Biology & Genetics, University of Florence, Via Romana, 17; 50125 Florence, Italy

Abstract. Molecules which store genetic information (DNA and RNA) are central to all life on Earth. The formation of these complex macromolecules, and ultimately life, required specific conditions, including the synthesis and polymerization of precursors (nucleotides), the protection and persistence of information polymers in a changing environment, and the expression of the "biological potential" of the molecules, i.e. their capacity to multiply and evolve. Determining how these steps occurred and how the earliest genetic molecules originated on Earth is a problem that is far from being resolved. Recent observations on the synthesis of polynucleotides on clay surfaces, the resistance of clay-adsorbed nucleic acid molecules to environmental degradation and the biological activity of clay-adsorbed DNA and RNA molecules suggest that mineral surfaces could have played a crucial role in the prebiotic formation of the biomolecules basic to life.

1. Introduction

All life forms on Earth today, and all life for which there is fossil evidence, are based on the DNA molecule, which is made up of nucleotides. It is the sequence of these subunits that supplies the language of life. These are the genes, which code for all information about a particular life form. The specific sequence of nucleotides is handed down from one generation to the next by DNA replication, thus ensuring the perpetuation of the "genetic information" on Earth. At present, we can say that the beginning of life coincided with the appearance of a nucleic acid-like polymer able to undergo evolution through processes of replication, mutation and natural selection. The formation of genetic material is thus the starting point for any discussion of the origins of life on our planet, and perhaps any other.

The building of primordial genetic polymers entails at least four fundamental steps: the synthesis and availability of precursors (e.g. nucleotides), the joining of these precursors into macromolecules (e.g. polynucleotides), the protection of polymers against degradation (e.g. by cosmic and UV radiation), thus ensuring their persistence in a changing environment, and—finally—the expression of the "biological" potential of the molecule, i.e. its capacity to self-replicate and evolve. Determining how these steps occurred and how the primordial genetic molecules originated on Earth is a very difficult problem that still must be resolved (Lazcano and Miller, 1996).

J. Seckbach et al. (eds.), Life in the Universe, 145–148.

With regard to the first two steps—the synthesis of precursors of biological macromolecules and their polymerization—classical research in this field has focused on processes in aqueous solutions, in the belief that building blocks of the biomolecules could be readily obtained by chemical reactions involving simple organic compounds present on Earth's surface. However, the fact that the right components were present in primeval habitats is not sufficient by itself to explain the appearance of complex molecules. The creation of these macromolecules required the polymerization of single components. As biological polymers are generally formed by dehydration, it is difficult to conceive that complex macromolecules could have originated by random collisions in the presence of a high concentration of water, like that of a primordial ocean. In these conditions hydrolysis is favoured, not polymerization (Pace, 1994).

This problem is exemplified by the properties of RNA. It is currently believed (Joyce, 2002) that in an era indicated as the "RNA World", the RNA molecule could have functioned both as genetic material and as an enzyme ("ribozyme") (Doudna and Cech, 2002). The presence of the 2'-OH group in ribose, which renders RNA catalytic, also makes the molecule particularly susceptible to hydrolysis. Therefore, the very complex structure of RNA and its intrinsic instability make it very difficult to imagine the origin of a hypothetical RNA World in free solution.

It has long been suggested that surface chemistry on clays or other minerals was involved in the prebiotic chemical evolution that culminated in the origin of life. In 1951, J.D. Bernal suggested that clay minerals could have bound organic molecules from the surrounding water, concentrating them and protecting biomolecules against destruction by high temperatures and strong radiation. In recent years, numerous observations have reinforced this hypothesis. Ferris *et al.* (1996) demonstrated the polymerization of oligonucleotides up to the length of a small ribozyme on montmorillonite clay. More recently, Huang and Ferris developed a new method for the synthesis of RNA oligomers in the presence of clays, in a "one step reaction" without the need of a primer (Huang and Ferris, 2003). Smith *et al.* (1998, 1999) provided a theory for the assembly of biopolymers on silica-rich minerals resembling zeolites. In addition, experimental data in the field of molecular microbial ecology have strengthened the hypothesis of a surface-mediated origin of life. Studies carried out on the "fate" of DNA in soil habitats have indicated that DNA originating from dead or living cells can persist for a long time in the environment and still maintain its biological activity as a result of its interaction with clay particles (Stotzky *et al.*, 1996). All these observations suggest that mineral surfaces could have played an important role in the formation and accumulation of genetic molecules in primeval terrestrial habitats (Franchi *et al.*, 1999), promoting their polymerization and allowing their persistence in an inhospitable environment like that of early Earth.

Nevertheless, for some sort of RNA World to develop, genetic polymers not only had to accumulate, they also had to interact with surrounding molecules. In other words, genetic polymers must have been able to acquire new specialized functions (catalysis, information, etc.) in order to self-organize spontaneously into the first self-replicating "living" systems.

For these reasons, we decided to investigate the physical-chemical and biological characteristics of nucleic acid-clay complexes in order to better understand their possible role in the origin of life.

2. Nucleic Acid-Clay Complexes

Nucleic acid-clay complexes were obtained by reacting DNA and RNA with two clay minerals, montmorillonite (M) and kaolinite (K), as described by Franchi *et al.* (1999). They were extensively analyzed by different techniques to determine the nature of the interaction of the nucleic acids with the clay particles. X-RD showed that DNA molecules do not intercalate the Al-Si layers of the clays, indicating that the adsorption occurs primarily on the external surface of the particles, as also suggested by electron microscopy analysis. FT-IR spectra of clay-DNA complexes showed a change of the nucleic acid conformation, with a transition from the B to the A form, as a consequence of its adsorption on the mineral particles. Mono- and divalent cations take part directly in the adsorption/binding of nucleic acids on clay particles, acting as a "bridge" between the negative charges on the mineral surface and those of the phosphate groups of the genetic polymer. Double-stranded DNA needs higher cation concentrations than single-stranded DNA to establish an interaction with clay. This suggests that the number of strands in nucleic acid molecules was an important selection factor for their persistence in prebiotic habitats, influencing their adsorption/release by clay minerals (Franchi *et al.*, 2003).

Studies on the resistance of nucleic acids (DNA and RNA) to degradation have indicated that both chromosomal and plasmid DNA bound to montmorillonite and kaolinite are protected from the activity of endonucleases (Gallori *et al.*, 1994). Similar results have been obtained for the RNase-A dependent degradation of 16S rRNA bound on clays. Adsorption of DNA on clay minerals also increases its resistance to UV radiation, as shown by a reduction of the number of pyrimidine dimers (T-T).

The transition from the B to the A form of the DNA double helix could partially account for the increased resistance of nucleic acid-clay complexes to environmental degradation: the transformation makes the molecule less available to biotic and abiotic degrading agents.

3. Biological Properties of Nucleic Acids Adsorbed on Clay

Adsorption/binding of the DNA molecule on mineral particles does not prevent its biological activity. Chromosomal and plasmid DNA bound to M and K retain the ability to transform competent bacterial cells for a long time, even after enzymatic digestion (Gallori *et al.*, 1994). Moreover, clay-bound DNA can be enzymatically replicated and amplified outside the cellular context by the Polymerase Chain Reaction (PCR) (Vettori *et al.*, 1996), confirming that biological information stored in DNA is not lost because of its adsorption to clay. Results of reverse transcriptase and amplification (RT-PCR) of 16S ribosomal RNA from the bacterium *Escherichia coli* indicate that the same is true for adsorbed RNA molecules. These observations support the hypothesis that nucleic acid complexes could have acted as a "storage" of genetic information in primeval habitats.

The possibility of further spontaneous organization was evaluated by studying the molecular reactivity of clay-adsorbed genetic polymers. Single-stranded RNA (Poly[A]) adsorbed on clay was able to recognize complementary molecules (Poly[U]) present in solution and to form double-stranded stretches (Poly[A]·Poly[U]), showing that clay-adsorbed RNA molecules are able to establish specific interactions with surrounding biopolymers.Moreover, the catalytic activity of "hammerhead" ribozymes, i.e. self-cleavage of the molecule via transesterification (2',3' cyclic phosphate) (Doudna and Cech, 2002), is currently being tested. Finally, the viroid PSTVd (Potato Spindle Tuber Viroid) is able to maintain its infectivity when adsorbed on montmorillonite and kaolinite, even after enzymatic digestion. This is particularly important since viroids are the only nucleic acid molecules (RNA) known to undergo replication without DNA intermediates and without coding for proteins. These characteristics, together with the frequent presence of active ribozymes inside these particular RNA molecules, suggest that they may be relics of the previously cited "RNA World" (Diener, 2001).

4. Conclusions

The experimental results summarized above, together with previous observations of this and other research groups, suggest that the formation of a close association between prebiotic genetic molecules (whatever they were) and mineral surfaces could have been a crucial step in the origin of life on Earth. Clay minerals may not only have promoted the formation and accumulation of genetic material in prebiotic environments, but also allowed its self-organization into complex molecular systems that could self-replicate.

5. References

Bernal, JD (1951): The physical basis of life. Routledge & Kegan Paul, London.

Diener, T.O. (2001) The viroid: biological oddity or evolutionary fossil?, *Advances in Virus Research* **57**, pp. 137–184.

Doudna, J.A. and Cech, T.R. (2002), The chemical repertoire of natural ribozymes, *Nature* **418**, pp. 222–228;

Ferris, J.P., Hill, A.R. Jr., Liu, R. and Orgel, L.E. (1996), Synthesis of long prebiotic oligomers on mineral surfaces, *Nature* **381**, pp. 59–61.

Franchi, M., Bramanti, E., Morassi Bonzi, L., Orioli, P.L., Vettori, C. and Gallori, E. (1999), Clay-nucleic acid complexes: characteristics and implications for the preservation of genetic material in primeval habitats, *Origins Life Evol Biosphere* **29**, pp. 297–315.

Franchi, M., Ferris, J.P. and Gallori, E. (2003), Cations as mediators of the adsorption of nucleic acids on clay surfaces in prebiotic environments, *Origins Life Evol Biosphere* **33**, pp. 1–16.

Gallori, E., Bazzicalupo, M., Dal Canto, L., Fani, R., Nannipieri, P., Vettori, C. and Stotzky G. (1994), Transformation of *Bacillus subtilis* by DNA bound on clay in non-sterile soil, *FEMS Microbiol Ecol* **15**, pp. 119–126.

Huang, W. and Ferris, J.P. (2003), Synthesis of 35-40mers of RNA oligomers from unblocked monomers. A simple approach to the RNA World, *Chemical Communications* pp. 1458–1459.

Joyce, G.F. (2002), The antiquity of RNA-based evolution, *Nature* **418**, pp. 214–221.

Lazcano, A. and Miller, S.L. (1996), The origin and early evolution of life: prebiotic chemistry, the pre-RNA world, and time. *Cell* **85**, pp. 793–798.

Smith, J.V. (1998), Biochemical evolution. I. Polymerization on internal, organophilic silica surfaces of dealuminated zeolites and feldspars, *Proc Natl Acad Sci USA* **95**, pp. 3370–3375.

Smith, J.V., Frederick, P.A. Jr., Parsons, I. and Lee, M.R. (1999), Biochemical evolution. III. Polymerization on organophilic silica-rich surfaces, crystal-chemical modeling, formation of first cells, and geological clues, *Proc Natl Acad Sci USA* **96**, pp. 3479–3485.

Stotzky, G., Gallori, E. and Khanna, M. (1996), In: A.D.L., van Elsas, Y.D. and de Brujin, F.J. (eds) *Molecular Microbial Ecology Manual*. Akkermans, , Dordrecht: Kluwer Academy Publishers, pp 1–28.

Vettori, C., Paffetti, D., Pietramellara, G., Stotzky, G. and Gallori, E. (1996), Amplification of bacterial DNA bound on clay minerals by the random amplified polymorphic DNA (RAPD) technique, *FEMS Microbiol. Ecol.* **20**, 251–260.

ADSORPTION AND SELF-ORGANIZATION OF SMALL MOLECULES ON INORGANIC SURFACES
Some Applications to the Origin of Life

DONALD G. FRASER
Department of Earth Sciences, University of Oxford, Parks Road, OXFORD OX1 3PR, U.K.

Abstract. Recent laboratory studies have shown that the polymerization of monomeric bases to form oligomeric nucleotide sequences, requires the presence of a solid catalyst such as the mineral montmorillonite. The formation in nature, at least of prebiotic precursor materials, may therefore also have required the presence of active mineral surfaces. The nature of reactions between organic precursors and mineral substrates is thus of considerable interest in studies of the origin of life. This paper reviews results obtained for the process of H_2O chemisorption on MgO and Mg_2SiO_4 surfaces from in situ synchrotron X-ray reflectivity studies, ^{1}H-^{29}Si cross polarisation NMR spectra and ab initio calculations. The results show that the exact nature of molecular attachment and reactions on oxide and silicate surfaces can now be characterized with precision. Understanding the details of heterogeneous catalysis on mineral surfaces is a key step in understanding the processes that led to the origin of life.

1. Introduction

Recent studies of the polymerisation of bases to form oligonucleotide sequences have shown that, at least for purine and pyrimidine, silicate catalysts are required in the form of added montmorillonite (Ferris et al. 2003). In the presence of this clay mineral, up to at least 50-mer nucleotide sequences have been produced synthetically. The activity and complexity of hydroxy-Fe-montmorillonite surfaces have been highlighted in other recent work (Liao and Fraser 2002) and an increasingly strong case can be made for the involvement of condensed phase surfaces in catalysing the production of many prebiotic complex organic chemicals. The existence of chirality in crystalline phases such as quartz and calcite may also be important in selectively determining the production of particular prebiotic enantiomers (e.g. Hazen et al. 2001). Adsorption of chiral molecules onto a non-chiral substrate can even lead to the formation of new chiral facets and channels in a non-chiral substrate (Switzer et al. 2003). Around 72 amino acids have been identified in meteorites (Botta et al. 2001). The amino acid, glycine, has also been observed spectroscopically in interstellar space (Kuan et al. 2003) and studies of infra-red absorption indicate that interstellar dust contains material of Mg/Si ratio identical to olivine (Snow 2000). The existence of organic precursors in space and in meteorites, coupled with observations that silicate catalysts are required to assist laboratory-induced polymerisation, suggests that interactions between small molecules and mineral surfaces play a key role in chemical evolution and the origin of life.

J. Seckbach et al. (eds.), Life in the Universe, 149–152.

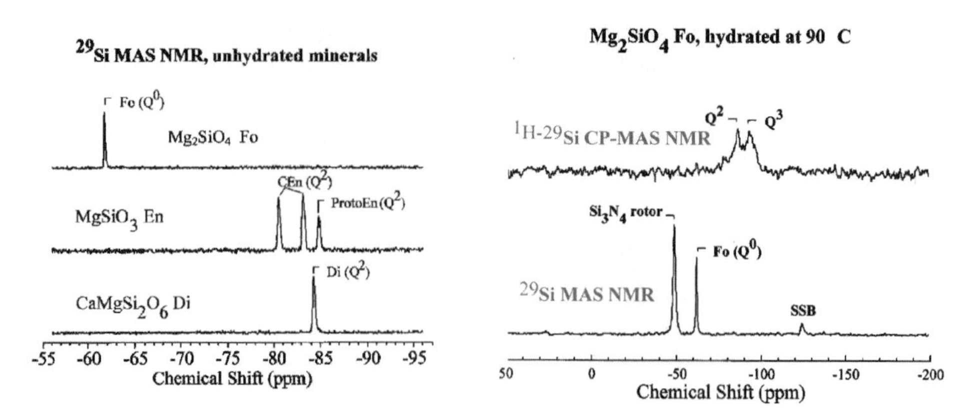

Figure 1. MAS NMR spectra of unhydrated and hydrated forsterite, Mg_2SiO_4.

2. Reactions at the Olivine Surface

In several recent papers we have reported experimental and ab initio theoretical calculations of the precise nature of the interaction of water molecules with inorganic surfaces (e.g. Mejias et al. 1999, Xue et al. 2002, Wogelius et al. 1998). Forsteritic (approx Fo_{90}) olivine is a common constituent of meteorites and makes up, in low and high pressure forms, some 45% of the Earth's mantle in composition. In situ, glancing incidence X-ray reflectivity studies of reactions of H_2O with polished Fo_{92} surfaces show the development of a hydrated reaction layer tens of nm thick (Wogelius et al. 1998).

To study the nature of this reacted surface more closely, MAS NMR measurements were carried out (Xue et al. 2002) on pure anhydrous and hydrated synthetic forsterite powders produced by sintering stoicheiometric amounts of natural abundance MgO and SiO_2 at 1bar 1,500°C. The ^{29}Si spectra of the anydrous powders were determined and are shown together with those of synthetic enstatite and diopside in Fig.1. The unground forsterite powders were reacted with pure H_2O at 90°C for two days and then washed and dried.

^{1}H-^{29}Si cross polarisation spectra select ^{29}Si resonances in Si atoms in immediate proximity to H atoms. The ^{1}H-^{29}Si cross polarisation spectra are also shown in Fig. 1 and give a remarkable result. The single sharp ^{29}Si peak at -62 ppm of anhydrous forsterite and characteristic of Si atoms in isolated Q^0 tetrahedra remains unchanged in the spectrum of the bulk hydrated sample. However by selecting only surface Si atoms in the hydration zone, the ^{1}H-^{29}Si cross polarisation spectrum shows polymerisation of the surface with only Q^2 and Q^3 groups present. Not only does exposure to water lead to formation of a hydrated surface, but the surface itself is active and polymerizes in response, forming a new surface composed of chain and sheet silicate groups.

3. Ab Initio Calculations

Theoretical and computational advances in the past decade have made possible accurate ab initio calculations of the energetics of systems of considerable complexity. In an attempt to

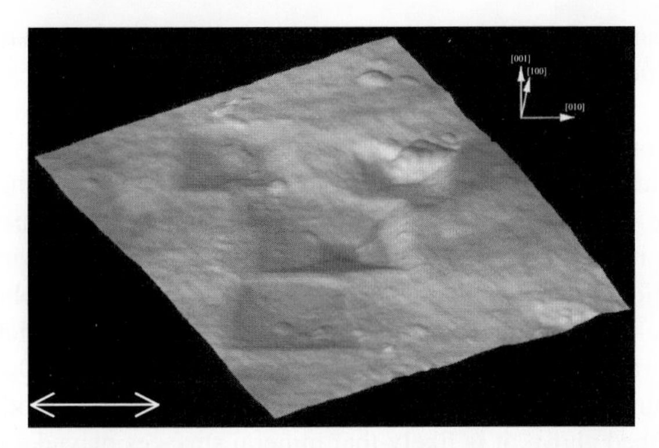

Figure 2. AFM image of an MgO (100) cleavage surface etched lightly in water for 24 hours. The dissolution etch pits caused by the detailes of the reaction of water molecules with specific crystallographic planes. Scale bar is approximately 1.2 μm. Note axes for orientation.

study the nature of the water attachment process in more detail, calculations were made of the hydration energies of MgO surfaces using density functional theory with plane-wave pseudopotentials under GGA as described elsewhere (Mejias et al. 1999). This method gives bond lengths accurate to 2% and energies within $10 kJ.mol^{-1}$ in the system $MgO-SiO_2$. The energetics of water molecule chemisorption on precisely defined crystallographic surfaces were calculated for the systems: MgO(100) – H_2O; MgO(111) – H_2O; MgO(110) – H_2O; MgO(130) – H_2O. The simple (100) cleavage surface of the rock-salt structure MgO is unstable to chemisorption with $\Delta E = +130 \text{ kJ.mol}^{-1}$. Monolayer chemisorption on (111), in contrast, is energetically favoured ($\Delta E = -20 \text{ kJ.mol}^{-1}$). Hydration energies of (110) and (130) steps or terraces are not distinguishable from zero.

An AFM image of a hydrated MgO surface is shown in Fig 2. The surface is a (100) cleavage surface exposed to pure water for 24 hours at approx 293K. This image shows that reaction of water molecules at the surface proceeds, not by dissolution on the (100) surface itself, but down (hk0) step edges and terraces on the surface as suggested by the calculations. H_2O chemisorption is kinetically determined and the presence of defects such as steps on mineral surfaces may play a critical role in determining active sites for small molecule adsorption and catalysis.

4. Conclusions

The identification of organic molecules, including amino acids, in space (Kuan et al. 2003), together with the presence of suitable silicate and other inorganic dust particles (Snow 2000) indicates that abiotic precursors can form in extra-terrestrial conditions. The presence of active mineral surfaces may turn out to be crucial in the organization of small molecules to the point of carrying enough information to self-replicate (Ferris 2003) and recent work (Liao and Fraser 2002) shows the activity and complexity of hydroxy-Fe-montmorillonite surfaces. The importance of silicate lattice cages and surfaces as catalysts for organic chemical reactions is well-known (e.g. Rozanska et al. 2003) and zeolites filling vesicles in hydrated basalts could also provide suitable and protected environments for the assembly, at least, of abiotic precursors.

What is not clear is whether or not initial prebiotic synthesis took place on Earth under near-surface conditions or occurred elsewhere. There is strong geological evidence for the existence of liquid water and sedimentary processes on the Earth at 3.85 Ga. The Earth's remaining and re-outgassed new atmosphere following the late heavy bombardment at around 4Ga must have contained large amounts of water vapour and mineral dust on the nano-particle scale. This mixture would have reached high altitudes near the top of the atmosphere where it would have been exposed to high levels of ultraviolet radiation and free-radical formation. Even this high-energy environment could have played an important role in seeding the surface with suitable organic precursors from which self-replicating molecules could form under lower temperature aquatic conditions (cf. Dobson et al. 2000). Such terrestrial sources are, of course, additional to any meteoritic introduction of organic material. As described above, it is likely that mineral surfaces played a central role in all these processes.

5. Acknowledgements

It is a pleasure to acknowledge the contributions of my co-workers at various times; in particular, Roy Wogelius, Keith Refson, Jose-Antonio Mejias, Andy Berry, Xian-yu Xue and Masami Kanzaki.

6. References

Botta, O., Glavin, D. P. Kminek, G. and Bada, J. L. (2001) Classification of Carbonaceous Meteorites Through Amino Acid Signatures? Lunar Planetary Science, XXX11.

Ferris, J.P., Huang, W., Joshi, P., Miyakawa, S., Pitsch, S. and Wang, K-J.(2003) Catalysis and the emergence of the RNA world. Geochim.Cosmochim. Acta, **67**, A96.

Dobson, C.M., Ellison, G.B., Tuck, A.F. and Vaida, V. (2000) Atmospheric aerosols as prebiotic chemical reactors. Proc. Nat. Acad. Sci., 97, 11864–11868.

Hazen, R.M., Filley, T.R. and Goodfriend, G.A. (2001) Selective adsorption of L- and D-amino acids on calcite: Implications for biochemical homochirality. Proc. Nat. Acad. Sci., **98**, 5487–5490.

Kuan,Y-J.,Charnley,S.B., Hui-Chun Huang, Tseng, W-L. and Kisiel, Z. (2003) Interstellar Glycine. The Astrophysical Journal, 593:848–867.

Liao, L. and Fraser D.G. (2002) The adsorption of As onto hydroxy-Fe-montmorillonite complexes. Geochim. Cosmochim. Acta. 66, A455.

Mejias, J.A., Berry, A.J., Refson, K. and Fraser, D.G. (1999) The Kinetics and Mechanism of MgO Dissolution. Chem. Phys. Letters. 314, 558–563.

Rozanska X, van Santen RA, Demuth T, Hutschka F and Hafner J (2003) A periodic DFT study of isobutene chemisorption in proton-exchanged zeolites: Dependence of reactivity on the zeolite framework structure. J Phys Chem B 107 (6): 1309–1315.

Snow, T.P. (2000) Composition of interstellar gas and dust. J. Geophys. Res. A10, 10239–11248.

Switzer, J.A., Kothari, H.M., Poizot, P., Nakanishi S. and Bohannan, E.W. (2003); Enantiospecific electrodeposition of a chiral catalyst. Nature 425, 490–493.

Wogelius R., Farquhar, M., Fraser D.G. and Tang C.C. (1998) Structural evolution of the mineral surface during dissolution probed with Synchrotron X-Ray techniques. In: B. Jamtveit and P. Meakin (eds) *Growth and Dissolution in Geosystems*, pp 1–17.

Xue, X, Kanzaki, M. and Fraser D.G. (2002) The dissolution mechanism of Mg_2SiO_4 Forsterite: Constraints from ^{29}Si and ^{1}H MAS NMR. Geochim.Cosmochim. Acta. 66, A853.

STUDIES ON COPPER CHROMICYANIDE AS PREBIOTIC CATALYST

KAMALUDDIN and SHAH RAJ ALI
Department of Chemistry, Indian Institute of Technology Roorkee, Roorkee—247 667, India

Abstract. Interaction of ribonucleotides, namely 5′-AMP, 5′-GMP, 5′-CMP, and 5′-UMP with copper chromicyanide has been studied. Maximum adsorption was observed at neutral pH. The adsorption isotherms were found to be Langmuirian in nature. Copper chromicyanide was found to be effective adsorbent and purine nucleotides showed more adsorption than pyrimidine nucleotides. Infrared spectral studies suggested that the adsorption occurs due to interaction of phosphate moiety, N-1, N-3, and N-7 of ribonucleotide molecule and outer divalent copper ion present in the lattice of copper chromicyanide. Copper chromicyanide also has been found to be efficient in catalyzing the conversion of cysteine into cystine. The results of the present study support the hypothesis that the metal cyanogen complexes could have played important role in concentrating and stabilizing the biomonomers on their surfaces during the course of chemical evolution and could have acted as prebiotic catalyst.

1. Introduction

It is now widely accepted that a crucial step in chemical evolution on the earth involved the polymerization of important biomonomers such as amino acids, nucleotides, and pentose sugars which were formed from simple molecules under prebiotic environment, but till now it is not well established about how the biomonomers might have concentrated from their dilute aqueous solutions in primeval seas. However, one of the suggestions is that clays and other minerals have provided surfaces onto which small molecules could have been concentrated and might have undergone a class of reactions such as condensation, oligomerization, and redox reactions producing polymeric material from which life has emerged under prebiotic environment (Ferris and Kamaluddin, 1989; Ferris, 1999).

It is generally accepted that the transition metal ions abundantly present in primeval seas, might have complexed with simple molecules available to them. It has been reported that cyanide ions were readily formed under prebiotic environment. Since cyanide ion is strong field ligand, it is reasonable to assume that cyanide ions might have formed a number of insoluble and soluble cyano complexes with transition metal ions abundantly present in primeval seas. Arrhenious has proposed the existence of ferro-ferricyanide in Anoxic Archean Hydrosphere. Water insoluble metal cyanogen complexes might have locally settled at the bottom of sea or at sea shore. We propose that these metal cyanogen complexes might have concentrated the biomonomers on their surface through adsorption processes and subsequently catalyzed a class of reactions of prebiotic relevance.

J. Seckbach et al. (eds.), Life in the Universe, 153–156.

Various metal ferrocyanides have been synthesized in our laboratory and their interaction with biomonomers such as amino acid, nitrogen bases and nucleotides have been studied suggesting their possible role in chemical evolution (Kamaluddin et al., 1990; Alam and Kamaluddin, 2000). They have been found to be efficient in catalyzing reactions of prebiotic relevance (Alam et al., 2000, 2002). We studied the interaction of 5′-ribonucleotides with copper chromicyanide to test its possible role in chemical evolution.

2. Experimental

Copper chromicyanide was synthesized from potassium chromicyanide by double decomposition method. Potassium chromicyanide was synthesized using Christensen's method (Bigelow, 1946). 167 ml, 0.1M potassium chromicyanide solution was slowly added to 500 ml, 0.1M copper nitrate solution with constant stirring. After 24 h reaction mixture was filtered, washed with water, dried, ground and sieved to 80-mesh size. Characterization of the copper chromicyanide was done by CHN analysis, TGA, IR and X-ray diffraction studies.

The adsorption of all the four ribonucleotides on copper chromicyanide was studied at pH 4.0, 7.0 and 9.0 by adding 5 ml of 2.8×10^{-4} M ribonucleotide solutions to 25 mg adsorbent each time. The suspensions were shaken initially for 1 h and then allowed to equilibrate at 30°C with intermittent shaking. After 24 h the suspensions were centrifuged at 8000 rpm and the supernatant liquid was decanted. The amount of adsorbed ribonucleotide was estimated from the difference between their concentrations before and after adsorption spectrophotometrically. The equilibrium concentration of ribonucleotide and the amount adsorbed were used to obtain adsorption isotherms.

3. Results and Discussion

For the chromicyanides of divalent metal ion, a general formula $M_3[Cr(CN)_6]_2.nH_2O$ has been reported, where M^{+2} represents an exchangeable divalent transition metal ion. The $[Cr(CN)_6]^{3-}$ ions possess an octahedral geometry in which Cr^{+3} is surrounded by six CN^- ligands and has electronic configuration of t_{2g}^3. One of the t_{2g} orbital has two electrons, second t_{2g} orbital has one unpaired electron whereas the third t_{2g} orbital remains empty because the electrons are filled against Hund's rule in the presence of cyanide ligand which is strong field in nature. Although CN^- ligands bond with Cr through σ donation, Cr donates π electrons present in its $d\pi$ orbital to antibonding $p\pi$ orbital of CN^- producing sufficient back bonding character. The transition metal chromicyanides generally exist in a polymeric lattice structure with $[Cr(CN)_6]^{3-}$ anions, in which another transition metal ions may be coordinated through the nitrogen end of the cyanide ligand.

3.1. INTERACTION OF RIBONUCLEOTIDES WITH COPPER CHROMICYANIDE

The preliminary adsorption studies were carried out over a wide pH range, and subsequent studies were performed at neutral pH, which exhibited the maximum adsorption. The adsorption data obtained at neutral pH and over a concentration range of adsorbate

(4×10^{-5}M to 2.8×10^{-4}M) followed Langmuir adsorption isotherms. The trend of adsorption was found as below:

$$5'\text{-GMP} > 5'\text{-AMP} > 5'\text{-CMP} > 5'\text{-UMP}$$

At lower pH, the nucleotide molecule has both the negative and positive charges and as the pH increased the negative charge on the nucleotide molecule is increased. The adsorption is presumably related to the involvement of N-1, N-3 and N-7 of the base residues as well as dissociated phosphate group. At neutral pH, 5′-monophosphate nucleotides exist in dianionic form because both the protons of phosphate group are dissociated at neutral pH. Pyrimidine nucleotides are able to form complex through its phosphate moiety only and purine nucleotides are able to form a bridging complex due to availability of the N-7 position present in addition to the phosphate moiety. Presence of N-7 as an additional interaction site in cases of purine nucleotides may be responsible for their greater adsorption with metal cations, therefore, 5′-AMP and 5′-GMP show greater adsorption than that of 5′-CMP and 5′-UMP.

The nature of interaction between ribonucleotides and copper chromicyanide was investigated in terms of infrared spectral studies of adsorption adducts. A shift towards higher wavelength in the characteristic frequencies of ribonucleotides was observed, indicating the interaction between ribonucleotides and copper chromicyanide. The typical strong bands in the region 950-1150 cm^{-1} are due to presence of ribose residue and change negligibly after adsorption. It suggests that ribose residue is not interacting with metal chromicyanides. A remarkable shift was observed in characteristic frequencies of purine nucleus and phosphate groups of ribonucleotides, which suggested probable involvement of N-1, N-3, N-7 and phosphate groups. Typical infrared frequencies of copper chromicyanide were almost unchanged suggesting that the ribonucleotide molecules do not enter into the coordination sphere of copper chromicyanide by replacing the cyanide ion. Further the insertion of ribonucleotide in the coordination sphere of copper chromicyanide.

3.2. DIMERIZATION OF CYSTEINE BY COPPER CHROMICYANIDE

The thiol group of cysteine was found to be readily oxidised to disulphide group by copper chromicyanide. 10 ml, 0.01M cysteine solution was reacted with 25 mg of copper chromicyanide. After 24 h, insoluble product was deposited which on analysis with ir, and uv showed disulphide bond formation. No reaction took place in the absence of the copper chromicyanide.

4. Conclusion

The results of the present study support the postulate that the metal cyanogen complexes could have provided their surface onto which the biomonomers might have concentrated from their dilute aqueous solutions during the course of chemical evolution. The biomonomers so concentrated have been protected from degradation and might have undergone a class of reactions such as condensation, oligomerisation and polymersation producing the biopolymers. Thus metal cyanogen complexes played an important role as prebiotic catalyst.

5. Acknowledgements

This research work was sponsored by Indian Space Research Organisation, Bangalore (India).

6. References

Alam, T., Gairola, P., Tarannum, H., Kamaluddin and Kumar, M.N.V. R. (2000) Conversion of Aniline to their Oligomers by Copper hexacyanoferrate(II), Indian J. Chemical Technology, **7**, 230–235.

Alam, T. and Kamaluddin (2000) Interaction of 2-Amino, 3-Amino and 4-Aminopyridines with Nickel and Cobalt Ferrocyanides, Colloids Surf, **162**, 89–97.

Alam, T., Tarannum, H., Ali, S.R. and Kamaluddin (2002) Adsorption and Oxidation of Aniline and Anisidine by Chromicyanide, J. Colloid Interface Science, **245**, 251–256.

Biegelow, J.H. (1946) Potassium hexacyanoferrate(III). In: Inorganic Synthesis, Vol. II, ed. W.C. Fernelius, Mc-Graw – Hill Book Company, Inc., Printed in the United States of America, 203.

Ferris, J.P. (1999) Prebiotic Synthesis on Minerals: Bridging the Prebiotic and RNA Worlds, Biol. Bull, **196**, 311–314.

Ferris, J.P. and Kamaluddin (1989) Oligomerization Reactions of Deoxyribonucleotides on Montmorillonite Clay: The Effect of Mononucleotide Structure on Phosphodiester Bond Formation, Origins Life Evol. Biosphere, **19**, 609–619.

Kamaluddin, Nath, M., Deopujari, S. W. and Sharma, A. (1990) Role of Transition metal Ferrocyanides(II) in Chemical Evolution, Origins Life Evol. Biosphere, **20**, 259–268.

PHOSPHATE IMMOBILIZATION BY PRIMITIVE CONDENSERS
Implications in the Availability of Soluble Phosphate in Abiotic Environments

FERNANDO DE SOUZA-BARROS[3], MARISA B. M. MONTE[1], ANA C. P. DUARTE[1], JOSÉ A. P. BONAPACE[2], MANOEL ROTHIER DO AMARAL JR.[3], RAPHAEL BRAZ LEVIGARD[4], YONDER A. CHING-SAN JR.[4], CRISTIANO S. COSTA[4], and ADALBERTO VIEYRA[4]

[1]Mineral Technology Centre (CETEM), MCT/BRAZIL; [2]Chemistry Institute / Federal University of Rio de Janeiro / UFRJ-BRAZIL); [3]Physics Institute / UFRJ; [4]Carlos Chagas Filho Biophysics Institute / UFRJ/BRAZIL.

1. Introduction

It is well known that co-precipitation of soluble Fe-oxide precursors with reactive phosphate is the dominant mechanism responsible for the low concentrations of this organic molecule in contemporary aqueous media. The formation of Fe-oxide precursors requires free oxygen molecules. This feature is well established and it is a common observation that reactive phosphate is more abundant in anoxic, i.e., non-oxidizing aqueous media (Van Cappellen and Ingall, 1996).

As noted by Stanley Miller and Harold Urey, the major overall chemical change in the geological scale is the transition from a reducing/neutral to the contemporary oxidizing atmosphere (Miller and Urey, 1959; Cloud, 1972; Kasting, 1993; Holland, 2002). This leads to the obvious conclusion that the major contemporary trapping mechanism could not have been a relevant process at the early stages of chemical evolution.

Since the pioneer proposals by Bernal and Ponnamperuma (Bernal, 1951; Ponnamperuma et al., 1982), minerals that can mediate phosphate binding have been considered in anoxic abiotic scenarios. A survey of these condensation reactions is beyond the scope of this report. We shall focus our attention to the mineral pyrite [FeS_2]. Studies of this mineral have been intensified with the original work of Wächterhäuser in the early 1990's (Wächtershäuser, 1992). His model of an autotrophic process converting ferrous ions and hydrogen sulphide into pyrite induced investigations in environmental conditions simulating those near the hydrothermal vents at the mid-ocean ridges.

Pyrite formations are common features in these hydrothermal sites (Seyfried, Kang Ding, and Berndt, 1991; Shock, 1992; Rona and Scott, 1993; MacLeod et al., 1994). It has been shown that pyrite reactive surfaces have a strong affinity to reactive phosphate (Bebié and Schoonen, 1999; Vieyra, et al., 2003). Estimates support the expectations that a significant volume of the oceans circulates in short time scale through the hot basalt and

J. Seckbach et al. (eds.), Life in the Universe, 157–160.

up through black smokers (Pilson, 1998). It is therefore reasonable to assume that soluble orthophosphate present in the primitive sea could complex with pyrite formations of the hydrothermal springs.

We report in this communication experimental results supporting the conjecture that in this particular scenario orthophosphate could be released from pyrite formations. The required mechanism would be based on possible fluctuations of pH-factor and ionic concentrations common to this particular environment. We propose that these mechanisms could be triggered by the dynamics of water streams through the hydrothermal vents.

2. Experimental

Details on the experimental settings for both the electrophoresis and the sorption/desorption determinations are given elsewhere (Monte et al., 2003; Vieyra et al., 2003). A surface treatment removed oxide, carbonate, and silicate impurities from the pyrite particles. The net negative-charge coverage of these particles decreases with the acidity of the aqueous medium. This is due to the attachment of protonic charges onto the pyrite surface. Thus, the effective net charge of pyrite particles can be modulated in anoxic media. Our electrophoresis results indicate that the ions SO_4^- – reported to exist in the emissions of hydrothermal vents – can also attach to the Pyrite surface. This is a specific sorption of sulphate ions – not electrostatic – for it is known that the net surface charge of Pyrite is negative.

The data shown in Figure 1 were obtained with suspensions of pyrite particles in an aqueous media simulating a primitive sea. The different pH-values used are shown in the figure. The arrow in this figure draws attention to the fact that the sampling procedure precluded observations at very short time intervals. It also highlights that an efficient desorption process takes place in mild alkaline medium. It is observed that the acidity of an aqueous medium simulating a primitive sea affects the orthophosphate affinity to pyrite. A change

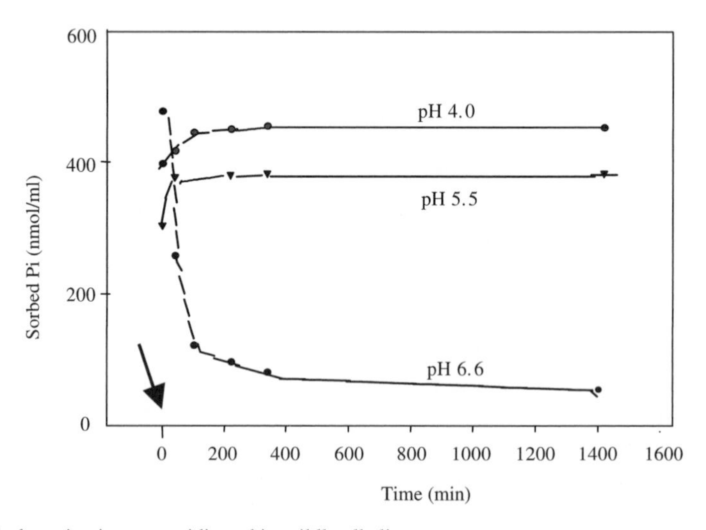

Figure 1. Pi adsorption in more acidic and in mildly alkaline seawater.

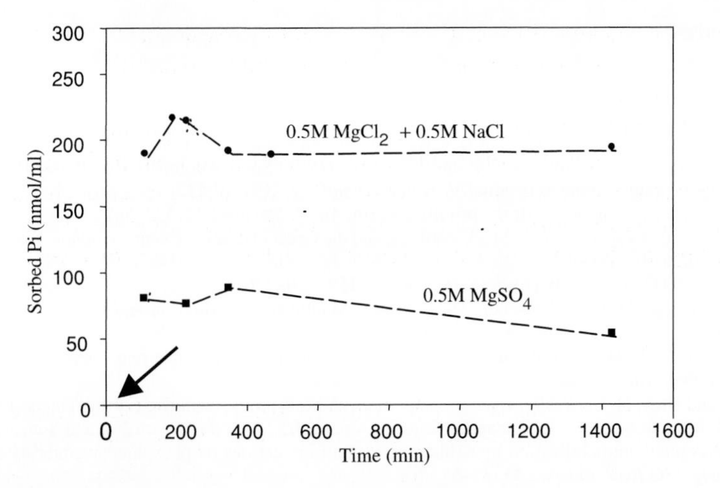

Figure 2. Phosphate sorption onto pyrite in the presence of different ions.

from an acidic to mild alkaline environment decreases the effective pyrite affinity to the phosphate.

Figure 2 clearly shows that sulphate ions can also affect the orthophosphate attachment to pyrite particles. In contrast, the results obtained with magnesium and sodium chloride are given as a comparative reference. The arrow indicates that the sampling procedure limits the time resolution of the assays.

3. Conclusion

We are proposing possible mechanisms that could have mediated the availability of reactive phosphate in special abiotic anoxic environments. In these environments, alternative phosphate condensation mechanisms would not have to compete with the dominant trapping process of modern eras: phosphate co-precipitation with Fe/Al-oxide precursors. We have shown that orthophosphate binding to pyrite can be modulated by alterations of the surrounding aqueous medium. It is suggested that short-duration perturbations affecting the acidity or the ionic composition of the surrounding media near the hydrothermal vents could have provided these specials niches with bursts of reactive phosphate. This opens new possibilities for models of molecular evolution that could use the well-known collective process of phosphate co-precipitation onto mineral surfaces. The onset of this process requires concentrations of reactive phosphate beyond its equilibrium value.

4. Acknowledgements

This work was supported by grants from the Brazilian Agencies CNPq, FINEP PRONEX, FAPERJ, and of the FUJB Foundation. The authors are grateful for very constructive comments of anonymous reviewers.

5. References

Bebié, J. and Schoonen, M.A.A. (1999) Pyrite and phosphate in anoxia and an origin-of-life hypothesis, *Earth Planet. Sci. Lett.* **171**, 1–5.

Bernal, J. D. (1951) *The physical basis of Life*, Routhledge and Kegan Paul, London.

Butkus, M.A., Grasso, D., Schulthess, C.P. and Wijnja, H. (1998) Surface complexation modelling of phosphate adsorption by water treatment residual, *J. Environ. Qual.* **27**, 1055–1063.

Cloud, P. (1972) A working model of the primitive Earth, *Am. J. Science* **272**, 537–548.

Holland, H. D. (2002) Volcanic gases, black smokers, and the Great Oxidation Event. *Geochim. Cosmochim. Acta,* **66**, 3811–3826.

Kasting, J. F. (1993) Earth's early atmosphere, *Science*, **259**, 920–926.

Kasting, J. F., Eggler D. H. and Raeburn, S. P. (1993) Mantle redox evolution and the oxidation state of the Archaean atmosphere, *J. Geol.* **101**, 245–257.

MacLeod, G., et al. (1994) Hydrothermal and oceanic pH conditions of possible relevance to the origins of life. *Orig. Life Evol. Biosph.*, **24**, 19–41.

Miller, S. L. and Urey, H. C. (1959) Organic compound synthesis on the primitive Earth. *Science* **130**, 245–251.

Monte, M. B. M., Duarte, A. C. P., Bonapace, J. A. P., Amaral Jr., M. R. do, Vieyra, A., and Souza-Barros. F. de, (2003) Phosphate immobilization by oxide precursors: implications on phosphate availability before life on Earth, *Orig. Life Evol. Biosph.*, **33**, 37–52.

Pilson, M. E. Q. (1998) *An introduction to the chemistry of the sea*, Prentice Hall, Upper Saddle River, New Jersey, p. 324.

Ponnamperuma, C., Shimoyama, A. and Friebele, E.: 1982, Clay and the origin of life, *Orig. Life* **12**, 9–40.

Rona, P. A. and Scott, S. D. (1993) A special issue on sea-floor hydrothermal mineralization: new perspectives. *Econom. Geology*. **88**, 1935–1976.

Seyfried Jr., W.E., Kang Ding, and Berndt, M.E. (1991) Phase equilibria constraints on the chemistry of hotspring fluids at mid-ocean ridges, Geochim. Cosmochim. *Acta* **55**, 3559–3580.

Shock, E. L. (1992) Chemical Environments of Submarine Hydrothermal Systems, *Orig. Life Evol. Biosph.*, **22**, 67–107.

Van Cappellen, P. and Ingall, E. D. (1996) Redox stabilization of the atmosphere and oceans by phosphorus-limited marine productivity, *Science* **271**, 493–496.

Vieyra, A., Tessis, A. C., Pontes-Buarque, M., Bonapace, J.A.P., Monte, M.B.M., Amorim, H. S. de, and Souza-Barros, F. de (2003), Catalysis of nucleotide hydrolysis by pyrite: possible role of reactive surfaces in prebiotic energy interconnection. *This proceedings.*

Wächtershäuser, G., (1992) Groundworks for an evolutionary biochemistry: the iron-sulphur world. *Prog. Biophys. molec. Biol.* **58**, 85–201.

ADSORPTION AND CATALYSIS OF NUCLEOTIDE HYDROLYSIS BY PYRITE IN MEDIA SIMULATING PRIMEVAL AQUEOUS ENVIRONMENTS

ADALBERTO VIEYRA[1], ANA CLAUDIA TESSIS[1,2], MILA PONTES-BUARQUE[1], JOSÉ A.P. BONAPACE[3], MARISA MONTE[4], HELIO SALIM DE AMORIM[5] and FERNANDO DE SOUZA-BARROS[5]

[1]Instituto de Biofísica Carlos Chagas Filho/UFRJ, Rio de Janeiro, Brazil, [2]Centro de Ciências Biológicas e da Saúde, Universidade Estácio de Sá, Rio de Janeiro, Brazil, [3]Instituto de Química/UFRJ, Rio de Janeiro, Brazil, [4]Centro de Tecnologia Mineral/MCT, Rio de Janeiro, Brazil and [5]Instituto de Física/UFRJ, Rio de Janeiro, Brazil

1. Introduction

Metal sulfides have been proposed as physical support for primitive bidimensional metabolism and chiral discriminators. There are controversial data about the possibly coupling between the exergonic pyrite syntheses and the endergonic amino acids syntheses in pyrite pulled metabolic-like processes. However, a role in adsorption and catalysis involving biomonomers – such as nucleotides – cannot be ruled out in an especial aqueous scenario. In an aqueous environment with a vast reservoir of reducing power in the form of, especially, iron (II) and sulfide within and beneath the crust. Our hypothesis is that ancient minerals such as pyrite could have played a role in concentrating mononucleotides and in determining their catalytic routes in later periods, when polynucleotides appeared and evolved (Tessis *et al.*, 1999). Pyrite – as well as other sulfides – could have participated in nucleotide adsorption and in primitive phosphoryl transfer reactions in the less drastic conditions that prevail away from the hot hydrothermal vents, the big sources of starting molecules. In this presentation, we shall briefly comment about our results on the following themes: 1) Adsorption of mononucleotides (AMP and ADP) onto untreated and acetate-treated samples of pyrite in which soluble Fe^{3+} ions were removed; 2) The influence – on nucleotide adsorption – of the previous adsorption onto the crystal of an oxo acid that could have appeared near the mineral and modified its surface properties (acetate can be formed onto the surface of mixed iron nickel sulfides at high temperature and pressure [Huber and Wächtershäuser, 1997]); 3) The role of the interface. In some models, the two-dimensional system represented by a mineral surface would be in a seawater environment having a passive role. The actual conditions prevailing when sulfur materials are in contact with salt solutions are quite complex. The very reactive iron-sulfide immersed in a solution resembling a primitive ocean generates a continuously changing interface with ionic gradients and formation of different oxide layers; 4) Hydrolysis of ATP adsorbed onto pyrite.

J. Seckbach et al. (eds.), Life in the Universe, 161–164.

2. Results and Discussion

In a recent paper (Pontes-Buarque *et al.*, 2001), assays with [^{14}C]acetate show that the oxo acid is firmly and almost completely adsorbed onto the surface of pyrite. When the solution containing the [^{14}C]acetate-coated pyrite samples was removed and replaced by artificial sea water, loss of pyrite-bound radioactivity to the medium after was barely detectable, indicating that acetate remains firmly attached to the mineral. Moreover there is no displacement of labeled acetate after removal of the supernatant and resuspension of [^{14}C]acetate-coated pyrite in unlabeled acetate solutions, indicating that interaction between the oxo acid and the surface can be considered irreversible. The pyrite/solution interface when acetate-coated pyrite is resuspended in artificial (primitive) seawater could have been important in the modulation of an early, starting metabolism onto Fe/S mineral surfaces. The interface lies between the bulk pyrite and the outer aqueous medium and has two layers. The inner layer, known as the Stern layer, is made of firmly attached ions such as acetate; the outer interface, characterized by the diffuse ionic gradients, is much thicker than the Stern layer. It is expected that an increase of both ionic or molecular species concentrations that can attach to the mineral surface results in the decrease of the Stern layer thickness and vice versa, thus modulating the interaction of ancient metabolites with the Fe/S minerals. It is clear that in this complex and interactive system, a large variety of processes able to continuously change the composition of the intermediate layers could had been changed, in several directions, extant metabolic reactions catalyzed by minerals. The acetate-induced modifications of the interface could have been modified, for example, the adsorption of mononucleotides onto Fe/S in primeval seas, in mildly alkaline environments such as found in the warm, less drastic scenarios away the hot hydrothermal vents. The enhancement of nucleotide adsorption due to the presence of acetate can be seen in two completely different adsorption isotherms of AMP onto pyrite in two extreme conditions: in a acetate-dominant solute and in acetate-free primitive sea water. In a concentrated acetate medium the adsorption is greatly enhanced at low AMP concentrations, contrasting with the one observed in an acetate-free primeval solution. Therefore, the adsorption mechanism is highly cooperative due to the presence of acetate. In contrast AMP adsorption is very low in the absence of acetate. Only when additions of nucleotide are in the millimolar range, a self-induced adsorption mechanism clearly sets in. In conclusion, there is a clear-cut different behavior due to a different coating of the mineral. It is clear that high nucleotide implies in a greater possibility of polymerization induced by acetate, the universal precursor of carbon compounds in living systems. Acetate also enhances ATP adsorption (Tessis *et al.*, 1999). Adsorption is faster and complete when the crystals are previously treated with an acetate solution, separated and then supplied with ATP-containing artificial sea water. Again, the effect of coating pyrite with acetate promotes an increase in the adsorption capacity of pyrite, favoring the interaction of the nucleotide with the mineral surface. This experiment was carried out in the presence of iron oxides onto the surface of pyrite. When the surface oxides are removed with hydrofluoric acid and the experiments are done in the absence of O_2, the levels of adsorbed ATP attained the same maximal values with and without pretreatment of pyrite with acetate. It is known that in the presence of O_2 pyrite is oxidized, first producing Fe^{2+}; Fe^{2+} is further oxidized to Fe^{3+}, and then $Fe(OH)_3$ forms. Acetate can form complexes with Fe^{2+} and these complexes are stable even at high temperatures. Thus it is plausible to postulate that acetate could have had another role besides that of providing a source of carbon after its formation in hydrothermal vents. The flow of sulfides to cooler and more superficial regions away from the main ridges was probably as high in primitive eras as they are in the present (Corliss, 1990). In these mild and distant regions, the presence of acetate strongly bound to the surface of sulfide minerals could have favored adsorption of nucleotides by preventing the formation of iron hydroxides, even in the presence of the dissolved O_2 formed by photolysis of water.

In contemporary living systems ATP hydrolysis is the key exergonic reaction and ADP is also an energy donor in Archaea (Kengen *et al.*, 1995). It was, therefore, interesting to test the hypothesis that in an ancient scenario in which ATP and pyrite were present, the mineral could have been able to catalyze phosphoryl transfer reactions involving adsorbed nucleotides (Tessis *et al.*, 1999). The acetate-coated pyrite suspended in artificial seawater promotes breakdown of the β,γ- and the α,β-phosphoanhydride bond of ATP adsorbed on its surface. The ATP adsorbed onto acetate-treated pyrite is sequentially cleaved by the mineral. ATP undergoes progressive hydrolysis to ADP and AMP (whereas the nucleotide concentration in artificial sea water remains unchanged in the absence of mineral). ATP and ADP hydrolysis are not the result of Fe^{2+}-catalyzed hydrolysis in solution since ATP was totally adsorbed and soluble Fe^{2+} ions were removed by the acetate washing. Pyrite-catalyzed hydrolysis of ATP and ADP and their reversal are second-order sequential reactions. The disappearance of ATP, the biphasic time course of ADP formation and the accumulation of AMP can be described by exponential equations obtained with a model in which ATP hydrolysis is complete and ADP coexists in equilibrium with AMP (free energy change of ADP hydrolysis as low as –3.5 kJ/mol instead of –35 kJ/mol for ATP hydrolysis). This difference may be due to differences in solvation when polyphosphate chains of different length are attached to the acetate-coated pyrite surface. This observation means that phosphorylation of AMP to ADP is thermodynamically favored onto the surface of pyrite minerals in which two-carbon organic fragments are attached. The reversibility of the ADP hydrolysis in iron-sulfide mineral surfaces might have been an evolutional pressure factor that allowed the utilization of this nucleotide by very primitive organisms (Kengen *et al.*, 1995). In addition, pyrite surface exhibits specific properties in adsorption and catalysis that could have been relevant in chemical evolution: 1) It needs divalent cations for the attachment of phosphoryl groups but not for carboxylates; 2) It catalyzes hydrolysis of phosphoanhydride bonds but not of ester-phosphates; 3) It can selectively change the free energy of hydrolysis of the same chemical bond. Since phosphates can also catalyze the synthesis of ADP from AMP (Tessis *et al.*, 1995), a varied assemblage of minerals in aqueous environments could have acted as adsorbants and catalysts of phosphorylations and energy conservation in primordial eras (Vieyra *et al.*, 1995).

3. Conclusions

The results presented support the view that surfaces of iron-sulfide minerals, implicated in several primitive processes (Russell *et al.*, 1994), could have participated in catalytic reactions involving phosphoanhydride bonds. The reactions occurring on those primitive interfaces, could have been modulated by the presence of hydrocarbon molecules formed in neighboring areas. More evolved energy-transducing systems could have taken over the hydrolytic properties of pyrite minerals and coupled them to energy-requiring processes.

4. Summary

There is an extensive literature on the possible roles of minerals in the chemical evolution of life. Minerals have considered in: a) processes that would discriminate molecular chirality; b) condensation reactions of biomolecular precursors; c) biochemical templates; d) prebiotic catalysis; e) autocatalytic metabolism. This is the case of metal sulfides, especially iron sulfides, including pyrite. In this paper we present results showing that pyrite adsorbs nucleotides and that this process is modulated by acetate – which is considered a primitive carbon donor – and by the size and composition of the interface of the mineral with the aqueous environment. Moreover, pyrite exhibits specificity towards phosphoanhydride linkage but not for ester phosphates.

5. References

Corliss, J. B. (1990) Hotsprings and the origin of life, *Nature* **347**, 624.

Huber, C. and Wächtershäuser, G. (I997) Activated acetic acid by carbon fixation on (Fe,Ni)S under primordial conditions, *Science* **276**, 245–247.

Kengen, S. W. M., Tuininga, J. E., de Bok, F. A. M. Stams, A. J. M. and de Vos, W. M. (1995) Purification and characterization of a novel ADP-dependent glucokinase from the hyperthermophilic archean Pyrococcus furiosus, *J. Biol. Chem.* **270**, 33453–33457.

Pontes-Buarque, M., Tessis, A. C., Bonapace, J. A. P., Cortés-Lopes, G., Souza-Barros, F. de and Vieyra, A. (2001) Modulation of adenosine 5′-monophosphate adsorption onto aqueous resident pyrite: potential mechanisms for prebiotic reactions, *Origins Life Evol. Biosphere,* **31**, 343–362.

Russell, M. J., Daniel, R. M., Hall, A. J. and Sherringham, J. A. (1994) A hydrothermally precipitated catalytic iron sulphide membrane as a first step toward life, *J. Mol. Evol.* **39**, 231–243.

Tessis, A. C., Salim de Amorim, H., Farina, M., de Souza-Barros, F. and Vieyra, A. (1995) Adsorption of 5′-AMP and catalytic synthesis of 5′-ADP onto phosphate surfaces: correlation of solid matrix structures. *Origins Life Evol. Biosphere* **25**, 351–373.

Tessis, A. C., Penteado-Fava, A., Pontes-Buarque, M., Amorim, H. S. de, Bonapace, J. A. P., Souza-Barros, F. de and Vieyra, A. (1999) Pyrite suspended in artificial seawater catalysis hydrolysis of adsorbed ATP: enhanced effect of acetate, *Origins Life Evol. Biosphere* **29**, 361–374.

Vieyra, A., Gueiros-Filho, F., Meyer-Fernandes, J. R., Costa-Sarmento, G. and Souza-Barros, F. de (1995) Reactions involving carbamyl phosphate in the presence of precipitated calcium phosphate with formation of pyrophosphate: a model for primitive energy-conservation pathways, *Origins Life Evol. Biosphere* **25**, 335–350.

VI. Cosmological and Other Space Science Aspects of Astrobiology

DUST AND PLANET FORMATION IN THE EARLY UNIVERSE

Clues from studies of Damped Ly α galaxies

GIOVANNI VLADILO
Osservatorio Astronomico di Trieste—I.N.A.F.Italy

1. Introduction

The process of planet formation is poorly understood, even for stars of the solar vicinity. The key factors that govern this process need to be identified if we wish to understand whether and how planet formation takes place in other galaxies and in other epochs. Interstellar dust may play a key role in this context. Proto-planetary disks form from the sedimentation of the left-over material of the parent interstellar nebula which gives birth to the central star. The dust of the nebula accumulates in the midplane of the disk and, according to accretion theories, leads to the formation of planetesimals (Lissauer 1993). The quantity of dust present in the nebula prior to the formation of the central star can affect the efficiency of planetesimal (and planet) formation. In this contribution, evidence is presented for a deficit of dust in galaxies observed during their early stages of evolution. The study concerns a particular class of quasar absorption-line systems, called Damped Lyman α (DLA) systems, associated with gas-rich galaxies observed at look-back times up to ~12 billion years ago. The deficit of dust in DLA galaxies is briefly discussed in the context of planet formation in the early universe. More details on DLA systems are presented in a separate contribution in these Proceedings.

2. Dust and Planetesimal Formation at Very Low Metallicity

The metallicity and dust content of DLA systems can be estimated from the abundances of volatile and refractory elements, such as Zn and Fe, respectively. In particular, one can derive the fraction of atoms of iron in dust form, f_{Fe}, which is essentially a dust-to-metal ratio by number (Vladilo 2002). In the course of chemical evolution the dust is expected to track the metals and, therefore, the dust-to-metal ratio should be roughly constant. Contrary to this expectation, we find that the dust-to-metal ratio increases with metallicity, with a severe deficiency of f_{Fe} at metallicity $\sim 10^{-2}$ the solar level (Vladilo 2004). Given the virtual absence of dust, we wonder whether circumstellar disks can still create planetesimals at very low metallicity. In fact, the absence of a microscopic solid component in the disk-feeding material could dramatically delay the planetesimal formation; if this delay is larger than the life time of the disk ($\sim 10^6$ / $\sim 10^7$ years), planet formation would be inhibited. For instance, for conditions typical of the solar nebula, e-folding sedimentation times can easily be as high as $\sim 10^6$ years for 1 μm grains, and several e-folding times are required to produce a thin layer in which the dust and gas density are comparable, a minimum requirement for planetesimal

J. Seckbach et al. (eds.), Life in the Universe, 167–168.

formation (Lissauer 1993). To reduce these times one needs to invoke collisional growth of pre-existing grains during their descent to the midplane of the disk (Weidenschilling 1980). The scarcity of pre-existing dust grains at low metallicity may prevent collisional growth and make the build-up of the critical dust density for planetesimal formation impossible within the life time of the disk. If we require at least ∼50% of iron atoms to be in dust form ($f_{Fe} > 0.5$) in order to reduce the sedimentation time, we derive a metallicity threshold from the trend between f_{Fe} and metallicity observed in DLA galaxies (Vladilo 2004). With few exceptions, we find that a metallicity [Fe/H] > -1.5 dex is required so that $f_{Fe} > 0.5$. Below this threshold, which should be interpreted in a statistical sense, we argue that planetesimal formation may be exceedingly delayed and planet formation inhibited.

3. Conclusions

The very low fraction of dust found in DLA galaxies of low metallicity suggests a possible scenario for inhibition of planetesimal formation: the scarcity of pre-existing dust grains may yield sedimentation times exceeding the life-time of proto-planetary disks. In order to reduce the sedimentation time, a minimum fraction of metals must be in dust form. Model computations are required to define this minimal dust fraction in terms of physical quantities and metallicity. Efforts to relate the efficiency of planetesimal formation to the dust-to-gas ratio are already under way (e.g. Youdin & Shu 2002). If we adopt $f_{Fe} > 0.5$ as the minimal dust fraction of iron atoms, we obtain a metallicity threshold [Fe/H] > -1.5 dex for planetesimal formation. With this criterion planets should be rare in DLA galaxies at redshift $z > 2$. The metallicity limit proposed here is less stringent than the threshold [Fe/H] > -0.3 dex proposed by Gonzalez et al. (2001) for the formation of habitable planets. Taken at face value, these limits indicate that the interval $-1.5 <$ [Fe/H] < -0.3 is characteristic of non-habitable planets. The discovery of a planet in the globular cluster M4 (Sigurdsson et al. 2003) may provide the first example of a non-habitable planet in this metallicity interval.

4. References

Gonzalez, G., Brownlee, D., and Ward, P. (2001) The Galactic Habitable Zone: Galactic Chemical Evolution, *Icarus* **152**, 185–200

Lissauer, J.J. (1993) Planet Formation, *Ann. Rev. Astron. Astrophys.* **31**, 129–174

Sigurdsson, S., Richer, H.B., Hansen, B.M., Stairs, I.H., Thorsett, S.E. (2003) A Young White Dwarf Companion to Pulsar B1 620–26: Evidence for Early Planet Formation, *Science*, **301**, 193–196

Vladilo, G. (2002) Chemical abundances of damped Ly alpha systems: A new method for estimating dust depletion effects, *Astron. Astrophys.* **391**, 407–415

Vladilo, G. (2004) The Early Build-up of Dust in Galaxies: A Study of Damped Lyman α Systems (work in preparation)

Weidenschilling, S.J. (1980) Dust to planetesimals—Settling and coagulation in the solar nebula, *Icarus* **44**, 172–189

Youdin, A.N., and Shu, F. H. (2002) Planetesimal Formation by Gravitational Instability, *Astrophys. J.* **580**, 494–505

QUASAR ABSORPTION-LINE SYSTEMS AND ASTROBIOLOGY

GIOVANNI VLADILO
Osservatorio Astronomico di Trieste—I.N.A.F. Italy

1. Introduction

Quasars are among the most powerful background sources for probing gas at cosmological distances by means of absorption-line spectroscopy. Quasar spectra show a large number of absorption lines, most of which are attributed to the Ly α transition of neutral hydrogen (HI) originating in layers of gas located in the direction of the quasar. Thanks to the cosmological expansion of the universe, each Ly α line falls at a different wavelength in the observer's rest frame according to the redshift of the layer. In most cases, the HI lines are weak and do not show accompanying metal lines at the same redshift. These weak lines are believed to originate in the intergalactic medium, an environment highly ionized and extremely metal poor. Our attention here is focused on a less frequent type of quasar absorbers, characterized by very strong Ly α profiles broadened by radiation damping and called damped Ly α absorptions (Wolfe et al. 1986). These absorptions are always accompanied by a complex of low-ionization metal lines at the same redshift, all together forming a Damped Ly α (DLA) system. There is general agreement that DLA systems originate in the interstellar medium (ISM) of intervening galaxies. DLAs are most easily identified at redshift $z > 2$, when the Ly α is redshifted to the optical band, but can be detected up to $z \sim 6$, the redshift of the most distant quasars. This redshift interval corresponds to an interval of look-back time between $\sim$10 and $\sim$12 billion years ago, according to the current values of the cosmic expansion parameters. Therefore we can say that DLA studies probe the ISM of galaxies observed at very large look-back times. As a consequence, the link between DLA studies and astrobiology is basically the same that exists between ISM studies and astrobiology, with the advantage that DLA observations probe a variety of galaxies, back to the earliest epochs of their evolution.

2. Studies of Dla Systems and Astrobiology

The connections between DLAs studies and astrobiology can be classified in a very schematic form according to the following points, which also summarize the links between ISM studies and astrobiology.

1. The ISM collects the elements produced and ejected by the stars and, in particular, the biogenic elements and those necessary for the formation of habitable planets; studies of DLAs can be used to measure the abundances of these elements during the early epochs of galactic evolution.

J. Seckbach et al. (eds.), Life in the Universe, 169–172.

2. The interstellar dust is an important component of the ISM, a basic ingredient in planet formation, according to planetesimal accretion models, and a catalyst of molecular formation; studies of DLAs can be used to probe the early build-up of dust in galaxies.
3. The ISM is the site where the first formation of organic molecules occur in the course of galactic evolution; DLAs can be used to investigate the early interstellar chemistry and trace the presence of complex organic material at different cosmic epochs.
4. The ISM probes the physical conditions of the environments where the stars and their planetary systems, if any, are embedded; DLAs can be used to investigate the habitability of these environments during the early stages of galactic evolution.

For each of these four points we now give a concrete example, in order to show that DLA studies can offer unique results relevant to the field of astrobiology.

2.1. THE FORMATION OF BIOGENIC ELEMENTS

As a first example we consider the early build-up of nitrogen. Measurements of nitrogen abundances in DLAs (Fig. 1, left panel) are particularly important for testing the early chemical evolution of galaxies. The abundances of nitrogen in DLAs are the lowest measured so far in any astrophysical site, with values as low as $\sim 10^{-3.5}$ the solar N/H number ratio (Centurión et al. 2003). Also the relative abundances are low; as an example, the N/Si ratio is $\sim 10^{-1.5}$ the corresponding solar value. This severe deficit of an important biogenic element may pose a real problem for building up pre-biotic material during the earliest stages of galactic evolution. For instance, it is hard to imagine how a nitrogen-rich atmosphere, favourable to the formation of aminoacids, could be created in these conditions.

Also other elements in DLAs have a low abundance, but they do not attain the extremely low values of nitrogen. For instance, the silicon abundance can be as low as $\sim 10^{-2.5}$ the solar Si/H ratio (Fig. 1, central panel). Quite interestingly, while the *absolute* abundances

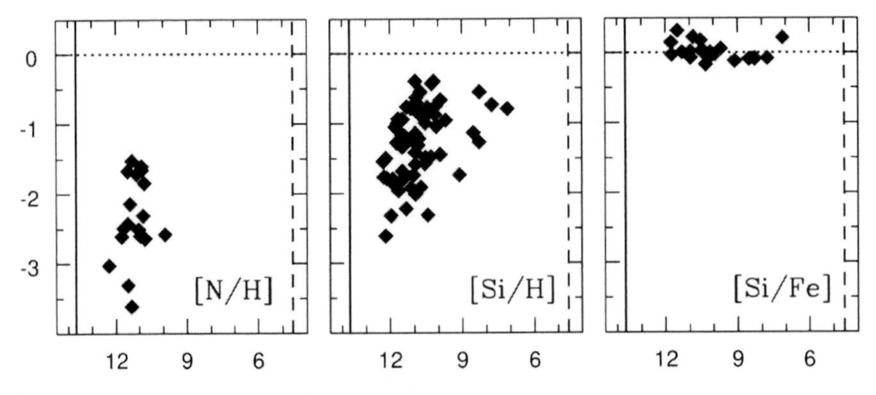

Figure 1. Element abundances in DLAs plotted versus look-back time (billion years ago). Abundances are given by number on a logarithmic scale; the zero point represents the Solar System value (dotted line). Most data are compiled from the literature. Also shown are the look-back time of the Big Bang (solid line) and of the Solar System formation (dashed line).

(relative to H) are below solar, the *relative* abundances are roughly solar in many cases, with the remarkable exception of nitrogen noted above. For instance, the Si/Fe ratio is approximately solar in DLAs (Fig. 1, right panel) when depletion effects are accounted for (Vladilo 2002). The rapid convergence in early cosmic epochs of relative abundances to present-day solar values may be considered an example of convergent evolution at cosmic level (see Chela-Flores, these Proceedings).

2.2. THE FORMATION OF DUST

The relative abundances of volatile and refractory elements in DLAs can be used to derive the fraction of metals in dust form (Vladilo 2002). This fraction is expected to be roughly constant with metallicity because the dust is expected to track the metals in the course of chemical evolution. Instead, recent studies indicate that in DLAs the dust fraction (or depletion) is correlated with the metallicity (Ledoux et al. 2003, Vladilo 2003). This implies that DLA galaxies have a severe deficit of dust in the earliest stages of evolution. This deficit may pose a serious problem in planet formation, according to planetesimal accretion models. In a separate contribution to these Proceedings, I propose that planet formation in DLA galaxies may be inhibited when the dust fraction of iron is below $\sim$50%, i.e. when the metallicity is below $\sim 10^{-1.5}$ the solar value. This would imply that planets are rare in DLAs at redshift $z > 2$.

2.3. THE FORMATION OF ORGANIC MOLECULES

Molecular hydrogen is detected in a small number of DLAs; even in these cases the molecular fraction is rather low, with typical values between $\sim 10^{-4}$ and $\sim 10^{-6}$ (Ledoux et al. 2003). Since H_2 is the starting point of the interstellar chemical reaction network, the general lack of H_2 suggests that the formation of organic molecules was inhibited at $z > 2$, i.e. interstellar organic molecules were not present until $\sim$10 Gyr ago, at least in DLA galaxies. On the other hand, the first detection of the extinction bump at 2175 Å at high redshift (Motta et al. 2002) provides an indication that complex organic material was already in place not much later. The 2175 Å bump is a well known emission feature of the interstellar extinction curve, seen in the Milky Way galaxy and other nearby galaxies. The identification of the carrier responsible for this feature is controversial. However, there is little doubt that it is related to the presence of complex, carbon-bearing material. The detection of the bump at redshift $z = 0.83$ represents the most distant record of this feature so far, and indicates that complex, carbon-rich interstellar material was already in place $\sim$7 Gyr ago.

2.4. INTERSTELLAR IONIZATION AT DIFFERENT COSMIC EPOCHS

Quasar absorption data indicate that the intergalactic radiation field at high redshift is dominated by the continuum emitted by quasars themselves. The integrated emission of quasars is characterized by a hard ionizing continuum and one may wonder if this can affect the physical conditions of high redshift galaxies. A recent study of neutral argon in DLA systems suggests that this may be the case (Vladilo et al. 2003). The abundance of argon is significantly lower than that of other α-capture elements observed in DLAs and detailed computations indicates that this is an effect of ionization rather than of nucleo-synthetic evolution. The analysis indicates that about 12 Gyr ago the ionizing continuum in DLAs started to be dominated by the integrated radiation of quasars. The quasar-dominated era must have lasted for at least two billion years. Such a hard ionizing radiation might have been hostile to the formation of habitable environments; on the other hand, it might have triggered chemical evolution in the interstellar gas.

3. Conclusions

Studies of DLA systems provide unique information on the early build-up of biogenic elements, dust and molecules and on the existence of habitable zones in the early universe. Some of the results mentioned above are examples of convergence at cosmic level of the type discussed by Chela-Flores in these Proceedings. For instance, the approximately solar values of Si/Fe ratios in DLAs, already attained at early cosmic epochs; or the presence of the 2175 Å extinction bump in a galaxy observed at a look-back time of 7 Gyr ago. On the other hand, other results can be used to set temporal limits to the existence of habitable zones in the universe. For instance, the severe deficit of nitrogen, dust and molecules in DLAs suggests that habitable zones are unlikely to have been formed earlier than 10 Gyr ago, at least in DLA galaxies. By adding the age of the earth to this limit, we obtain an estimate of 5.5 Gyr ago for the formation of the oldest intelligent civilization of terrestrial type. However, care must be taken in generalizing the DLA results to all types of galaxies since the exact nature of DLA galaxies is still under debate (e.g., see Calura et al. 2003). Future studies of quasar absorbers in different spectral bands will be able to set more stringent time limits and, hopefully, establish unexpected links with other topics of interest in astrobiology.

4. References

Calura, F, Matteucci, F., and Vladilo, G. (2003) Chemical evolution and nature of damped Lyman α systems, *Mon.Not.Roy.Astron.Soc.*, **340**, 59–72.

Centurión, M., Molaro, P., Vladilo, G., Péroux, C., Levshakov, S. A., and D'Odorico, V. (2003) Early stages of nitrogen enrichment in galaxies: Clues from measurements in damped Lyman alpha systems, *Astron. Astrophys.* **403**, 55–72.

Ledoux, C., Petitjean, P., and Srianand, R. (2003) The VLT-UVES survey for molecular hydrogen in high-redshift damped Lyman-alpha systems, *Mon.Not.Roy.Astron.Soc.*, in press (astro-ph/0302582).

Motta,V., Mediavilla, E., Muñoz, J.A., Falco, E., Kochanek, C.S., Arribas, S., García-Lorenzo, B., Oscoz, A., and Serra-Ricart, M. (2002) Detection of the 2175Å Extinction Feature at $z = 0.83$, *Astrophys. J.*, **574**, 719–725.

Vladilo, G. (2002) Chemical abundances of damped Ly alpha systems: A new method for estimating dust depletion effects, *Astron. Astrophys.* **391**, 407–415.

Vladilo, G. (2003) Evolution of the dust content of Damped Ly α systems, in Astrophysics of Dust, Estes Park, Colorado, May 26–30, 2003 (poster presentation).

Vladilo, G., Centurión, M., D'Odorico, V., and Péroux, C. (2003) Ar I as a tracer of ionization evolution, *Astron. Astrophys.* **402**, 487–497.

Wolfe, A. M., Turnshek, D. A., Smith, H. E., and Cohen, R. D. (1986) Damped Lyman-alpha absorption by disk galaxies with large redshifts. I—The Lick survey, *Astrophys. J. Suppl.*, **61**, 249–304.

A NEW SEARCH FOR DYSON SPHERES IN THE MILKY WAY

DANTE MINNITI, FRANCISCA CAPPONI, ALDO VALCARCE,
and JOSÉ GALLARDO
Depto. de Astronomía, P. Univ. Católica, Casilla 306, Santiago 22, Chile

1. Introduction

Our civilization consumes more and more energy as it progresses. Consider that the total ouput from the Sun is about 4×10^{26} W, of which the illuminated Earth intercepts a small fraction (10^{-9}). With our accelerated energy consumption rate, we would soon use that whole illuminated Earth energy. Given the cosmic timescale (see G. V. Coyne in this proceedings), there is plenty of room for other civilizations to have been born well before ours. These more advanced civilizations would require unthinkable amounts of energy. Freeman Dyson (1960) proposed that advanced civilizations would have the means to use all the energy of their parent stars by building A.U. size shells around them. Using the laws of thermodynamics he predicted how these "Dyson spheres" could be observed, even without knowing the nature of these objects. Whole or partial Dyson spheres would dim the original stellar light, and radiate in the IR (at about 10 μm).

In this paper we only address the question: Can we find candidate Dyson spheres in the Milky Way? We do not try to answer other questions like: Will an advanced civilization build a Dyson sphere?; How can a Dyson sphere be built?; Why would they build a Dyson sphere? How stable would it be? etc. However, most of which follows is also valid for partial Dyson spheres, Niven's rings, or similar configurations. We will assume that Dyson spheres are rare, and that therefore one has to search through thousands of stars. Prime fields to monitor large numbers of stars are the disk of the Galaxy in the solar neighborhood, and the dense regions of the Galactic bulge.

2. Searching for Dyson Spheres in Our Galaxy

The main-sequence stars of the solar neighborhood can be observed with great detail and exquisite sensitivity. The problems are that there are relatively few stars available, and that they are spread all over the sky. The most systematic searches for late-type main sequence stars that are too faint for their spectral types, and that have IR excess have been carried out by Jugaku and Nishimura (1997, 2000). They used the mid-IR data from IRAS in combination with near-IR photometry with 1.5 m class telescopes. They obtained photometry for 180 F-G-K main sequence stars within 25 pc of the Sun, using the K-[12 mm] index to search for Dyson spheres. If the waste heat of the Dyson sphere is 1% of the radiation energy

J. Seckbach et al. (eds.), Life in the Universe, 173–176.

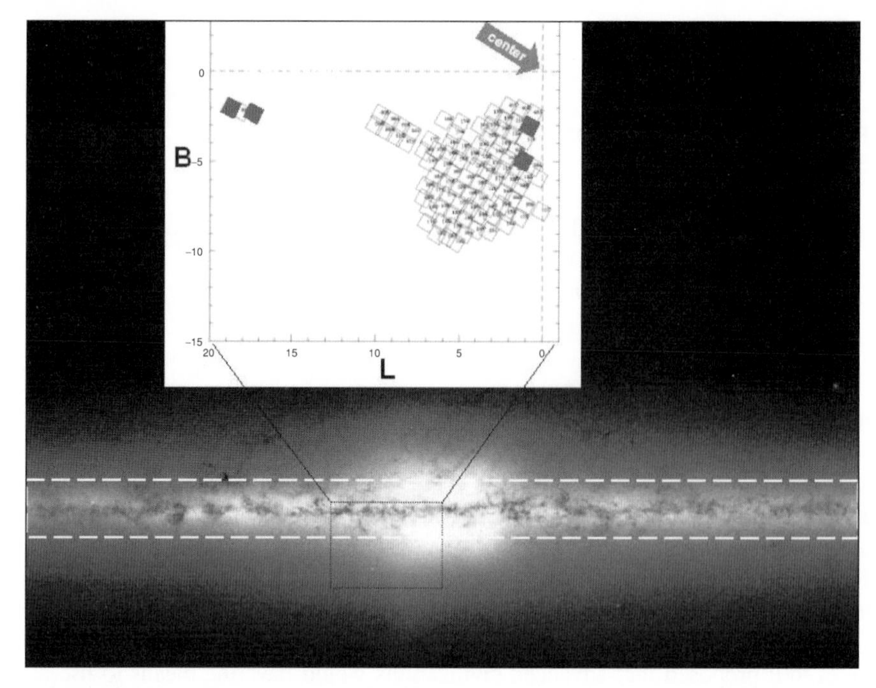

Figure 1. 2MASS face-on view of the inner 90 × 30 sqdeg of the Milky Way. The dashed lines limit the MSX observations of the Milky Way disk, and the insert shows the MACHO fields, including the fields F301, F303, F121 and F118 studied here (full squares from left to right).

of the star, they expected a color excess >1 mag. These searches have been unsuccessful so far (Jugaku and Nishimura 2000).

The large numbers of stars present in the inner MW disk and the Galactic bulge are metal rich, which would favor the presence of planets, and old, which would give more chance to advanced civilizations to develop. We will direct our search to these fields. We should then search for long-lived stars, such as late-type main-sequence stars that are fainter than normal in the optical and near-IR, and that have infrared excesses in the mid-IR bands.There should be no overlap with other types of known sources in the Milky Way. But there are still three serious observational problems for a massive search in the inner fields of the Milky Way: (1) source associations may be ambiguous in crowded fields due to different spatial resolutions of different datasets, (2) the presence of unresolved companions can mimic IR excesses, and (3) variable sources can mimic IR excesses when data acquired in different epochs are combined.

We will combine three large databases in order to search for Dyson Spheres in the inner Milky Way: the MACHO database, the 2MASS database, and the MSX database. The Two Micron All Sky Survey (2MASS) covered the whole sky in the near-IR bands. The 2MASS database is public at http://irsa.ipac.caltech.edu/, containing accurate JHK single epoch photometry down to $K = 14 - 15$ depending on crowding. The Midcourse Space Experiment (MSX) database contains single epoch mid IR photometry of sources throughout the whole Galactic plane (Cohen et al. 2000). The calibrated MSX photometry is also public at http://irsa.ipac.caltech.edu/. Of the six MSX bands, we will use the A-band

TABLE 1. Numbers of Optical and IR Sources in the MACHO Fields

Field	RA(2000)	DEC(2000)	L	B	NMACHO	N2MASS	NMSX	NVAR
F118	17:56:31	−29:46:47	0.8	−3.1	6.4e + 5	4.1e + 4	204	1700
F121	18:04:50	−30:22:52	1.2	−4.9	5.9e + 5	2.0e + 4	52	600
F301	17:32:38	−13:31:44	18.8	−2.0	5.6e + 5	2.6e + 4	27	300
F303	18:30:51	−14:57:52	17.3	−2.3	4.1e + 5	2.3e + 4	21	300

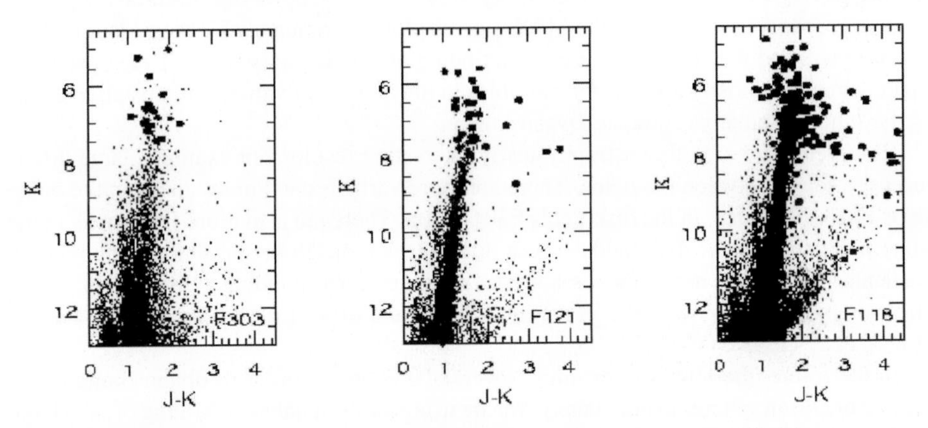

Figure 2. 2MASS infrared color-magnitude diagrams for three different fields. The MSX IR excess sources are plotted as full circles, and the MACHO constant stars as black dots.

at 8 μm, which is the most sensitive one, reaching 10 times deeper than IRAS. The main problem with this data is the poor spatial resolution. Even though it is still better than IRAS, the 3″ to 8″ resolution can lead to mismatches in the crowded bulge fields The MACHO bulge database contains multi-epoch optical photometry about 50 million stars in 96 bulge fields each of 0.5 square degree. These stars were observed for several seasons from 1993 to 1999, and a typical light curve contains more than 1000 points. The MACHO data are calibrated into the standard V and R photometric system (Alcock et al. 1999), and are publicly available at: http://www.mcmaster.ca/macho.html. We have started our search by looking at 2 inner bulge MACHO fields and 2 inner disk MACHO fields of the Milky Way. The locations of the four fields are shown in Figure 1. Table 1 lists their positions in equatorial and Galactic coordinates, along with the total number of sources found in the MACHO, 2MASS and MSX databases. They contain in total about 300 MSX sources, about 10^5 2MASS JHK sources, of which about 3000 are variable stars, and about 2×10^6 MACHO sources.

The stellar positions of the MACHO and 2MASS data match to better than 1″, while we consider matches within 7″ for the MSX sources. An important potential bias is that when more than one source is present within this matching radius, we chose the brightest optical source as the counterpart of the IR source. After the match is done, we search for stars that are >1 mag fainter in VRJHK that lie in the disk main sequence, that have normal VRJHK colors, and that have >1 mag excess at 8 μm. Another limitation is that the MSX data is not deep enough. Typical Solar-type stars have absolute A-band magnitudes of about 3 (Cohen et al. 2000). For the MSX limiting magnitude of A = 7.5, only nearby main sequence stars (within about 100 pc) would be detected, and main sequence stars with A-band excess would be detected only out to about 500 pc.

3. Summary

We have matched optical sources of the MACHO database with the near-IR 2MASS and the mid-IR MSX sources in the inner Milky Way. Figure 2 shows the IR color-magnitude diagrams of the four fields, with the MSX sources plotted as squares. We have identified the variable stars and the stars with IR excess, finding about 300 stars with >1 mag excess at 8 μm, which are mostly long period variables in the bulge, with dusty circumstellar shells. We have not found any candidate Dyson sphere.

However, there are other astrophysically interesting results. For example, large differences are found between the fields. There are more variables and more giants in the bulge fields F118, F121 than in the disk fields F301, F303. There are also more IR sources in the bulge fields than in the disk fields. These differences indicate the presence of population gradients. We also see neatly the variable star sequence corresponding to first ascent giants. These are semiregular variables in the inner bulge and disk of the Milky Way studied by Minniti et al. (1998).

In the future, the aim is to examine several 10^6 stars, in order to obtain limits to the number of Dyson spheres in our Galaxy. We are using public databases. Mining of such large databases is specially suited for searches like this, and the availability of the Astro-physical Virtual Observatory would be important for these searches.

Even though at first sight searching for Dyson spheres may seem far-fetched, if we are seriously contemplating questions like the existence of life in the Universe or the number of advanced civilizations in the Milky Way, we must consider all possibilities.

4. References

Alcock, C., Allsman, R., Alves, D., Axelrod, T., Becker, A., Bennett, D., Cook, K., Freeman, K. C., Griest, K., Marshall, S.L., Minniti, D., Peterson. B., Pratt, M., Quinn, P., Rodgers, A., Stubbs, C., Sutherland, W., Tomaney, A., Vandehei, and T., Welch, D. (1999) Calibration of the MACHO Photometric Database, *PASP*, **111**, 1539.

Cohen, M., Hammersley, P. L., Egan, M. P. (2000) Radiometric Validation of the Micourse Space Experiment, *The Astrophysical Journal*, **120**, 3362.

Dyson, F. (1960) Search for Artificial Stellar Sources of Infrared Radiation, *Science*, **131**, 1967.

Jugaku, J., and Nishimura, S. (1997) A Search for Dyson Spheres Around Late Type Stars in the Solar Neighborhood, in *Astronomical and Biochemical Origins, and the Search for Life in the Universe*, IAU Coll. No. 161, (Editrice Compositori: Bologna), p. 707.

Jugaku, J., and Nishimura, S. (2000) A Search for Dyson Spheres Around Late Type Stars in the Solar Neighborhood, in *A New Era in Bioastronomy*, G. Lemarchant and K. Mitch (eds.), ASP Conf. Series **213** (ASP: San Francisco), p. 581.

Minniti, D., Alcock, C., Allsman, R., Alves, D., Axelrod, T., Becker, A., Bennett, D., Cook, K., Freeman, K., Griest, K., Marshall, S. L., Peterson. B., Pratt, M., Quinn, P., Rodgers, A., Stubbs, C., Sutherland, W, Tomaney, A., Vandehei, T., and Welch, D. (1998) Pulsating Variable Stars in the MACHO Bulge Database: The Semiregular Variables, in: *Pulsating Stars: Recent Developments in Theory and Observations* (University Academy Press: Tokio), p. 5 (astro-ph/9712048).

SPACE WEATHER AND SPACE CLIMATE
Life Inhibitors or Catalysts?

MAURO MESSEROTTI
INAF-Trieste Astronomical Observatory, Loc. Basovizza n. 302, 34012 Trieste, Italy and Department of Physics, Trieste University, Trieste, Italy

Abstract. Today the Sun exhibits a stable radiation output, which is expected to endure on a long time scale and characterizes the *Space Climate* (SpC). On a short time scale the solar activity perturbs the heliosphere by originating radiation outbursts, highly energetic particles and plasmoids, which characterize the *Space Weather* (SpW). A similar phenomenology can occur in solar-like stars and affect the exoplanetary environments. In this work we speculate on the possible mutual role of SpW and SpC on life birth and evolution, stressing the inadequacy of the basic concept of *Habitability Zone* and the relevance of SpW and SpC to *life-genicity* and *life-sustainability*.

1. Introduction

The Sun is a yellow dwarf star (G2V class) presently in a stable Hydrogen burning phase, variable on a second order scale and with low magneticity, which drives the solar activity cycle. Solar activity is a complex of phenomena, that are variable on spatial, time and energy scales and occur in the photosphere (sunspots), chromosphere (flares), corona (Coronal Mass Ejections, CME) and solar wind (fast plasma streams), as heating, particle acceleration, waves and shocks, emission of radiation, plasmoid formation, triggered by fluid motions and interacting magnetic fields at different spatial scales. Short- and long-term evolution of solar irradiance and activity has been determining the energy input to the planets and the physical state of the planetary environments. Such an evolution played a key role in favoring the birth of life on the Earth, an attitude of the solar-planetary environment which we define *life-genicity*, and has been playing a fundamental role in keeping favorable conditions to the preservation and evolution of life, which we define *life-sustainability*. By extending these concepts to exoplanetary systems around solar-like stars, we elaborate on the need to extend the concepts of Habitability Zone (HZ) by explicitly incorporating the related Stellar Space Meteorology.

2. Space Meteorology of the Stellar-Planetary Environment

An unperturbed stellar-planetary environment is a complex physical system composed of coupled physical subsystems such as: a) the Interstellar Wind (ISW), a diluted magnetized plasma accelerated by neighboring stars which feeds the Interstellar Medium (ISM),

J. Seckbach et al. (eds.), Life in the Universe, 177–180.

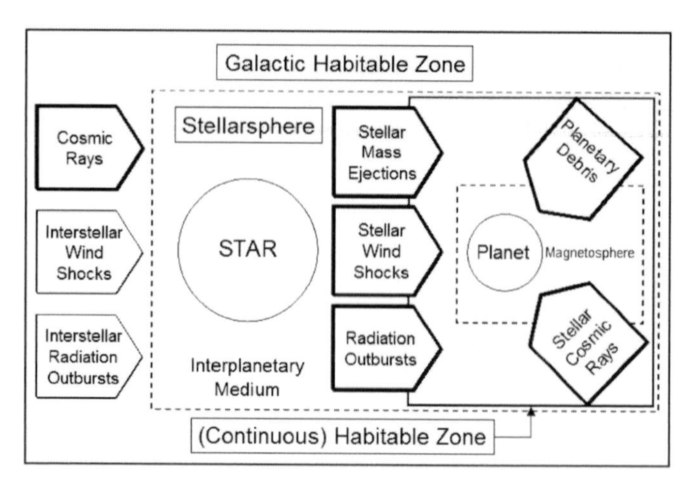

Figure 1. Schematic location of the Galactic and the Continuous Habitable Zones.

permeates the local galactic environment and vehiculates the interstellar magnetic field; b) the central star, a magnetized plasma in an organized state; c) the Stellar Wind (SW), a magnetized plasma accelerated by the central star that, together with dust particles, originates the Interplanetary Medium (IPM), permeates the planetary environment and vehiculates the Interplanetary Magnetic Field (IPMF); d) planet(s), gaseous or condensed mass(es) of organized matter with or without a Planetary Magnetic Field (PMF). The ISW flows externally to the IPMF and confines the SW in a region called *stellarsphere*, compressed towards the ISW source and elongated in the opposite direction. Analogously the SW acts on the PMF and originates the planetary magnetosphere(s). Such physical subsystems are characterized by different physical conditions defined by the state of the neighboring stars, the central star and the dynamical history of the planetary system. Various *outer perturbations* can affect the stellar-planetary environment (Figure 1), e.g. cosmic rays, high energy particles accelerated in violent astrophysical processes, shocks in the ISW and radiation outbursts as observed in supernovae explosions. *Inner perturbations* come from the central star, which can be characterized by a certain level of magnetic activity as starspots. Stellar flares, signatures of impulsive magnetic energy release, determine radiation outbursts in the γ, X, Extreme UltraViolet (EUV) and UV domains. Large mass ejections from the stellar corona (plasma clouds) can be accelerated towards the planet(s) and originate stellar wind shocks. Stellar cosmic rays are generated by the star in energetic activity phenomena. Planetary debris (meteoroids, comets and asteroids) hit the planet(s), whose size and trajectories are defined by the planetary population diversity and orbital dynamics. Similarly to the solar case, *Stellar Space Meteorology* (SSpM) is aimed to observing and modeling the state of the stellarsphere and its perturbations on a short (*Stellar Space Weather*, SSpW) to a long time scale (*Stellar Space Climatology*, SSpC).

3. Stellar Space Meteorology and Planetary Response Drivers

The *drivers of SSpM* (Figure 1) are intrinsic to: a) the local galactic neighborhood (LGN); b) the central star typology; c) the planetary system environment. Specifically, the *LGN star population* determines: – the flow and energy spectrum of galactic cosmic rays; – the occurrence frequency and morphology of ISW shocks; – the occurrence frequency and intensity of interstellar radiation outbursts in the γ, UV and EUV bands, and of stellar cosmic rays flow. Similarly, the *central star* is characterized by: – its evolutionary state and global irradiance level; – magneticity, which determines the surface activity level on a short time scale; – variability, which biases the long-term irradiance variations; – wind, which modulates the galactic and stellar cosmic rays, and sweeps the planetary environments. Stellar mass ejections, wind shocks and radiation outbursts are dependent on evolutionary stage, magneticity and activity level. In fact, the main activity features observed in solar-like stars on a non-systematic basis are: – very large photospheric starspots; – high energy stellar flares; – huge stellar coronal mass ejections; – stellar activity cycles. As such stellar features can occur at scales order of magnitudes larger than the solar homologous ones (Schrijver and Zwaan, 2000), the stellar activity level and the variety of stellar activity phenomena represent fundamental features to consider when characterizing stellar-planetary environments for life origination and preservation. The *planetary system environment* is furthermore populated by a variety of planetary debris. The *planetary response drivers* to SSpM are respectively mass, radius, density, orbital dynamics, surface morphology, atmosphere (that can filter and modulate the SpM radiation effects and determines the thermodynamics driven by the stellar radiation) and magnetosphere (key factor in shielding and modulating the SpW particular effects).

4. Relevance of Stellar Space Meteorology to Habitability Zones

At a galactic spatial scale the *Galactic Habitability Zone* (GHC) (Gonzalez *et al.*, 2001) is defined as an environment in a galaxy favorable to life, populated by suitable neighboring stars (e.g. no nearby supernovae), solar-like stars (characterized by a "normal", non-cataclysmic evolution), and a suitable interstellar medium (with organics). At a planetary spatial scale the *Habitability Zone* (HZ) (Kasting, 1996) is defined as a set of stable planetary orbits to allow for liquid water to exist on a planet. This requirement is a necessary condition, but is it also a sufficient condition for life-genicity? The *Continuous* HZ (Kasting, 1996) is defined in terms of a stellar environment *favorable to life in time*. Is this condition sufficient for life-sustainability?

Let us assume a possible *scenario for the origin of life* (e.g. Baross, 2002): a) a primordial soup is originated; b) the environment is fed with energy and some unidentified synergetic mechanism(s) operates (the "twilight zone"); c) an RNA world is formed; d) RNA and protein biosynthesis occurs; e) the environment is progressively populated by cells with DNA, RNA and proteins. The concept of *life-genicity* implies stellar-planetary *conditions favorable to originate life*, which means a central star with *"suitable" stellar activity* (respectively "moderate" irradiance, wind, radiation outbursts, particle emissions, coronal mass ejections) and a *"suitable" planetary configuration* (adequate atmosphere and magnetosphere, moderate debris bombardment). The *role of SSpM* in life-genicity has to be considered in terms of both *SSpC* (presently the solar irradiance variations are lower than 0.1% but new solar models imply an initial irradiance higher than the present one) (Tehrany et al., 2002 and references therein), which defines the *peak energy input* to the planet, and *SSpW* (stellar activity modulates and planetary magnetosphere shield cosmic rays; planetary atmosphere modulates radiation and particles at sea level), which determines the *peak biological suitability of the planetary environment*. How much relevant are SSpC and SSpW to the "Twilight Zone"? How to quantify the attributes "suitable", "moderate" and "adequate"? How to define the "peak biological suitability"? When does SSpM act as *life catalyst* and when like *life inhibitor*? The *existence of life* can be schematically described respectively in terms of: a) *existence of* cells with DNA, RNA and proteins; b) *evolution to* complex life forms; c) *evolution of* complex life forms; d) *persistence of* complex life forms. The concept of *life-sustainability* involves stellar-planetary *conditions favorable to preserve life*, i.e. *"suitable" stellar activity* and a *"suitable" planetary configuration* (adequate biosphere and magnetosphere, moderate debris bombardment). The *role of SSpM* in life-sustainability is determined by the *SSpC* (long-term evolution of stellar irradiance and existence and intensity of an activity cycle), which *modulates in time the planetary energy input*, and by the *SSpW* (cosmic rays modulation and shielding, modulation of radiation and particles at sea level), which determines the *long-term biological harshness of the planetary environment*. How much relevant are SSpC and SSpW to the "long-term biological harshness"? How to define the "biological harshness"?

5. Conclusions

New environmental aspects are worthwhile investigating in the framework of Stellar Astrophysics, Exoplanetology and Astrobiology. Solar Space Weather and Space Climate have been playing a fundamental role in life-genicity and life-sustainability on the Earth and acted as life catalysts. Hence the definitions of Galactic and (Continuous) Habitability Zones must be extended to account for Stellar Space Meteorology, based on a refinement of Solar SpM, as Stellar SpM is beyond the present instrumental capabilities. A detailed analysis in the framework of Astrobiology can provide new information to quantify the relevant biological characteristics of a planetary environment such as life-genicity and peak biological suitability as well as life-sustainability and long-term biological harshness. This will help in understanding when SSpM acts as life catalyst or inhibitor in exoplanetary environments, unobservable at a high level of detail.

6. References

Baross, J.A. (2002) The Definition of Life and the Origin of Life on Earth, Bull. Am. Astron. Soc. **34**, p. 1213.

Gonzalez, G., Brownlee, D. and Ward, P. (2001) The Galactic Habitable Zone: Galactic Chemical Evolution, Icarus **152**, 1, pp. 185–200.

Kasting, J.F. (1996) Habitability of Planets, Astrobiology Workshop: Leadership in Astrobiology, pp. A10–A11.

Schrijver, C.J. and Zwaan, C. (2000) *Solar and Stellar Magnetic Activity*, Cambridge Astrophysics Series **34**, Cambridge University Press, Cambridge, UK.

Tehrany, M.G., Lammer, H., Selsis, F., Ribas, G., Guinan, E.F., Hanslmeier, A. (2002) The particle and radiation environment of the early Sun, In: A. Wilson (ed.) Proc. *Solar variability: from core to outer frontiers*, ESA SP-506, **1**, pp. 209–212.

VII. Planetary Exploration in our Solar System: The Interstellar Medium, Micro-Meteorites and Comets

SPONTANEOUS GENERATION OF AMINO ACID STRUCTURES IN THE INTERSTELLAR MEDIUM

UWE J. MEIERHENRICH
Dept. Physical Chemistry, University of Bremen
Fachbereich 2, Leobener Straße, 28359 Bremen, Germany

Abstract. In dense interstellar clouds dust particles accrete ice mantles. As seen in infrared (IR) observations, this ice layer consists mainly of water ice, but also of carbon and nitrogen containing molecules. We deposited a gas mixture consisting of H_2O, CO_2, CO, CH_3OH, and NH_3 onto an aluminium surface at 12 K under high vacuum, 10^{-7} mbar. During deposition the molecules were subjected to ultraviolet radiation with main intensity at Lyman-α. After warm-up, the refractory material was extracted from the aluminium block, hydrolysed for 24 h at 110 °C with 6 M HCl, derivatized and finally analysed by enantioselective gas chromatography coupled to a mass spectrometer. We were able to identify 16 amino acids in the room temperature products of irradiation. The results were confirmed by parallel experiments using ^{13}C-labelled ices in order to exclude contamination. A first 'group' of the identified amino acids was suggested to serve as the precursors of peptides and proteins. A second 'group' namely the diamino carboxylic acids is assumed to contribute to the development of the first genetic material, the peptide nucleic acid PNA. Beside the two groups of amino acids, N-heterocyclic organic molecules were identified that resemble the molecular building block of biological cofactors. The obtained results support the assumption that the photochemical products could be preserved in interstellar objects, and in term be delivered to the Earth during the heavy bombardment which ended about 3.8 Gyr ago, where they triggered the appearance of live.

1. Introduction

In order to understand availability and distribution of molecular building blocks of biological systems during defined phases of the Chemical Evolution we studied and simulated interstellar/circumstellar processes in the laboratory. Based on the knowledge of intensity, energy, and polarization of interstellar/circumstellar electromagnetic radiation (Bailey *et al.*, 1998), and on the occurrence and abundance of volatile compounds in interstellar clouds/circumstellar disks like H_2O, CO_2, CO, CH_3OH, and NH_3 interstellar/circumstellar photochemical processes were simulated in the laboratory. Scientific objective of these experiments was to synthesize interstellar ices under most realistic conditions in order to test the enantioselective GC-MS instrumentation developed for the Rosetta-Lander cometary sampling and composition experiment COSAC at the Max-Planck-Institut für Aeronomie in Katlenburg-Lindau, Germany. The experiments were not optimised in order to identify any organic compounds under simulated interstellar/circumstellar conditions.

J. Seckbach et al. (eds.), Life in the Universe, 183–186.

2. Simulation of Interstellar/Circumstellar Ices

At the Raymond and Beverly Sackler Laboratory for Astrophysics at Leiden Observatory, the Netherlands, J. Mayo Greenberg's (†) student G. M. Muñoz Caro deposited the volatile compounds H_2O, CO_2, CO, CH_3OH, and NH_3 (molecular composition 2:1:1:1:1) onto an aluminium block at a representative temperature of 12 K during UV-irradiation in a specially developed space simulation vacuum chamber. After 24 h irradiation a residue of yellowish color remained on the aluminum block (Muñoz Caro *et al.,* 2001). After a stepwise warm-up procedure and the vacuum venting process this residue was kept under inert gas until analysis.

3. Enantioselective GC-MS Analysis of Amino Acids

At the Centre de Biophysique Moléculaire in Orléans, France, the irradiated samples of interstellar/circumstellar ice analogues were analysed in the group of André Brack and Bernard Barbier. Therefore, a new and sensitive analytical procedure was developed that consists of the sample's extraction with 100 μL water, the 24 h hydrolysis at 110 °C with 6 M HCl, the chemical transformation of amino acids into the N-ethoxycarbonyl ethyl ester derivatives (Abe *et al.*, 1996), and the enantioselective capillary GC-MS analysis (Huang *et al.*, 1993) using the Agilent 6890/5973 GC-MSD system.

4. Amino Acid Identification in Simulated Interstellar/Circumstellar Ices

The application of the new analytical procedure allowed us the identification of 16 amino acids in the simulated interstellar/circumstellar ice samples (Muñoz Caro *et al.,* 2002). The gas chromatogram is given in Figure 1.

Among the 16 amino acids are L-Ala, Gly, L-Val, L-Pro, L-Asp and L-Ser, i.e. six α-amino acids being the molecular constituents of proteins. The results suggest, that the structures of these α-amino acids can be synthesized under interstellar/circumstellar conditions and be delivered via asteroids, comets and interplanetary dust particles to the early Earth, where they might have triggered the appearance of live.

Beside the proteinacous α-amino acids, six diamino acids were identified in the interstellar/circumstellar ice analogues. These structures are the molecular constituents of the backbone of the peptide nucleic acid PNA (Egholm *et al.*, 1992). PNA is considered to be a potential early genetic material on Earth, where it might have preceded RNA and DNA (Nielsen, 1993 and Nelson *et al.*, 2000). PNA hybridizes to complementary oligonucleotides obeying the Watson-Crick hydrogen-bonding rules (Egholm *et al.*, 1993). Its backbone consists of diamino acids linked via peptide bonds. The nucleic bases can be attached with the help of various molecular spacer groups to the PNA backbone.

As expected, the identified amino acids showed racemic occurrence. In future experiments, the applied light source that emitted unpolarized light will be substituted by a circularly polarized synchrotron beam in the French synchrotron facility LURE, Paris. Circularly polarized electromagnetic radiation has been identified in the Orion star formation region by J. Bailey and his group (1998). With these experiment we try to mimic

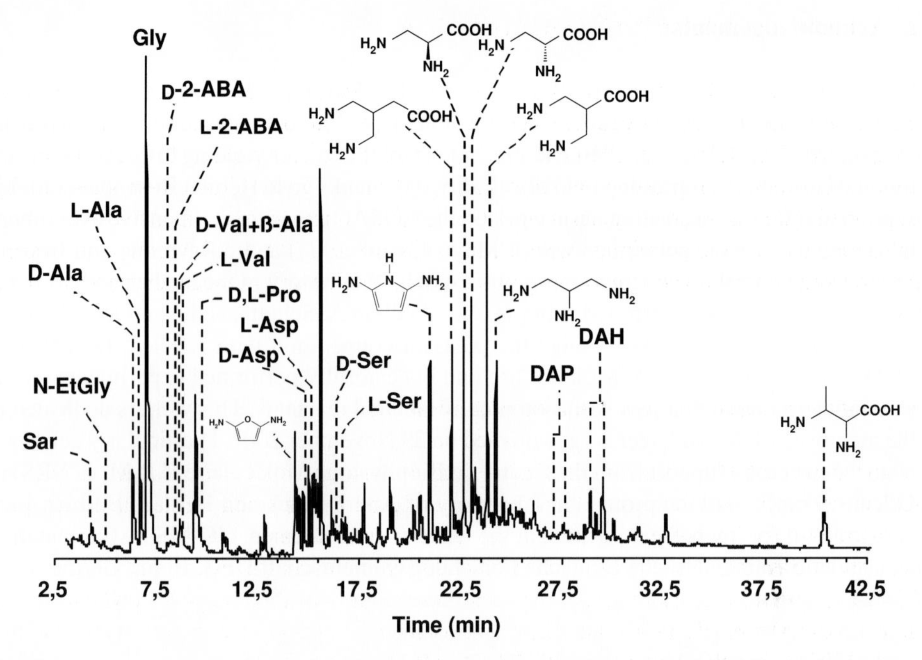

Figure 1. Amino acids identified in interstellar/circumstellar ice analogues by enantioselective gas chromatography. Varian-Chrompack Chirasil-L-Val capillary column 12 m × 0.25 mm inner diameter, film thickness 0.12 μm, splitless injection, 1.5 mL/min constant flow of He carrier gas, oven temperature programmed for 3 min at 70 °C, 5 °C/min, and 17.5 min at 180 °C, detection of total ion current TIC. DAP, diaminopentanoic acid; DAH, diaminohexanoic acid.

interstellar/circumstellar conditions even more realistically and to introduce an enantiomeric excess into the produced amino acids by absolute asymmetric photochemistry (Griesbeck and Meierhenrich, 2002).

5. Discussion

Amino acid structures, the molecular building blocks of proteins *and* diamino acid structures, the molecular constituents of the PNA backbone, were identified in simulated interstellar/circumstellar ice analogues. The results suggest that both early proteins and early genetic material was synthesized from building blocks that had been delivered from interstellar/circumstellar space to the early Earth. Organic molecules like amino acids might survive an impact onto the early Earth (Greenberg *et al.*, 1994) as the results from the Stone artificial meteorite experiments indicate (Brack, 2003) The organic molecules might have played an important role on the appearance of primitive life on Earth. The identification of amino acids in interstellar ice analogues is suggested to be linked with the prebiotic development of proteins, genetic material and biological cofactors on Earth. The COSAC experiment with its enantioselective GC-MS instrumentation (Meierhenrich *et al.*, 1999; Thiemann and Meierhenrich, 2001) is envisaged to verify the analyses of organic molecules in cometary ices *in situ* and in continuation to elucidate possible origins of life.

6. Acknowledgements

I would like to thank Wolfram H.-P. Thiemann, Dept. Physical Chemistry at the University of Bremen, Germany, most sincerely for uncountable enlightening discussions based on his experience in the field and his tremendous support for the studies yielding to this work. He introduced me into the intriguing field of chirality. My thanks go to Helmut Rosenbauer for his support and for his fascinating basic work on the COSAC experiment. Intensive experimental cooperations were performed with J. Mayo Greenberg (†) at the Raymond and Beverly Sackler Laboratory for Astrophysics at the Leiden Observatory in the Netherlands, and his group composed of Guillermo M. Muñoz Caro, Willem A. Schutte, and Almudena Arcones Segovia. I particularly acknowledge the pleasant cooperation with J. Mayo Greenberg's Ph.D. student Guillermo M. Muñoz Caro and his carefully performed experiments on the simulation of interstellar ices with isotopically labelled reactants. This work is dedicated to the memory of J. Mayo Greenberg, who died on 29 November 2001. I particularly acknowledge the generous support from the Centre de Biophysique Moléculaire C.B.M. (CNRS) in Orléans, France, and the prominent laboratory of André Brack and Bernard Barbier. I am very grateful for my present position at the University of Bremen, GC-MS instrumentation and students founded by the Deutsche Forschungsgemeinschaft DFG, Bonn, Germany.

7. References

Abe, I., Fujimoto, N., Nishiyama, T., Terada, K. and Nakahara, T. (1996) Rapid analysis of amino acid enantiomers by chiral-phase capillary gas chromatography, *J. Chromatogr. A* **722**, pp. 221–227.

Bailey, J., Chrysostomou, A., Hough, J.H., Gledhill, T.M., McCall, A., Clark, S., Ménard, F. and Tamura, M. (1998) Circular Polarization in Star-Formation Region: Implications for Biomolecular Homochirality, *Science* **281**, pp. 672–674.

Brack, A. (2003) Search for life on Mars: The Beagle 2 Lander and the Stone Experiment, in this volume.

Egholm, M., Burchardt, O., Nielsen, P.E. and Berg, R.H. (1992) Peptide Nucleic Acids (PNA). Oligonucleotide Analogues with an Achiral Peptide Backbone, *J. Am. Chem. Soc.* **114**, pp. 1895–1897.

Egholm, M., Burchardt, O., Christensen, L., Behrens, C., Freier, S.M., Driver, D.A., Berg, R.H., Kim, S.K., Norden, B. and Nielsen, P.E. (1993) PNA hybridizes to complementary oligonucleotides obeying the Watson-Crick hydrogen-bonding rules, *Nature* **365**, pp. 566–568.

Greenberg, J.M., Kouchi, A., Niessen, W., Irth, H., van Paradijs, J., de Groot, M. and Hermsen, W. (1994) Interstellar dust, chirality, comets and the origins of life: life from dead stars? *J. Biol. Phys.* **20**, pp. 61–70.

Griesbeck A.G. and Meierhenrich U.J. (2002) Asymmetric Photochemistry and Photochirogenesis, *Angew. Chem. Int. Ed. Engl.* **41**, pp. 3147–3154.

Huang, Z.-H., Wang, J., Gage, D.A., Watson, J.T. and Sweeley, C.C. (1993) Characterization of N-ethoxycarbonyl ethyl esters of amino acids by mass spectrometry, *J. Chromatogr. A* **635**, pp. 271–281.

Meierhenrich, U., Thiemann, W.H.-P., and Rosenbauer H. (1999) Molecular Parity Violation via Comets? *Chirality* **11**, pp. 575–582.

Muñoz Caro, G.M., Ruiterkamp, R., Schutte, W.A., Greenberg, J.M. and Mennella, V. (2001) UV photodestruction of CH bonds and the evolution of the 3.4 micron feature carrier. I. The case of aliphatic and aromatic molecular species, *Astron. Astrophys.* **367**, pp. 347.

Muñoz Caro, G.M., Meierhenrich, U.J., Schutte, W.A., Barbier, B., Arcones Segovia, A., Rosenbauer, H., Thiemann, W.H.P., Brack, A. and Greenberg, J.M. (2002) Amino acids from ultraviolet irradiation of interstellar ice analogues, *Nature* **416**, pp. 403–406.

Nelson, K.E., Levy, M. and Miller, S.L. (2000) Peptide nucleic acids rather than RNA may have been the first genetic molecule, *Proc. Natl. Acad. Science* **97**, pp. 3868–3871.

Nielsen, P.E. (1993) Peptide nucleic acid (PNA): A model structure for the primordial genetic material? *Orig. Life Evol. Biosphere* **23**, pp. 323–327.

Thiemann, W.H.-P. and Meierhenrich, U. (2001) ESA Mission ROSETTA Will Probe for Chirality of Cometary Amino Acids, *Orig. Life Evol. Biosphere* **31**, pp. 199–210.

EXPERIMENTAL STUDY OF THE DEGRADATION OF COMPLEX ORGANIC MOLECULES. APPLICATION TO THE ORIGIN OF EXTENDED SOURCES IN COMETARY ATMOSPHERES

N. FRAY, Y. BENILAN, H. COTTIN, M.-C. GAZEAU and F. RAULIN
LISA, Universités de Paris 7 et 12, UMR CNRS 7583,Universités Paris 7 and 12, CMC, 61 Av. du Gal de Gaulle, 94010 Créteil Cedex, France

1. Introduction

Most of the molecules observed in the cometary environment are directly produced by sublimation from the nucleus, and the main fraction of the observed radicals is the result of the photodissociation of gaseous "parent" molecules already observed. However some of these species (especially CO, H_2CO, CN...) have a spatial distribution which can not be explained by these processes. They are produced by an unknown "extended source" in the coma. If one could infer the nature of the material involved in this phenomenon, this should allow to constraint the chemical composition of cometary grains and nucleus which is of prime interest for the exobiology studies.

In this paper, we report the study of the origin of the formaldehyde (H_2CO) and CN radical extended sources. The H_2CO density profiles has been derived in comet 1P/Halley from the *Giotto* NMS measurements (*Meier et al.*, 1993). These observations have demonstrated that H_2CO is not produced only by nucleus sublimation but rather by an "extended source". The case of CN radicals is similar but not identical since it is a radical, and an important fraction is produced by the photodissociation of HCN (*Fray et al.*, submitted b).

In both cases, it has been proposed that these species could be produced by the degradation of large organic molecules present on cometary grains. These compounds could be decomposed into gaseous molecules by UV irradiation or heating when the comet reaches perihelion. Polyoxymethylene [formaldehyde polymer: $(-CH_2-O-)_n$, also called POM] could produce gaseous H_2CO (*Meier et al.*, 1993) and CN radicals could be produced by the decomposition of hexamethylenetetramine ($C_6H_{12}N_4$, also called HMT) (*Bernstein et al.*, 1995) and/or of HCN polymers (*Huebner et al.*, 1989). Unfortunately the lack of quantitative data relative to the decomposition of such compounds has prevented to confirm this hypothesis so far. Thus, we have developed a specific experimental setup to measure the kinetics of the decomposition of large solid organic molecules into gaseous molecules by thermal or photo processes.

J. Seckbach et al. (eds.), Life in the Universe, 187–190.

2. Experimental Setup

The experimental setup designed for the study of photodegradation is mainly composed of a UV lamp and a pyrex reactor. It has been presented in details in *Cottin et al.* (2000). The UV lamp can emit at 122, 147 and 193 nm depending on the filling gas. For each wavelength, the UV flux is measured by chemical actinometers and is about 10^{15} photons per second. The photochemical reactor, in which solid molecules are deposited in order to be irradiated, is equipped with two vacuum stopcocks, one leading to the analysis system and another one to a turbomolecular pump. A vacuum, better than 10^{-4} mbar, can be achieved in the reactor. Generally the reactor is directly connected to a FTIR spectrometer. Successive spectra can be collected to follow the production of gaseous molecule as a function of time. Then, combining the known UV flux and the kinetic measurements, we can derive the production quantum yields of each gaseous molecule.

In order to study thermal degradation, we have developed two different reactors. Between 210 and 350 K, we use a pyrex reactor equipped with a double wall, which allows the circulation of a thermostated fluid. For higher temperatures, up to 600 K, we use a metallic reactor in which a heating resistance is included. According to the temperature range of degradation of the studied molecule, we choose one reactor or the other. The pyrex reactor has been used to study the thermal degradation of POM. But, as HMT and HCN polymers are stable at room temperature, the study of their thermal degradation has been performed only at high temperature in the metallic reactor.

3. Decomposition of H_2CO Polymers. Application to the H_2CO Extended Source

The photodegradation of POM has been performed by *Cottin et al.* (2000). They have shown that several oxygenated compounds (H_2CO, HCOOH, CO, CO_2 and CH_3OH) are produced by UV irradiation of POM at 122, 147 and 193 nm. The production quantum yield of H_2CO is roughly equal to 1 up to 147 nm and decreases for longer wavelengths. By thermal heating, POM produces only gaseous H_2CO. This production has been measured between 255 and 330 K for two different types of POM coming from two different suppliers: Prolabo and Aldrich (*Fray et al.*, submitted a). Our measurements are presented in Figure 1, which

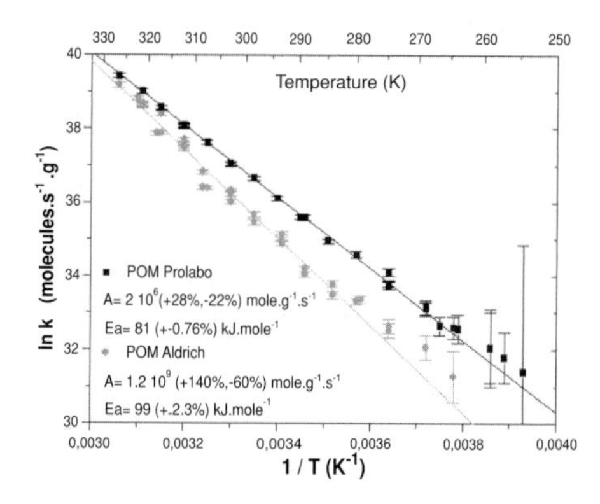

Figure 1. Logarithim of the H_2CO production rates as a fuction of the inverse of the temperature for the "Prolabo POM" in black and for the "Aldrich POM" in grey.

shows the logarithm of the production (in molecules . g^{-1}. s^{-1}) as a function of the inverse of the temperature.

These measurements are well fitted by a straight line; this shows that the production of gaseous formaldehyde from solid POM follows the Arrhenius law (i.e. $k(T) = A.e^{-Ea/RT}$). Hence, these fits allow us to determine Ea, the activation barrier and A, the frequency factor for both types of POM (see Figure 1).

The value of the activation energy is quite important; it shows that the kinetics of gaseous H_2CO production is highly sensitive to the temperature. The production of gaseous formaldehyde is different for both type of POM; at 310 K it is 1.8 times higher for the "Prolabo POM" than for the "Aldrich POM". So far, the origin of this discrepancy is not understood but could be due to different general structure of the polymer or to different ending of their chains. These quantitative results on the thermal and photo degradation of POM have been incorporated in a model of the cometary environment to explain the H_2CO extended source.

This model of the coma (*Cottin et al.*, in press) allows us to fit very well the H_2CO density profiles which have been obtained from *Giotto* NMS measurements (*Meier et al.*, 1993) taking into account the production of gaseous H_2CO by degradation of solid POM present on grains. New results which have been obtained taking into account these new measurements of the thermal degradation of POM, are presented in *Fray et al.* (submitted a). Depending on the parameters that we used for the cometary grains, the percentage of POM by mass needed to reproduce the observations is ranging from 1.3 to 15.5%. These values are in agreement with previous estimations. Thus the degradation of POM is so far the best quantitative explanation of the H_2CO extended source in comets. And thereafter the presence of POM in cometary nucleus is highly probable.

4. Degradation of HMT and HCN Polymers. Application to the CN Extended Source

HCN has been detected in small amount after UV irradiation of HMT at 147 nm. NH_3 and some heavier compounds are also produced if 0.1 mbar of water is added during irradiation (*Cottin et al.*, 2002). Nevertheless, HMT seems to be quite resistant to photolysis. Whereas *Iwakami et al.* (1968) have reported the thermal decomposition of HMT by pyrolysis, we found that HMT sublimates when slowly heated. Thus, HMT does not seem to be a good candidate to explain the CN extended source.

After UV irradiation at 122 and 147 nm of HCN polymers, we have detected HCN and C_2H_2 by IR spectroscopy. And, as we show in Figure 2, by heating the sample, we have observed the signature of HCN, NH_3, HNCO, CO and CO_2 at every temperature between 430 and 670 K. The major products are NH_3 and HCN. CO_2 does not seem to be a degradation product of HCN polymer and its presence in spectra is certainly due to instability of the purge of the FTIR spectrometer. HNCO and CO could be produced by reaction between degradation products with H_2O which is trapped in HCN polymers. We are currently measuring the production kinetics of each product as a function of the temperature. Nevertheless these first results show that HCN polymers produce gaseous compounds in conditions relevant of the cometary grains, and thus could also be plausible parents for the cometary CN radicals, which cannot be directly detected in our experiments yet. Indeed, to date, it not clear whether HCN is directly released from the polymer, or if it is rather CN that

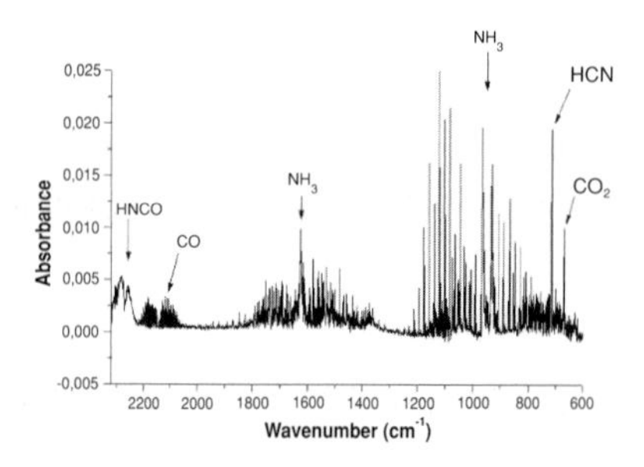

Figure 2. IR spectra of the gaseous ocmpounds produced by heating of HCN polymers at 450 K.

is produced and promptly reacts to form HCN. Direct detection of CN radical by LIF (Laser Induced Fluorescence) and quantify-cation by CRDS (Cavity Ring-Down Spectroscopy) are planned to investigate this point.

5. Conclusion

To explain H_2CO and CN extended source, we have studied the photo and thermal de-composition of POM, HMT and HCN polymers. We show that the degradation of POM in solid state on grains is to date the best explanation of the H_2CO extended source. Thus, the presence of POM in cometary nucleus is highly probable. To explain the CN origin in cometary atmospheres, HCN polymers seem to be a better candidate than HMT. New experiments are planned to confirm this hypothesis.

6. References

Bernstein, M.P., S.A. Sandford, L.J. Allamandola, S. Chang, and M.A. Scharberg (1995) Organic Compounds Produced By Photolysis of Realistic Interstellar and Cometary Ice Analogs Containing Methanol. *The Astrophysical Journal*, **454**, 327–344.

Cottin, H., S. Bachir, F. Raulin, and M.C. Gazeau (2002) Photodegradation of HMT by VUV and its relevance for CN and HCN extended sources in comets. *Advances in Space Research.*, **30(6)**, 1481–1488.

Cottin, H., M.C. Gazeau, J.F. Doussin, and F. Raulin (2000) An experimental study of the photodegradation of polyoxymethylene at 122, 147 and 193 nm. *Journal of photochemistry and photobiology A: Chemistry*, **135**, 53–64.

Cottin, H., M.C. Gazeau, and F. Raulin (1999) Cometary organic chemistry: a review from observations, numerical and experimental simulations. *Planetary and Space Science*, **47(8–9)**, 1141–1162.

Cottin, H., Y. Benilan, M.C. Gazeau and F. Raulin (in press) Origin of cometary extended sources from degradation of refractory organics on grain: polyoxymethylene as formaldehyde parent molecule. *Icarus.*

Fray, N., Y. Benilan, H. Cottin, and M.-C. Gazeau (submitted-a) New experimental results on the degradation of polyoxymethylene. Application to the origin of the formaldehyde extended source in comets. *Journal of Geophysical Research Planets.*

Fray, N., Y. Benilan, H. Cottin, M.-C. Gazeau, and J. Crovisier (submitted-b) CN extended source: a review of observations and modelisations. *Planetary and Space Science.*

Huebner, W.F., D.C. Boice, and A. Korth (1989) Halley's polymeric organic molecules. *Advances in Space Research*, **9**, 29–34.

Iwakami, Y., M. Takazono, and T. Tsuchiya (1968) Thermal decomposition of Hexamethylene Tetramine. *Bulletin of the Chemical Society of Japan*, **41**, 813–817.

Meier, R., P. Eberhardt, D. Krankowsky, and R.R. Hodges (1993) The extended formaldehyde source in comet P/Halley. *Astronomy and Astrophysics*, **277**, 677–691.

FATE OF GLYCINE DURING COLLAPSE OF INTERSTELLAR CLOUDS AND STAR FORMATION

SANDIP K. CHAKRABARTI[1,2], SONALI CHAKRABARTI[2,3] and KINSUK ACHARYYA[2]
[1]S.N. Bose National Center For Basic Sciences, JD—Block, Sector—III, Salt Lake, Kolkata 700098, India; [2]Indian Astrobiology Network(IAN),Center For Space Physics, Chalantika 43, Garia Station Rd, Kolkata 700084, India; [3]Maharaja Manindra Chandra College, Kolkata 700003, India.

1. Introduction

Many bio-molecules have now been observed in meteorites which showered the Earth. However, even though about 124 organic molecules have been observed in star forming regions, it is doubtful if any amino acids have been detected. The problem is probably not that these bio-molecules are not present, but that they are of very small in abundance so that the present detection techniques are sufficiently sensitive. In our view, for any reaction rate, given a large laboratory such as a molecular cloud and a sufficient time such as about ten million years (the time scale of collapse) it is not unlikely that complex bio-molecules could form during the star formation itself. The problems lies in (a) to identify the pathways to produce these molecule, given that the ice chemistry and grain-chemistry are very important, (b) to use appropriate reaction rates for each pathways and finally (c) to use an appropriate hydrodynamic evolution of the collapse which govern the temperature and density of the collapsing matter.

In our work we follow an unconventional route to check if the complex molecules could be formed or not. Instead of waiting for chemists to come up with the right pathways and experimentalists to come up with the right reaction rates, we assume standard pathways from textbooks and use reaction rates as parameters. By choice of reasonable parameters we conclude that simple amino acids like glycine and alanine may be produced in space and they may even be detectable. We think that more complex molecules, such as urea, adenine etc. may also form, though the abundance of adenine is even lower, and possibly may not be detectable in near future. This exercise does trivialize a complex part of the formation of life—in fact it makes the presence of bio-molecules (and therefore life) to be more generic than thought previously.

2. Our Approach

Our approach may be briefly written down as follows: (a) We collect a large number of chemical species which we think could be responsible for life formation, (b) We collect the

J. Seckbach et al. (eds.), Life in the Universe, 191–194.

reactions which may take place among them, and also in presence of radiations, cosmic rays etc. and compile the reaction rates as functions of density and temperature. (c) We define a model of the collapsing interstellar cloud, i.e., we pre-define the variation of the density, temperature and velocity of a collapsing cloud. (d) We allow the chemical reactions to take place at each time step during infall. We expect that part of the matter would form outflows and jets as in any rotating and collapsing system (such as young stellar objects, or YSOs) and therefore, we expect that some of the matter would be re-cycled and there will be scope to produce more complex molecules.

We simplify the problem presently (a) by choosing a spherical freely falling collapse, (b) by ignoring the effect of the energy release on the collapsing cloud, (c) by assuming isothermal infall till the rotation becomes important and (d) by ignoring the effect of recycling mentioned above. Furthermore, we assume two parameter family (α, β) of reaction rates mentioned in Chakrabarti and Chakrabarti (2000a). In this case, while forming a complex molecule, we assume that it is made up of successive two-body reactions (wherever permissible) and the two body reaction rate of the first step is α. However, as the molecules grow in size, the reaction rate is increased by a factor of β. This will take care of the increase in reaction rates as the size grows. In future we plan to relax these constraints.

A very important improvement of our code has been done: In Chakrabarti and Chakrabarti (2000ab) we used the $H + H \rightarrow H_2$ rate from UMIST database. But presently, we assume compute the H_2 formation on grain surfaces by a very rigorous method. Briefly, we separated the existence of mainly three types of grains and used the master equation and rate equations (Biham et al. 2001; Herbst, 1992) to compute mass fraction of H_2 as matter collapses. We then use this H_2 to find out the mass fractions of more complex molecules such as glycine, alanine, urea etc.

The result of our computation clearly depends on the initial abundance of the chemical special present in the interstellar cloud. For instance, the chemical composition of the solar neighbourhood and that near a carbon star would be very much different and therefore the final product would vary. In the present analysis we use the solar neighbourhood abundance as the initial condition.

3. Results

Figs. 1(a–c) show the variation of the mass fraction (along Y-axis) of complex molecules such as glycine, alanine, urea and adenine with the radial distance (along X-axis) drawn in logarithmic scale. We use $\alpha = 10^{-11}$ in Fig. 1a, 10^{-12} in Fig. 1b and 10^{-13} in Fig. 1c. We assume $\beta = 1$ in the first two-body reaction in a chain reaction, $\beta = 4$ for the two-body reaction second in the chain (see, Chakrabarti and Chakrabarti, 2000a), $\beta = 9$ for the next two body reaction in the chain and so on.

A comparison of the mass fractions suggest that in the case of adenine, the abundance progressively goes down as the reaction rate is decreased. Glycine and alanine have peak abundances in Fig. 1a at around $\log(r) = 15.5$, but the peaks are not achieved (or achieved at a lower radius) for Fig. 1(b-c) while the matter is collapsing. In case of urea, the monotonicity with decreasing α is not maintained.

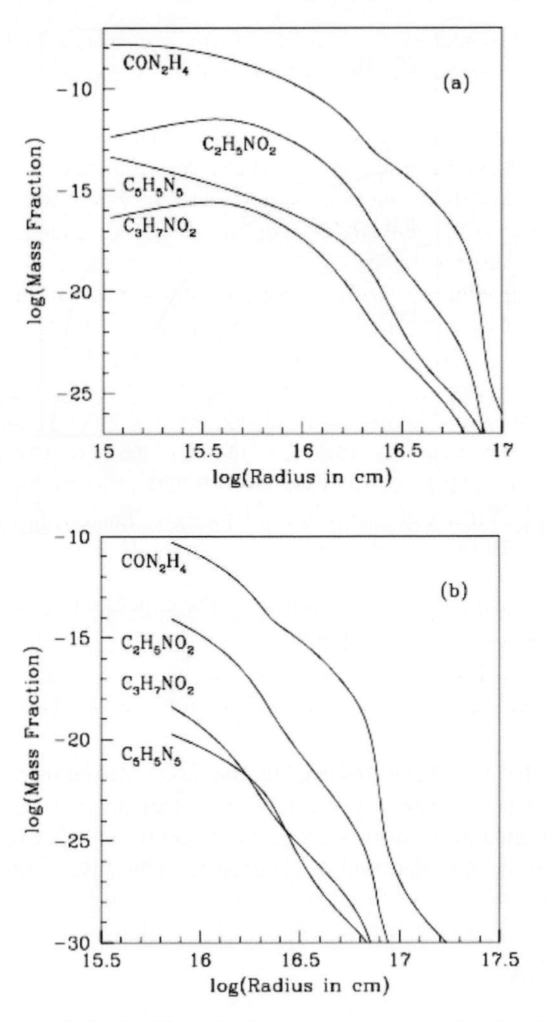

Figure 1a–b. Mass fraction of complex bio-molecules as functions of logarithmic radial distance (cm). See, text for details of the parameters used.

4. Observational Status

Recently, there are reports that glycine has been seen in the interstellar clouds (Kuan et al. 2003). Similarly there are contradicting claims (Hollis et al. 2003). The abundance that is observed is similar to what we predict from our model calculations. Given the uncertainly of the rate constants at the present moment nothing better could be done, but as we have used the rates in the reasonable parameter range our prediction is robust and must agree with the observations. Of course, the challenge would be to be able to explain all the observed species with chirality exactly. As our code grows with the inclusion of more complexity, we should be able to explain the whole set of lines, since our approach is nothing but a

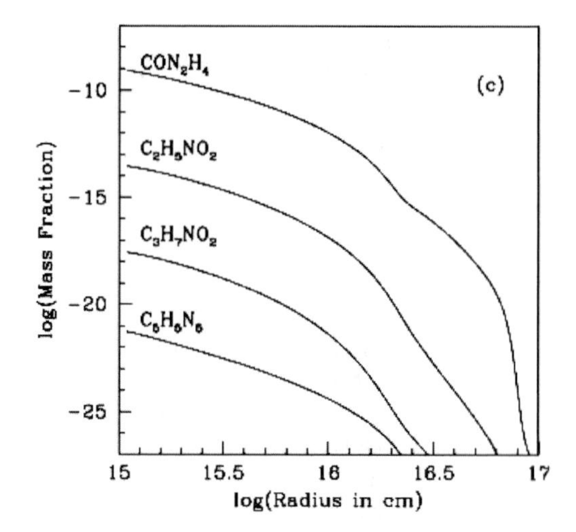

Figure 1c. Mass fraction of complex bio-molecules as functions of logarithmic radial distance (cm). See, text for details of the parameters used.

numerical experiment. Meanwhile, we appeal to the experimentalists to supply us with pathways and rates as accurately as possible.

5. Summary

Simple Amino acids have been detected in meteorites coming from space, but their existence in star-forming region is still a mystery. We show that if we employ reasonable reaction rates, then they may be formed in star-forming regions during the collapse of interstellar clouds but whether or not they can be detected would depend on the advancement in observational techniques.

6. Acknowledgments

This project is supported in part by a grant from Indian Space Research Organization, RESPOND programme.

7. References

Biham, O., Furman, I., Pirronello, V. and Vidali, G. (2001), Master Equation for Hydrogen Recombination on Grain Surfaces, Astrophys. J., **553**, 595.

Chakrabarti, S.K. and Chakrabarti, S. (2000a), Adenine Abundance in a Collapsing Molecular Cloud, Ind. J. Phys. **74B**, 97.

Chakrabarti, S. and Chakrabarti, S.K. (2000b), Can DNA bases be produced during molecular cloud collapse?, Astron. Astrophys. **354**, L6.

Herbst, E. (1992), The Production of Large Molecules in Dense Interstellar Clouds in *Chemistry and Spectroscopy of Interstellar Molecules*, Univ of Tokyo Press.

Hollis, J.M., Pedelty, J.A., Snyder, L.E., Jewell, P.R., Lovas, F.J., Palmer, P. and Liu, S.-Y. (2003), A Sensitive Very Large Array Search for Small-Scale Glycine Emission Toward OMC-1, Astrophys. J., **588**, 353.

Kuan, Yi-J., Charlney, S.B., Huang, H.-C., Tseng, W.-L. and Kisiel, Z. (2003), Interstellar Glycine, Astro-phys. J., **593**, 848.

FORMATION OF SIMPLEST BIO-MOLECULES DURING COLLAPSE OF AN INTERSTELLAR CLOUD

KINSUK ACHARYA[1], SANDIP K. CHAKRABARTI[1,2] and SONALI CHAKRABARTI[1,3]
[1]Indian Astrobiology Network (IAN), Center For Space Physics, Chalantika 43, Garia Station Rd. Kolkata 700084, India;
[2]S.N. Bose National Center For Basic Sciences, JD—Block, Sector—III, Salt Lake, Kolkata 700098, India; [3]Maharaja Manindra Chandra College, Kolkata 700003, India.

1. Introduction

Our earth is not a special place in this Universe. However, *it is* a planet with favorable conditions where complex bio-molecules could form and life could evolve. Chemical analysis of various carbonaceous chondrite like Murchison meteorite shows that plenty of organic molecules including eight types of biologically significant amino acids are present in them. Also, the spectral analysis of interstellar lines shows 124 types of molecules. Among them, nearly 80 species are organic and are present in the dense interstellar medium. All these motivate us to study the formation of bio-molecules in space. Our procedure is to couple hydrodynamic equations with chemical evolution of the interstellar medium (Chakrabarti and Chakrabarti, 2000) and the check if the complex molecules are produced in the process.

2. Brief Introduction to Hydrodynamic Study

A generic interstellar cloud may have two distinct regions. One is diffused having density $\rho \sim 1\text{–}10^3$ per cm^3 and temperature 80 K and the other is the dense molecular cloud having density $\rho \sim 10^3\text{–}10^6$ and temperature 10 K. Typical size of a molecular cloud is $\sim$10–10^4 pc and the average lifetime is $\sim$10–20 Myr. In the isothermal phase of the cloud collapse, density $\rho \propto r^{-2}$ (Chandrasekhar 1939) and the velocity is constant. When opacity becomes high enough to trap radiations, the cloud collapses adiabatically with $\propto r^{-3/2}$. In presence of rotation, centrifugal barrier forms at $r = r_c$, where centrifugal force balances gravity. Density falls off as $\propto r^{-1/2}$ in this region (Hartmann, 1998). Following Shu, Adams and Lizano (1987), we compute the density, temperature and velocity distribution inside the cloud and follow the chemical evolution at the same time.

J. Seckbach et al. (eds.), Life in the Universe, 195–199.

3. Chemical Evolution

At t = 0, the chemical content of a cloud is dominated by hydrogen and very little amount of other constituents. The abundance of various species (in mass fraction) of a molecular cloud is given by, H:He:C:N:O:Na:Mg:Si:P:S:Cl:Fe = 0.64:0.35897:5.6(−4):1.9(−4): 1.81(−3):2.96(−8):4.63(−8):5.4(−8):5.79(−8):4.12(−7):9.0(−8):1.08(−8), where, 10^a is written as 1.0(a).

We now present a few paths to complex molecules. The production of organic molecules in dense interstellar clouds can occur via gas phase. Exothermic ion molecular reactions are most probable (endothermic reactions are not possible in the cold conditions like interstellar medium). The initiating process is the cosmic ray-induced ionization of H_2 (Herbst, 1992) to yield mainly H_2^+, H_2^+ Cosmic Ray $\rightarrow H_2^+ + e^- +$ Cosmic Ray

This H_2^+ ion reacts with H_2 and produces H_3^+, which then reacts with carbon to produce the simplest hydro-carbon, $C + H_3^+ \rightarrow CH^+ + H_2$; $CH^+ + H_2 \rightarrow CH_2^+ + H$; $CH_2^+ + H_2 \rightarrow CH_3^+ + H$; $CH_3^+ + H_2 \rightarrow CH_5^+ + h\nu$; $CH_5^+ + e^- \rightarrow CH_4 + H$; again, this could similarly produce $C_3H_3, C_2H_5, C_6H_4^+, C_2H_5, C_3H_7, C_3H_7O, C_2H_5O$ etc. which then can produce, $N + C_3H_3^+ \rightarrow HC_3NH^+ + H$.

Similarly, ammonia NH_3 can be produced, which then reacts, $CH_3^+ + NH_3 \rightarrow CH_3 NH_3 + h\nu$; $CH_3^+ + HCN \rightarrow CH_3CNH^+ + h\nu$, which then produce methyl amine and acetonitrile via dissociative recombination reactions. The oxygen-containing organic molecules such as alcohol, acetaldehyde, dimethyl ether etc. can also be produced from the following reactions, $CH_3^+ + H_2O \rightarrow CH_3OH_2^+ + h\nu$; $H_3O^+ + C_2H_2 \rightarrow CH_3CHOH^+ + h\nu$; $CH_3^+ + CH_3OH \rightarrow (CH_3)_2OH^+ + h\nu$.

4. Effects of Grains on the Rate of Chemical Evolution

Grains are very important constituents of interstellar medium. The chemistry on the surface of these grains are to be included in the chemical evolution. Main effect of these grains is that they influence the production of hydrogenated species such as H_2O, NH_3, CH_4 etc. due to high mobility of atomic hydrogen on the cold surfaces. Therefore, we need to take into account the rates of such grain surface reactions as given below in the reaction network:

$H + H \rightarrow H_2$, $H_2 + H_2 \rightarrow H_3^+ + H$, $H + OH \rightarrow H_2O$, $C^+ + H_2 \rightarrow CH_2^+ + h\nu$, $CH_2^+ + H_2 \rightarrow CH_3^+ + H$, $CH_3^+ + H_2 \rightarrow CH_5^+ + h\nu$, $CH_5 + e^- \rightarrow CH_4 + H$. Tunneling reactions with H_2 could occur once $H_2 >> H$. Production of bigger species does not get influenced very much.

To simplify our computation, we plot in Fig. 1 a quantity representing characteristic size of the grain as a function of the grain size (x-axis) and note that there are three humps (Weingartner and Draine, 2001; hereafter WD01). Thus we assume that three types of grains are important. Second, in Fig. 2 we plot the product of the area of each grain times the number density of each grain and find that the smaller grains have the largest surfaces, and therefore are the most important ones as far as the grain chemistry is concerned. For the time being, we use the Master equation approach for the smallest grains as suggested by Biham et al. (2001) and the rate equation approach for the larger two types of grains. This gives us the mass fraction of H_2 molecules formed due to grain surfaces as a function of the

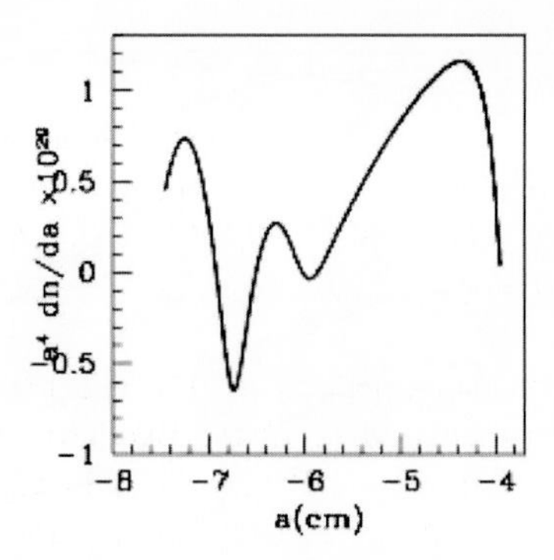

Figure 1. Plot of a^4 dn/da as a function of the size of the grain (WD01).

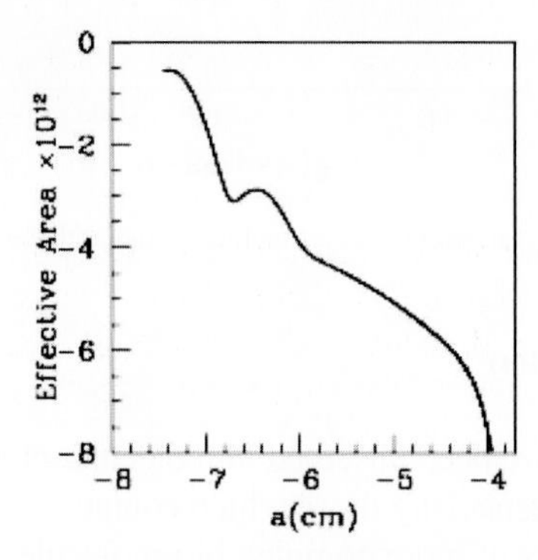

Figure 2. Effective surface area as a function of size of the grain.

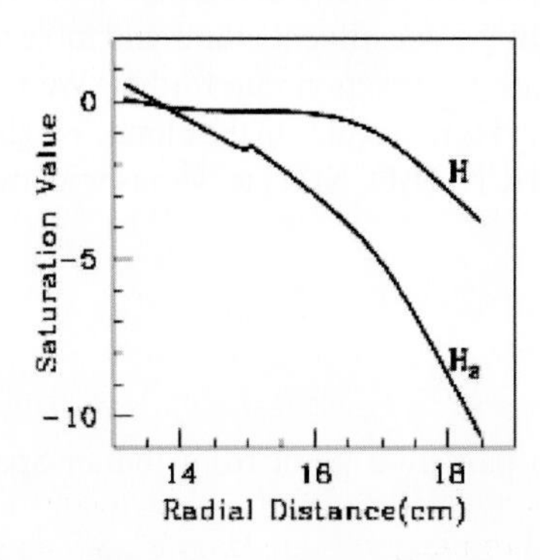

Figure 3. Plot of the saturation value of H_2 and H as a function of radial distance.

number radius of grains. In Fig. 3 we show the expectation value of the number as functions of the radius of H and H_2 after saturation occurs through desorption and adsorption.

Finally, in Fig. 4, we plot the mass fraction of the simple molecules including H and H_2 as a function of the logarithmic radial distance of the cloud. These simple molecules are relevant to form more complex bio-molecules in these clouds.

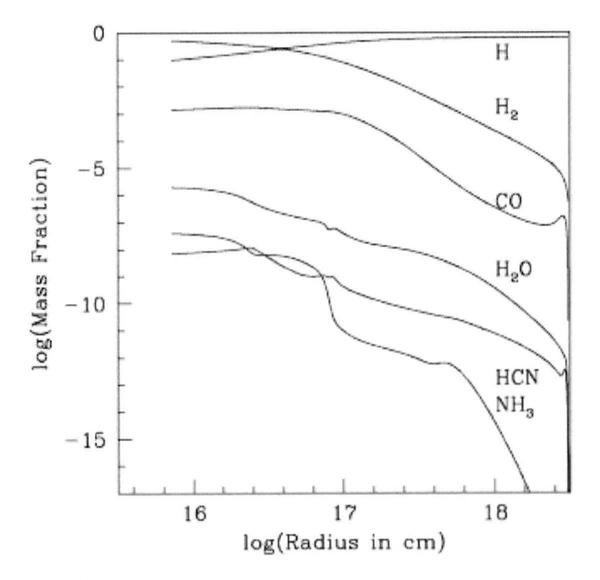

Figure 4. Mass fraction of simplest molecules as a function of radial distance of the cloud.

5. Discussion and Conclusion

More than 120 molecules have been observed in molecular cloud, more than half of which are organic. Some of them, especially those which contain C and N are important because they could be the pre-cursors of more complex bio-molecules. In this paper, we explored the possibilities of formation of these molecules during interstellar cloud collapse and star formation. During collapse, the density and temperature of the gas increases, thereby increasing the reaction rates of the constituent atoms and molecules. Presence of grains and metallic catalysts may increase the reaction rate further. We therefore discussed in details the effects of grains in forming H_2 molecules in the cloud. We showed how the mass fraction of simple molecules such as HCN, H_2O, NH_3 etc. varies with radial distance of a collapsing cloud.

6. Acknowledgments

This project is supported in part by a grant from Indian Space Research Organization, RESPOND programme.

7. References

Biham, O., Furman, I., Pirronello, V. and Vidali, G. (2001), Master Equation for Hydrogen Recombination on Grain Surfaces, Astrophys. J., **553**, 595.

Chandrasekhar, S., (1939), *An Introduction to Stellar Structure*, Chicago: Univ. of Chicago Press.

Chakrabarti, S. and Chakrabarti, S.K. (2000b), Can DNA bases be produced during molecular cloud collapse?, Astron. Astrophys. **354**, L6.

Hartmann, L., (1998), Accretion in Star Fomation, (Cambridge Univ.).
Herbst, E. (1992), The Production of Large Molecules in Dense Interstellar Clouds in *Chemistry and Spectroscopy of Interstellar Molecules*, Univ of Tokyo Press.
Shu, F.H., Adams, F.C., and Lizano, S., (1987), *Star formation in molecular clouds—Observation and Theory* Ann. Rev. Astron. Astrophys. **25**, 23.
Weingartner, J. C. and Draine, B.T., (2001) Dust Grain-Size Distributions and Extinction in the Milky Way, Large Magellanic Cloud, and Small Magellanic Cloud, Astrophys. J., **548**, 296.

CHEMICAL ABUNDANCES OF COMETARY METEOROIDS FROM METEOR SPECTROSCOPY

Implications to the Earth Enrichment

JOSEP Mª TRIGO-RODRÍGUEZ[1], JORDI LLORCA[2,3] and JOAN ORÓ[4]

[1] *Institute of Geophysics & Planetary Physics, UCLA, Los Angeles CA 90095-1567, USA.* [2] *Institut d'Estudis Espacials de Catalunya, Spain.*
[3] *Departament de Química Inorgànica, Universitat de Barcelona, Spain.*
[4] *Fundació Joan Oró, Spain.*

1. Introduction

The overabundance of comets and chondritic bodies during the first stages of the primitive Solar System produced a constant rain of material over our planet, which might have direct implications to the origin of life on it (Oró 1961; Chyba and Sagan, 1992). Besides direct impacts with comets and chondritic asteroids, other important source of Earth enrichment in volatile and organic compounds has been the constant entry of meteoroids from dense meteoroid streams produced during fragmentation processes of their parent bodies. Owing to the large abundance of comets and asteroids in the primordial stages of the solar system (Delsemme, 2000), frequent encounters between minor bodies and planets should be expected. During a close encounter, the induced changes in the orbital elements usually acts by reducing the relative velocity between the Earth and the cometary or asteroidal fragments. Such process reduces the entry velocities of these particles, therefore decreasing the effects of ablation on these meteoroids, some of them very rich in organic and volatile content (Rietmeijer, 2002). Until recently, meteoroids have not been considered as an important source of exogenous matter in the estimations of interplanetary mass reaching the Earth. Probably the main reason has been the assumption that the ablated material is destroyed during entry. Fortunately, the advances in the knowledge of the ablation processes provide us important evidences that this assumption, in general, is not true (Rietmeijer, 2002).

2. Determinating the Mass Influx

Precise information on the mass influx reaching the Earth from meteor storms is really scarce. Fortunately, some authors recovered recently valuable information on these streams in order to study the mass distribution of the particles and the flux densities. The only way to deduce the flux number density of old meteor outbursts is through Zenital Hourly Rates (ZHRs) historical visual records arrived to us from literature (Jenniskens, 1995). The ZHR

J. Seckbach et al. (eds.), Life in the Universe, 201–204.

is the number of shower meteors seen in one hour under a clear sky, the radiant at the zenith and the faintest visible star equal to +6.5. These standardized parameters were in the past not reported by observers and, in consequence, the deduced visual ZHRs are only rough estimations (Jenniskens, 1995). Fortunately, in the past century additional photographic and radar observations were made during storms, which were capable to improve the quality of the flux determinations in some cases (Trigo-Rodríguez et al., 2001). ZHR are obtained usually from the following equation:

$$ZHR = \frac{C \cdot N \cdot r^{6.5-Lm}}{T \cdot (\sin\theta)^{1.4}} \quad (1)$$

Where C is a correction factor for the perception of the observer relative to an average observer, N is the number of shower meteors recorded in T hours of effective time, Lm is the limiting stellar magnitude, θ is the elevation of the shower radiant, and r the population index. To deduce the mass influx for different meteoroid streams we used a very simple model taking into account the population index and hourly activity observed in the different apparitions. The population index r corresponds to the ratio of the number of meteors in magnitude class M to those in class M-1 for all magnitude ranges in the different observational intervals.

$$r = \frac{N(m)}{N(m-1)} \quad (2)$$

We used the population index in order to reconstruct the magnitude distribution for all meteoroids. The next step was to derive the ZHRs taking into account the probability of perception for each meteor magnitude for a sky limiting magnitude of +6.5.

$$N_m = r^{m-M_o} \quad (3)$$

The main procedure is quite simple. Our software is able to compute ZHR values from the proposed magnitude distribution. The magnitude origin of the distribution (M_o) is taken as a free parameter to obtain the ZHR values. When M_o is equal to a determinate magnitude class, it implies that the number of meteors of such magnitude is one ($N_m = 1$). In order to deduce reliable ZHRs we have searched in the literature proper r and M_o values (Jenniskens, 1995; Trigo-Rodriguez et al., 2001; IMO Visual Data Base). The ZHR in each case was obtained by correcting the observed magnitude distribution for a standard probability function (Roggemans, 1989). To compute the global mass reaching the atmosphere, when the magnitude distribution produces the ZHR level given in the literature, the program integrates the number of meteors for each magnitude class by multiplying this number by the mass for each class, derived from (Verniani, 1973):

$$0.92 \cdot \log m = 24.214 - 3.91 \cdot \log V_g - 0.4 \cdot M_v \quad (4)$$

where m is the meteoroid mass in grams, V_g is the geocentric velocity given in $\text{cm}\cdot\text{s}^{-1}$ and M_v is the magnitude of the meteor. Using these simple principles our software calculates the mass contribution of all magnitude classes from the meteors observed from an observing location. The final mass entering into the terrestrial atmosphere will be finally calculated considering the effective surface of the terrestrial atmosphere seen from one observing

TABLE 1. Incoming meteoroid mass from different meteoroid streams during the annual encounters, outburst or storms. The spatial number density for particles producing meteoroids of +6.5 magnitude ($\rho_{6.5}$) is the number of meteoroids included in a cube with an edge of 1000 km.

Meteoroid Stream	Encounter	r	Duration (hours)	Global Mass (kg)	Max. ZHR (meteors/hour)	$\rho_{6.5}$ (n/10^9 km^3)
Geminids	Annual shower	2.5	1	$0,7 \pm 0,2$	$\approx$150	50 ± 15
Giacobinids	1946 storm	2.6	6	12100 ± 3100	12000 ± 3000	84000 ± 22000
	1985 outburst	2.6	5	160 ± 48	700 ± 100	4500 ± 1500
	Annual shower	2.5	1	0.018 ± 0.006	30	10 ± 2
	1966 storm	2.9	3	$8,5 \pm 2.5$	15000 ± 3000	$\approx$10000
Leonids	1998 outburst	1.2/2.0	40	1700 ± 600	300	100 ± 5
	1999 storm	2.2	3	4.6 ± 1.5	3700 ± 100	5400 ± 1200
	2002 storm	2.2	3	5.1 ± 1.7	4500 ± 100	6600 ± 1900
Perseids	Annual shower	2.5	1	0.015 ± 0.004	$\approx$100	100 ± 15
	1991 outburst	2.5	4	1.5 ± 0.5	400 ± 50	500 ± 150
June Bootids	1916 outburst	1.7	4	100 ± 30	300 ± 80	150 ± 50
	1998 outburst	2.2	7	145 ± 35	300 ± 50	150 ± 50

place. The considered effective area was 128 times smaller that the covered by the terrestrial atmosphere, this factor being the same as we used to estimate the global mass entering into the Earth. The involved error for well-studied storms was in the order of a factor 1σ. The results of these models are given in Table 1.

Table 1 provides us the amount of cometary material that settle in the terrestrial atmosphere. The incoming mass of meteoroids reaching yearly the Earth from cometary, asteroidal and sporadic meteoroids, integrated over the 10^{-9} to 10^3 kg range of masses, is around $1.2 \pm 0.4 \times 10^6$ kg. From this data and assuming that the cometary flux on Earth followed a similar trend that the Moon cratering record, we obtained Figure 1.

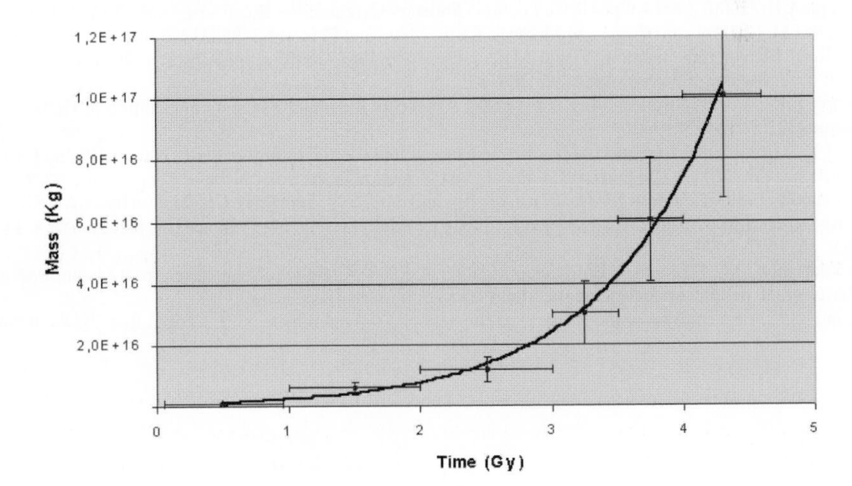

Figure 1. Total mass of meteoroids reaching the Earth.

3. Chemical Abundances from Meteor Spectra

In order to gain a better insight into the temperatures reached and the elements delivered during the ablation process, meteor spectroscopy has provided us with interesting new data. We have analysed recently the spectra of fireballs produced principally by cometary meteoroids (Trigo-Rodríguez et al., 2003). From these spectra we have derived the relative abundance of Na, Mg, Ca, Si, Ti, Cr, Mn, Fe, Co and Ni in the parent meteoroids by averaging the composition of the radiating gas along their respective fireball path produced during atmospheric entry, following the methodology developed by Borovicka (1993). In general we have found important chemical differences between these cometary meteoroids and the 1P/Halley dust analysed *in situ* by Giotto spacecraft (Jessberger et al., 1988), suggesting important differences between comets, as was already pointed out by Greenberg (2000). An interesting conclusion derived from these data is that 1P/Halley meteoroids analysed by Giotto cannot be used as reference sample of cometary dust. Also, the deduced abundance of the major rock-forming elements Si, Mg, and Fe are in accordance to the hierarchical dust accretion model (Rietmeijer, 2002).

4. Acknowledgments

J.M. Trigo-Rodriguez is grateful to the Spanish State Secretary of Education and Universities for a postdoctoral grant.

5. References

Borovicka J. (1993) A fireball spectrum analysis, Astronomy & Astrophysics 279, 627–645.
Chyba C., C. Sagan (1992) Endogenous production, exogenous delivery and impact-shock synthesis of organic molecules: an inventory for the origins of life, Nature 335, 125–132.
Delsemme A.H. (2000) 1999 Kuiper Prize Lecture Cometary Origin of the Biosphere, Icarus 146, 313–325.
Greenberg J.M. (2000) From Comets to Meteors, Earth, Moon and Planets 82–83, 313–324.
International Meteor Organization, Visual Meteor Database 1988–2000. Jenniskens P. (1995) Meteor stream activity. 2: Meteor outbursts, A&A295, 206–235.
Jessberger E.K., A. Christoforidis, A. Kissel (1988) Aspects of the major element composition of Halley's dust, Nature 322, 691–695.
Oró J. (1961) Comets and formation of biochemical compounds on the primitive Earth, Nature 190, 389–390.
Rietmeijer F. J. M., (2002) The Earliest Chemical Dust Evolution in the Solar Nebula, Chemie Erde 62–1, 1–45.
Roggemans P. (1989) *Handbook for visual meteor organization,* Sky Publishing Co., Massachusetts, USA.
Trigo-Rodríguez J.M., J. Llorca J., Fabregat J. (2001) Leonid fluxes: 1994–1998 activity patterns, Met. Planet. Sci., 36, 1597–1604.
Trigo-Rodríguez J.M., J. Llorca, J. Borovicka, J. Fabregat (2003), Chemical Abundances from Meteor Spectra: I. Ratios of the main chemical elements, Met. Planet. Sci. 38–8, 1283–1294.
Verniani F (1973) An analysis of the physical parameters of 5759 radio meteors, J. Geoph. Res. 78, 8429–8462.

VIII. Earth Analogues of Extraterrestrial Ecosystems

VIABLE HALOBACTERIA FROM ANCIENT OCEANS—AND IN OUTER SPACE?

H. STAN-LOTTER, C. RADAX, S. LEUKO, A. LEGAT, C. GRUBER, M. PFAFFENHUEMER, H. WIELAND and G. WEIDLER
Institute of Genetics and General Biology, University of Salzburg, Hellbrunnerstr. 34, A-5020 Salzburg, Austria

1. Permo-Triassic Salt Sediments

About 250 million years ago the continents were close together and formed Pangaea, a supercontinent, which persisted for about 100 million years and then fragmented. The landmasses at that time were located predominantly in the southern hemisphere. The climate was arid and dry; the average temperature is thought to have been several degrees higher than at present. This was one of the time periods in the history of the Earth, when huge salt sediments formed. A total of about 1.3 million cubic kilometers of salt were deposited during the late Permian and early Triassic period alone (Zharkov 1981). The thickness of the salt sediments can reach 1000 to 2000 meters. When Pangaea broke up, land masses were drifting in latitudinal and Northern direction. Mountain ranges such as the Alps, the Carpathians and the Himalayas were pushed up due to the forces of plate tectonics. The salt deposits in Austria originated in the Alpine basin, which extended from Innsbruck to Vienna. Some salt mines in the Alps are still in operation, and these were the sources of our samples. In the Alpine basin and in the Central European basin (Zechstein sea), no more salt sedimentation took place after the Triassic period; however, in other locations, e.g. in Poland, significant salt deposits were still formed until about 20 million years ago. Dating of the salt deposits by sulfur-isotope analysis (ratios of $^{32}S/^{34}S$ as measured by mass spectrometry), in connection with information from stratigraphy, indicated a Permo-Triassic age for the Alpine and Zechstein deposits, which was independently confirmed by the identification of pollen grains from extinct plants in the sediments (Klaus 1974).

Today, salt sedimentation is taking place in surface waters with high salt contents, e.g. natural salt lakes or solar evaporation ponds. At concentrations above 15% NaCl, extremely halophilic archaebacteria (also termed haloarchaea) are the predominant microorganisms. Most haloarchaeal type strains have been isolated from such hyper-saline environments or derived materials, e.g. heavily salted foods like fish and meats.

J. Seckbach et al. (eds.), Life in the Universe, 207–210.

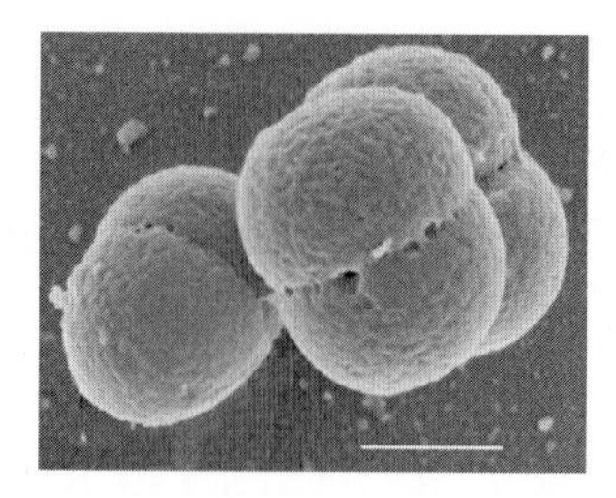

Figure 1. High resolution field emission scanning electron micrograph of *Halococcus dombrowskii* H4 grown in liquid culture. The bar represents 500 nm.

2. Microorganisms and Signature Sequences from Halite

We obtained two types of rock salt samples from the salt mines at Bad Ischl and Altaussee, Austria: lumps of 1–2 kg of weight, which were produced by blasting for the construction of new tunnels, and bore cores of 5 cm diameter from exploratory deep drilling operations. The samples were generally from depths between 400 and 700 m below surface. Samples were dissolved under sterile conditions and, following addition of nutrients, investigated for presence and growth of haloarchaea as described previously. To date, three novel haloarchaeal species from Alpine Permo-Triassic rock salt have been described, which include several strains of *Halococcus salifodinae*, one strain of *Hc. dombrowskii* (see Fig. 1) and one *Halobacterium* strain (Denner *et al.* 1994, Stan-Lotter *et al.* 1999, 2002, 2003).

The phylogenetic tree (Fig. 2) suggests that the coccoid rock salt isolates belong to the genus *Halococcus*, but they are sufficiently different to form separate species. The rod-shaped isolate strain A1 is related to known *Halobacterium* species and even more closely

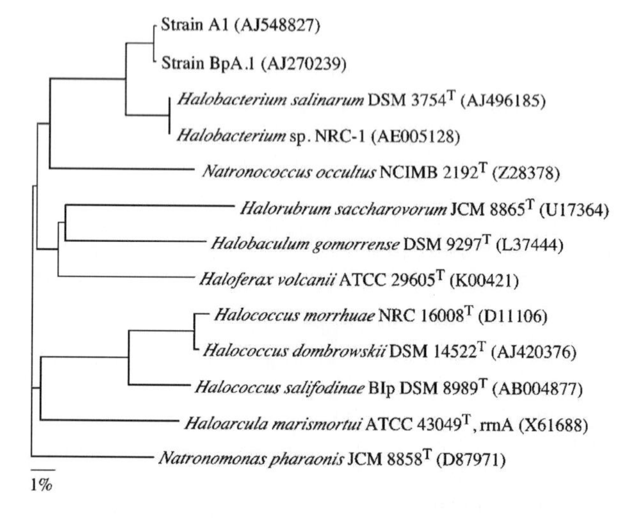

Figure 2. Dendrogram showing the phylogenetic relationship of the haloarchaeal isolates from rock salt (*Halococcus salifodinae* BIp, *Hc. dombrowskii* H4, strains A1 and BpA.1) to several genera of the *Halobacteriaceae*. The tree is based on an alignment of 16S rRNA gene sequences. Sequence accession numbers are in brackets. Bar represents a 1% sequence difference.

to as yet uncharacterized isolates from other ancient salt deposits, such as the Wieliczka mine in Poland or the salt mine at Boulby in Britain (McGenity *et al.* 2000).

Terry McGenity and Karl Stetter had investigated halococcal isolates from salt mines in Winsford, England, and Berchtesgaden, Germany, respectively. Their isolates Br3 and BG2/2 showed many similarities to our *Halococcus salifodinae* strain BIp, including pigmentation and morphology. Therefore, all coccoid strains were analyzed in detail. They were found to possess identical 16S rRNA gene sequences, very similar whole cell protein patterns, similar G + C contents, enzyme activities, phospholipid composition, and other properties. From these results we concluded that in geographically separated halite deposits of similar age, identical species of halococci are present (Stan-Lotter *et al.* 1999). The results are consistent with the notion that a large hypersaline sea, which covered parts of Europe, was populated by haloarchaea, which became trapped upon sedimentation, and whose progeny is present today in Alpine sediments, and also in Zechstein evaporites.

The amplification of 16S rRNA genes by the polymerase chain reaction has become the standard method for obtaining material for subsequent nucleotide sequencing. For this technique it is not necessary to cultivate the microorganisms, since the genes can be amplified by using DNA prepared from the sample of interest. We obtained amplification products from dissolved rock salt with archaeal or bacterial primers; the amplified archaeal rDNA was more prominent than bacterial rDNA. Following subcloning a total of fiftyfour 16S rRNA genes were sequenced (Radax *et al.* 2001); subsequently, 123 further sequences were obtained and analysed. The results suggested the presence of at least 12 clusters of sequences with similarities of 95%, or less, to known haloarchaeal genes, which indicated the presence of novel strains in the rock salt. Some sequences were similar to those of samples from other ancient salt deposits, such as the Wieliczka mine in Poland, or British salt mines. A few sequences were 92–98% similar to those of extant haloarchaea, for instance, to *Halorubrum vacuolatum* and *Halobacterium salinarum.*

3. Extraterrestrial Halite and Environments

The SCN meteorites stem from Mars (Treiman *et al.* 2000) and contain traces of halite, as does the carbonaceous Murchison meteorite. The Monahans meteorite, which fell in Texas in 1998, contained macroscopic crystals of halite, in addition to potassium chloride and water inclusions (Zolensky *et al.* 1999). The Galileo spacecraft collected evidence that support the existence of a liquid salty ocean on the Jovian moon Europa (McCord *et al.* 1998).

These results are intriguing, since they suggest that the formation of halite with liquid inclusions could date back billions of years and occurred probably early in the formation of the solar system (Whitby *et al.* 2000). Could halophilic life have originated in outer space and perhaps travelled with meteorites; could haloarchaea have persisted in environments as they are found today on Mars or on Europa?

We have started to investigate if our haloarchaeal isolates from Permian salt would survive Martian and other extraterrestrial conditions. We exposed several haloarchaea to a Martian atmosphere in the simulation chamber at the Austrian Academy of Science in Graz and obtained recovery rates in the order of 0.1 to 1% (Stan-Lotter *et al.* 2003); the addition of protective substances, such as glycerol, or divalent cations, increased survival rates. The presence of surviving cells can be visualized by fluorescent dyes, such as the LIVE-DEAD

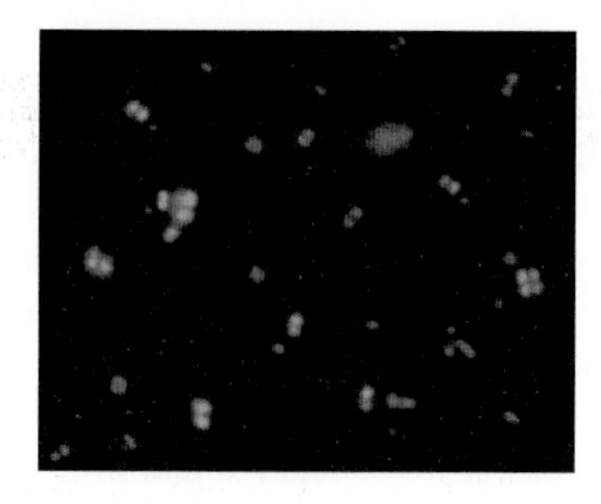

Figure 3. *Halococcus dombrowskii* cells following staining with the LIVE-DEAD kit (Molecular Probes). Green: viable cells; red: nonviable cells.

kit, which allows distinction between membrane-damaged, nonviable (red) cells, and cells with intact membranes (green; Fig. 3).

4. Summary

Viable extremely halophilic archaea, representing novel strains, were isolated repeatedly from salt sediments, which originated from Permo-Triassic oceans. The strains appear to be capable of long-term survival in dry environments. Since extraterrestrial halite has been discovered, it seems feasible to include into the search for life on other planets or moons specifically a search for halophilic microorganisms. The response of haloarchaea to simulated Martian conditions is being investigated.

5. Acknowledgments

This work was supported by the Austrian Science Foundation (FWF), projects P13995-MOB and P16260-B07. We thank M. Mayr, Salinen Austria, for the rock salt samples.

6. References

Denner, E.B.M., McGenity, T.J., Busse, H.-J., Grant, W.D., Wanner, G. and Stan-Lotter, H. (1994) Int. J. Syst. Bacteriol. 44:774 –780.

Klaus, W. (1974) Carinthia II, 164, Jahrg 84: 79–85.

McCord, T.B., Mansen, G.B., Fanale, F.P., Carlson, R.W., Matson, D.L., *et al.* (1998) Science 280:1242–1245.

McGenity, T.J., Gemmell, R.T., Grant, W.D., and Stan-Lotter, H. (2000) Environ. Microbiol. 2:243–250.

Radax, C., Gruber, C. and Stan-Lotter, H. (2001) Extremophiles 5:221–228.

Stan-Lotter, H., McGenity, T.J., Legat, A., Denner, E.B.M., Glaser, K., Stetter, K.O. and Wanner, G. (1999) Microbiology 145:3565–3574.

Stan-Lotter, H., Pfaffenhuemer, M., Legat, A., Busse, H.J., Radax, C. and Gruber, C. (2002) Int. J. System. Evol. Microbiol. 52: 1807–1814.

Stan-Lotter, H., Radax, C. Gruber, C., Legat, A., Pfaffenhuemer, M., Wieland, H., Leuko, S., Weidler, G., Kömle, N. and Kargl, G. (2003) Int. J. Astrobiol. 1:271–284.

Treiman, A.H., Gleason, J.D. and Bogard, D.D. (2000) Planet. Space Science 48:1213–1230.

Whitby, J., Burgess, R., Turner, G., Gilmour, J. and Bridges, J. (2000) Science 288:1819–1821.

Zharkov, M.A. 1981. History of Paleozoic Salt Accumulation. Springer Verlag, Berlin.

Zolensky, M.E., Bodnar, R.J., Gibson, E.K., Nyquist, L.E., Reese, Y., Shih, C.Y. and Wiesman, H. (1999) Science 285:1377–1379.

MARS-LIKE SOILS IN THE YUNGAY AREA, THE DRIEST CORE OF THE ATACAMA DESERT IN NORTHERN CHILE

RAFAEL NAVARRO-GONZÁLEZ[1], FRED A. RAINEY[2], PAOLA MOLINA[1], DANIELLE R. BAGALEY[2], BECKY J. HOLLEN[2], JOSÉ DE LA ROSA[1], ALANNA M. SMALL[2], RICHARD C. QUINN[3], FRANK J. GRUNTHANER[4], LUIS CÁCERES[5], BENITO GOMEZ-SILVA[6], ARNAUD BUCH[7], ROBERT STERNBERG[7], PATRICE COLL[7], FRANCOIS RAULIN[7], and CHRISTOPHER P. MCKAY[8]

[1]Laboratorio de Química de Plasmas y Estudios Planetarios, Instituto de Ciencias Nucleares, Universidad Nacional Autónoma de México, Circuito Exterior, Ciudad Universitaria, Apartado Postal 70-543, México D.F. 04510, México; [2]Department of Biological Sciences, 202 Life Sciences Building, Louisiana State University, Baton Rouge, LA 70803, USA; [3]SETI Institute, NASA Ames Research Center, Moffett Field, CA 94035-1000, USA; [4]Jet Propulsion Laboratory, Pasadena, CA, 91109, USA; [5]Instituto del Desierto y Departamento de Ingeniería Química, Facultad de Ingeniería, Universidad de Antofagasta, PO BOX 170, Antofagasta, Chile; [6]Instituto del Desierto y Unidad de Bioquímica, Departamento Biomédico, Facultad Ciencias de la Salud, Universidad de Antofagasta, PO BOX 170, Antofagasta, Chile; [7]Laboratoire Inter-Universitaire des Systèmes Atmosphériques, UMR CNRS 7583, Universités Paris 12 & Paris 7, CMC, 61 Avenue du Général de Gaulle F 94010 Créteil Cedex, France ; and [8]Space Science Division, NASA-Ames Research Center, Moffett Field, CA 94035-1000, USA

1. Introduction

The data obtained from the Viking lander's analyses of soils on Mars were unexpected. First, was the finding that when soil samples were exposed to water vapor in the gas exchange experiment (GE) there was rapid release of molecular oxygen, at levels of 70–770 nmole g^{-1} (Oyama and Berdahl, 1977). The next puzzling result was that organic material in the labeled release experiment (LR) was consumed as would be expected if life would have been present (Levin and Straat, 1977). Lastly, there were no organic materials at levels of part-per-billion (ppb), as measured by pyrolysis-gas chromatography-mass spectrometry (pyr-GC-MS) detected (Biemann *et al.*, 1977); which were in apparent contradiction with the presence of life as detected by the LR experiment. The reactivity of the martian soil is currently believed to result from the presence of one or more inorganic oxidants (*e.g.*,

J. Seckbach et al. (eds.), Life in the Universe, 211–216.

superoxides, peroxides, or peroxynitrates) at the part-per-million (ppm) level. The absence of organics in the soils results from their oxidation by such oxidants and/or direct ultraviolet radiation damage (McKay *et al.*, 1998).

We described here recent findings on the chemical and microbiological properties of soils in the Atacama Desert (Navarro-González *et al.*, 2003), an environment that serves as a model for Mars. The Atacama is an extreme, arid, temperate desert that extends from 20°S to 30°S along the Pacific coast of South America (Miller, 1976; Arroyo *et al.*, 1988; McKay et al., 2003). The extreme aridity is due to the combined effects of a high pressure system located on the western Pacific Ocean, the drying effect of the cold north-flowing Humboldt ocean current, the oceanic cloud barrier effect of the Cordillera de la Costa, and the rain shadow effect of the Cordillera de Los Andes intercepting precipitation from the Intertropical Convergence (Arroyo *et al.*, 1988; Navarro-González *et al.*, 2003). The Copiapó river (27°S) marks the southern limit of the extremely arid desert. The area north of Copiapó receives moisture from an occasional fog or a shower event every few decades (Arroyo *et al.*, 1988). The region south of Copiapó starts to receive precipitation from the occasional winter incursions of the polar front (Arroyo *et al.*, 1988). Proxy temperature records indicate increased precipitation from El Niño events occurred from 10 to 16 kyr ago; however, rains did not penetrate the absolute desert region (Betancourt *et al.*, 2000). In addition, a 106 kyr paleoclimate record from a Salar de Atacama drill core also indicates episodic wet periods (Bobst, *et al.*, 2001). Geological and soil mineralogical evidence suggest that the extreme arid conditions have persisted in the southern Atacama for 10–15 Myrs (Ericksen, 1983), making it one of the oldest, if not the oldest desert on Earth.

2. Sampling Collection Sites

Approximately 500 g representing a composite of 6 individual nearby sites (~2 m in radius) of the upper 10 cm soil layer were collected using sterile polyethylene scoops and stored in sterile polyethylene (WhirlpakTM) bags. The samples were kept at ambient temperature until analysis. The soil samples were collected in the driest core of the Atacama Desert, named the Yungay area. This area was sampled in the following sites: **AT01-03**, **AT02-03A** (S 24° 4′ 9.6″, W 69° 51′ 58.8″), **AT02-03B** (S 24° 4′ 11.1″, W 69° 51′ 58.1″), **AT02-03C** (S 24° 4′ 8.9″, W 69° 51′ 54.5″), **AT02-03D** (S 24° 4′ 7.1″, W 69° 51′ 57.7″), **AT02-03E** (S 24° 4′ 8.3″, W 69° 52′ 50.1″),**AT01-12** (S 24° 6′ 10.2″, W 70° 1′ 9.7″), **AT03-33** (S 24° 4′ 6.8', W 69° 51′ 58.1″), **AT03-34** (S 24° 4′ 6.2″, W 69° 51′ 48.4″), **AT03-35** (S 24° 4′ 0.4″, W 69° 51″ 49.7″), **AT03-36** (S 24° 3′ 50.2″, W 69° 51′ 51.2″), **AT03-37** (S 24° 3′ 44.0″, W 69° 51′ 53.3″), **AT03-38** (S 24° 3′ 38.8″, W 69° 52′ 5.3″), **AT03-39** (S 24° 3′ 33.0″, W 69° 52′ 11.3″). For comparison a less arid site with vegetation was also studied: **AT01-22**, **AT02-22** (S 28° 7′ 4.5″, W 69° 55′ 8″).

The samples from the less arid site provided a control in that we used exactly the same methods. The “wet” sites provided positive controls for the dry sites and show that the relative lack of detection in the dry areas was not a failure of the methods. Blank tests were also run simultaneously with all methods in which no soil was added to each assay.

3. Results and Discussion

3.1. PYR-GC-MS

Analysis of samples by pyr-GC-MS at 750° C under an inert atmosphere revealed that the most arid zone of the Atacama, the Yungay area, is depleted of most organic molecules. Only two peaks in the chromatograms corresponding to organic molecules (formic acid and benzene) are detectable. Formic acid and benzene are present at concentrations of ~1 μmole g^{-1}, and ~0.01 μmole g^{-1}, respectively. The ratio between formic acid and benzene has a high value (≥12 units) indicating that the organic matter present in the region is oxidized, possibly composed of refractory organics such as aliphatic and aromatic mono- and polycarboxylic acids.

3.2. CHEMICAL DERIVATIZATION-GC-MS

Refractory organics present in the soil were extracted at 60° C in ethanol by sonication (Buch et al., 2003). Then the supernatant was filtered on a 10 μm MoBiTec filter and the solvent evaporated to dryness under nitrogen at 40° C. The dry residue was then exposed to N,N-methyl-tert-butyl(dimethyl-silyl)trifluoroacetamide in dimethyl-formamide using pyrene as an internal standard. The resultant volatile compounds were analyzed by gas chromatography coupled to mass spectrometry using a polar polydimethylsiloxane column. The results of these analyses indicated that there are no free amino acids ($<10^{-4}$ μmole g^{-1}) in the Yungay area. The only organic compounds detected were aliphatic monocarboxylic (~10^{-3} μmole g^{-1}) and dicarboxylic acids ($<10^{-4}$ μmole g^{-1}), and aromatic monocarboxylic acids (~10^{-3} μmole g^{-1}). These results support Benner' view of the missing organics by the Viking landers due to the lower pyr-GC-MS temperature, e.g., 500° C (Benner *et al.*, 2000). However, under our experimental pyr-GC-MS conditions (750° C), these organic compounds would have been degraded to formic acid and benzene at detectable levels.

3.3. DETECTION OF HETEROTROPHIC BACTERIA AND DNA

Soil samples from the Yungay area were studied for the presence of viable heterotrophic microorganisms by serial dilution plating on a number of artificial culture media with both low (1/10 and 1/100) strength Plate Count Agar (PCA) and high nutrient single strength PCA media. The samples contain levels of heterotrophic bacteria below the detection limits of dilution plating. In many cases no bacterial colonies were observed at any dilutions on any of the nutrient media used for plating of these samples. Extensive plating with up to 100 replicates of samples AT01-03 and AT02-03 and a sample from a similar site in the same region (AT01-12) provided less than ten bacterial colonies in total. These data indicate that the Yungay area contains extremely low levels of heterotrophic bacteria. It was considered that there would be microorganisms entering this environment from the atmosphere, however, they were not detected in the soil samples analyzed. Air samples were collected at site AT02-03 in order to determine the load of culturable microorganisms entering this environment from the atmosphere. However, no culturable heterotrophic bacteria were obtained from the air samples collected in the Yungay area indicating the lack of a local source of airborne bacteria. Since only a small percentage of soil microorganisms can be cultured, DNA was extracted from the soil to construct 16S rRNA gene sequence clone libraries for the soil samples of the Yungay area. However, there was surprisingly no recoverable DNA in the Yungay soil samples studied. Contrary to results obtained in other regions of the Atacama, no DNA was recovered from these core region soil samples using a number of established DNA extraction procedures. The lack of recoverable DNA substantiates the absence of culturable bacteria in many of the core region soils.

3.4. EXPERIMENTS ON THE TOXICITY OF THE SOIL

The pH of the soils is in the range 5.5 to 8.6 indicating that extreme soil pH values are not the cause of the low microbial numbers found in the most arid zone, the Yungay area. Further studies were conducted to determine if the soil from the Yungay area was toxic for the growth of microorganisms. The soil from AT02–03A (most arid zone) was mixed with the soil from the less arid site, AT02–22 (which contains about 10^6 colony forming units per gram, CFU/g), in the following ratios 1:2, 1:1, and 2:1; and then plated on 1/10 strength PCA to determine the CFU/g number. The CFU/g values for these soil mixtures were not reduced by more than the expected dilution factor, suggesting that the soil AT02–03A is not toxic.

3.5. LR EXPERIMENT

In order to understand the reactivity of the soils, we performed a modified version of the Viking LR experiment (Levin and Straat, 1977) with a GC-MS detection system. In one procedure desert soil was incubated for several days in an aqueous solution containing ^{13}C-labeled sodium formate. A significant fraction (3–12 μmole) of the formic acid added (~50 μmole) was decomposed in the Yungay area (samples AT02–03A to AT02–03E) even where no culturable heterotrophic organisms were detected. This oxidation of formate to $^{13}CO_2$ could be attributable to either abiotic or biological activity or both. However, formate was selected as it is thought to be the only substrate oxidized in the Viking LR experiment (Levin and Straat, 1977). To distinguish between any abiotic *vs* biological activity, another set of experiments were performed in which desert soil was incubated for several days in an aqueous mixture of two ^{13}C-labeled chiral substrates: sodium alanine and glucose. Different combinations of these enantiomers were used so that any microorganisms present in the soil could (L-alanine + D-glucose) or could not (D-alanine + L-glucose) carry out metabolism. The LR response of the enantiomeric mixtures of alanine and glucose showed equal quantities of $^{13}CO_2$ (~0.4 μmole) released from both D-alanine + L-glucose and L-alanine + D-glucose mixtures in the Yungay area and were three orders of magnitude higher than those measured in the blank experiments; therefore, any biological explanation for the reactivity of the soil in this part of the desert can be ruled out. Degradation of the ^{13}C-labeled molecules was observed in the presence of hydrogen peroxide and sodium peroxide in control experiments while no reactivity was observed in the presence of nitrates.

3.6. SEARCH FOR OXIDANTS IN THE SOIL

The E_h of several Atacama Desert samples were oxidizing with values ranging from 365 to 635 mV. We performed chemical assays for superoxides and hydrogen peroxide because these are the most plausible oxidants and are those suggested as explanations for the reactivity seen by the Viking landers (McKay *et al.*, 1998). They were determined by measuring the absorbance of crystal violet at 592 nm at pH 4, formed by the oxidation of leuco crystal violet by H_2O_2 and/or O_2^{2-} in the presence of the enzyme horseradish peroxidase (Zhang and Wong, 1994). Our results rule out these oxidants as the cause of the reactivity seen at the Yungay area because the concentrations are too low (0.05–0.14 ppm) to explain our results. Nitrates present in the soil were reduced to nitrite using powdered cadmium. The nitrite was then determined by diazotizing sulfanilamide and coupling with N-(1 naphthyl)-ethylenediamine dihydrochloride to form a highly colored azo dye which was measured colorimetrically (Navarro-González and Castillo-Rojas, 1995). Nitrates were found to be present in the soil in high levels (10–140 ppm) but they alone are not oxidizing enough to account for the reactivity seen in our samples. Nitrates may lead to the formation of peroxonitrite (NOO_2^-) and this has been suggested as a possible martian oxidant (Plumb *et al.*, 1989). However the nitrate concentrations needed are in the percent level, much higher than in the Atacama. Thus, while our results show the presence of a strong oxidant in the soils in the Yungay area, the nature of the oxidant remains unexplained (Navarro-González *et al.*, 2003). Photochemical reactions initiated by sunlight continually produce oxidants in the lower atmosphere and surface. However, in most soils biological production of reduced organic material completely dominates the net redox state of soils. If biological production is less than the photochemical production of oxidation then the soil will become oxidizing. The transition from biologically dominated soils to photochemically dominated soils appears to be abrupt. Whichever process dominates will shift the redox state in one direction or another. In the Atacama there is a gradual decline in biological activity as conditions became drier, yet near the extreme arid region there is an abrupt transition to very low bacterial levels and low organic content. A gradual decline in biological activity is observed as conditions become more arid, and in the vicinity of the core arid region there is an abrupt transition to very low bacterial levels and low organic content.

4. Conclusion

It is improbable that the high UV flux would have caused the oxidizing conditions found at the site which is only 1 km above sea level. The Atacama desert's location and therefore it's extreme aridity must inhibit the biological production of reductants and could in fact increase the survival of photochemically produced oxidants. Microbiological findings suggest that in the core region of the Atacama, we have found the limit of microbial survival in extremely desiccated environments. Since photochemical processes dominate in the core region of the Atacama Desert, we find almost no microorganisms, low levels of organic material, and the organic material present appears to have been subject to oxidation. The labeled-release experiments point to the presence of, as yet unidentified, oxidants in the soil.

Many properties of these Atacama soils are analogous to the soils of Mars based on the current knowledge. These terrestrial soils provide an accessible resource for testing new instrumentation and experiments for use in future Mars exploration(Navarro-González *et al.*, 2003).

5. Acknowledgements

We acknowledge support from NASA ASTEP and BSRP, the National Autonomous University of Mexico (DGAPA-IN119999 and IN101903), the National Council of Science and Technology of Mexico (CONACYT No. 32531-T and F323-M9211), NASA-Ames/ LSU Cooperative Agreement (NCC 2-5469), the National Science Foundation (Award DEB 971427), the University of Antofagasta, and the National Center of Scientific Research of France.

6. References

Arroyo, M.T.K., Squeo, F.A., Armesto, J.J., and Villagran, C. (1988) Effects of aridity on plant diversity in the northern Chile Andes: results of a natural experiment. *Annals of the Missouri Botanical Garden*, **75**, 55–78.

Benner, S.A., Devine, K.G., Matveeva, L.N., and Powell, D.H. (2000) The Missing organic Molecules on Mars. *Proc. Natl. Acad. Sci.* **97**, 2425–2430.

Betancourt, J.L, Latorre, C., Rech, J.A., Quade, J., and Rylander, K.A. (2000) A 22,000-Year Record of Monsoonal Precipitation from Northern Chile's Atacama Desert. *Science* **289**, 1542–1546.

Biemann, K., Oro, J., Toulmin III, P., Orgel, L.E., Nier, A.O., Anderson, D.M., Simmonds, P.G., Flory, D., Diaz, A.V., Rushneck, D.R., Biller, J.E., and LaFleur, A.L. (1977) The Search for Organic Substances and Inorganic Volatile Compounds in the Surface of Mars. *J. Geophys. Res.* **30**, 4641–4658.

Bobst, A.L., Lowenstein, T.K., Jordan, T.E., Godfrey, L.V., Ku, T.-L., and Luo, S. (2001) A 106ka Paleoclimate Record from Drill Core of the Salar de Atacama, Northern Chile. *Paleogeogr. Paleoclimatol. Paleoecol.* **173**, 21–42.

Buch, A., Sternberg, R., Meunier, D., Rodier, C., Laurent, C., Raulin, F., Vidal-Madjar, C. (2003) Solvent Extraction of Organic Molecules of Exobiological interest for in situ analysis of the Martian Soil. J. Chromatogr. (in press).

Ericksen, G.E. (1983) The Chilean Nitrate Deposits. *Amer. Scientist* **71**, 366–374.

Levin, G.V. and Straat, P.A. (1977) Recent Results from the Viking Labeled release Experiment on Mars. *J. Geophys. Res.* **82**, 4663–4667.

McKay, C.P., Grunthaner, F.J., Lane, A.L., Herring, M., Bartman, R.K., Ksendzov, A., Manning, C.M., Lamb, J.L., Williams, R.M., Ricco, A.J., Butler, M.A., Murray, B.C., Quinn, R.C., Zent, A.P., Klein, H.P. and Levin, G.V. (1998) The Mars Oxidant Experiment (MOx) for Mars'96. *Planet Space Sci.* **46**, 769–777.

McKay, C.P., Friedmann, E.I., Gómez-Silva, B., Cáceres-Villanueva, L., Andersen, D.T., and Landheim, R. (2003) Temperature and moisture Conditions for Life in the Extreme Arid Region of the Atacama: Four Years of observations Including the El Niño of 1997–1998. *Astrobiology* **3**, 393–406.

Miller, A. (1976) The Climate of Chile, In: W. Schwerdtfeger (ed.) *Climates of Central and South America.* Elsevier Scientific Publishing Company, Amsterdam, pp. 113–145.

Navarro-González, R. and Castillo-Rojas, S. (1995) Lightning strikes. A simple undergraduate experiment, demonstrating the lightning-induced synthesis of NO_x in the atmosphere. Educ. Chem. 32, 161–162.

Navarro-González, R., Rainey, F.A., Molina, P., Bagaley, D.R., Hollen, B.J., de la Rosa, J., Small, A.M., Quinn, R.C., Grunthaner, F.J., Cáceres, L., Gomez-Silva, B. and McKay, C.P. (2003) Mars-Like Soils in the Atacama Desert, Chile, and the Dry Limit of Microbial Life. *Science* **302**, 1018–1021.

Oyama, V.I., and Berdahl, B.J. (1977) The Viking Gas Exchange Experiment Results from Chryse and Utopia Surface Samples. *J. Geophys. Res.* **82**, 4669–4676.

Plumb, R.C., Tantayonon, R., Libby, M., and Xu, W.W. (1989) Chemical Model fro Viking Biology Experiments: Implications for the Composition of the martian Regolith. *Nature* **338**, 633–635.

Zhang, L.S., and Wong, G.T.F. (1994) Spectrophotometric determination of H_2O_2 in Marine Waters with Leuco Crystal Violet. *Talanta* **41**, 2137–2145.

THE DISCOVERY OF ORGANICS IN SUB-BASEMENT FOSSIL SOILS DRILLED IN THE NORTH PACIFIC (ODP LEG 197): THEIR MODEL FORMATION AND IMPLICATIONS FOR ASTROBIOLOGY RESEARCH

R. BONACCORSI[1] and R.L. MANCINELLI[2]
[1]Dip. di Scienze Geologiche, Ambientali e Marine, University of Trieste Via Weiss, 2-34127, TS – Italy, and [2]SETI Institute, NASA Ames Research Center, Mail Stop 239-4, Moffett Field, CA 94035, USA.

1. Introduction

Although the recovery of sub-basement red paleosoil (or fossil soil) dates back to the 1980's (e.g., Holmes, 1995), the search for organics preserved in material retrieved from the deep earth' subsurface has been systematically initiated during the Ocean Drilling Program (ODP) Leg 197 (Emperor Seamounts, north Pacific Transect) (Tarduno et al., 2002). We address the astrobiology-relevant suggestion that preserved organics from extremely deep fossil soils in isolated diagenetic settings makes them a suitable test beds to develop hypotheses for future Deep Earth biosphere research and potential excellent Mars analogs. These soil sequences are rare in geologic collections (review in Holmes, 1995) because they are difficult to access (they are deeply buried −300 to −350 meters below volcanic basement) and sample (only by deep drilling).

Two independent lines of evidence for the isolation of these rare paleosoils are outlined here. They are: a) a model proposed for the atmosphere-ocean decoupled subsurface (Bonaccorsi, in press) and geochemical proxies for differentiation of the buried soils from their exposed counterparts, e.g., Hawaiian oxisoil (Bonaccorsi, 2002).

1.1. WHY ARE THE SUB-BASEMENT FOSSIL SOILS RELEVANT TO ASTROBIOLOGY?

Identifying organics throughout earth's subsurface materials has implications relevant to astrobiological research. In fact, the detection of organics buried deep beneath the surface of a planet is a fundamental step to constrain the presence and evolution of life on that planet. This would be particularly relevant to Mars where ***i)*** low atmospheric pressure (i.e., 4–10 mbar) and very low surface temperatures (down to −125°C) prevent the stability of liquid water; and ***ii)*** high near-surface UV flux interacting with possible oxidants are likely to affect any near-surface and present-day life and the stability of organic molecules.

Nevertheless, it has been suggested that previously existing life (e.g., chemosynthetic microbes) could have been preserved with depth on Mars (Mancinelli, 2000). This is especially true if conditions completely different from the present-day Mars surface and

J. Seckbach et al. (eds.), Life in the Universe, 217–220.

atmosphere (decoupled conditions) occurred at some deeper locations on this planet. Hence, we need to identify new suitable terrestrial analogs for surface-decoupled deep settings on Mars and search for pristine organics deeply buried beneath subsurface materials, and decoupled from the Earth surface (such as the sub-basement fossil soil from ODP Leg 197).

Integrating microbial ecological and geochemical studies on the ODP soil samples to develop a model for understanding what types of organic material may serve as potential biomarkers for a future exploration and deep to near-surface drilling missions (i.e., soils sampling during landed missions) on Mars (Mancinelli, 2000).

2. Background and Study Area

The stratigraphy and photographs of the soil units cored at Site 1205 and Site 1206 and the sites location map are available online at <http://www-odp.tamu.edu/publications/197_IR/197ir.htm>. Early Eocene red paleosols were cored deeply beneath volcanic basement at Site 1205 (Nintoku Seamount, ~41°20′N; ~170°23′E) and Site 1206 (Koko Seamount, ~34°56′N; ~172°9′E).

These Fe oxides/oxyhydroxide-rich soil interbeds represent the weathering product (tropical conditions) of mafic igneous rocks and basalts (Breccia and Plagioclase-Olivine basalts) and also contain hematite, magnetite, various clay minerals and palagonite (Holmes, 1995; Tarduno et al., 2002). Importantly, they contain very low, but detectable concentrations of total organic carbon (i.e., TOC = 0.12 – 0.01%, ± 0.02%, N = 36) (Tarduno et al., 2002; Bonaccorsi, in press) and ultra-low nitrogen (i.e., 0.01 to 0.006, N-tot wt%). The single fossil soil unit (Core 197-1206A-40R) has stable isotope values more negative (i.e., δ^{13}C-org. = ~ −25‰ to ~ −26‰ and δ^{15}N-tot = −9.5‰ to +2.5‰) than those of exposed Hawaiian counterparts (i.e., δ^{13}C-org = ~ −17‰ to ~ −23‰; and δ^{15}N -tot = 0‰ to +8.5‰; Bonaccorsi, in prep. Unpubl. Data). This stable isotope signature would suggest mixed sources of organic carbon (i.e., plant and primary/secondary bacterial), while values of δ^{15}N would indicate nitrogen fixation, nitrification, and denitrification processes that might have been related to some past/present microbial-induced activity (Bonaccorsi, 2002).

3. The Model

The model proposed for the formation of an atmosphere-ocean decoupled subsurface soil (Bonaccorsi, in press) consists of: a) Sub-aerial formation; b) burial and isolation from the atmosphere; and c) complete subsidence and decoupling from the ocean. It is expected that for each of those phases different organic traces were produced, preserved/altered and different microbial ecologies selected by changing environmental conditions.

According to this model we have:

1. sub-aerial formation of soil. Red-brown soil horizons were formed on the top of subsiding islands by subaerial weathering *of* lava flow tops or basaltic ashes during subaerial growth phases of the Nintoku and Koko Seamounts (emergent submarine volcano) and an interval of time characterized by lower eruption rates (Holmes,

1995; Tarduno et al., 2002). More specifically, the massive to mottled red-brown claystones from the Pacific are typical of tropical-subtropical soils and their clay mineral assemblage (smectite, kaolinite, gibbsite) indicates pedogenic oxic horizon formed at elevations of at least several tens of meters in lowland/upland areas such as those present in the Hawaiian islands today (Holmes, 1995).

2. Burial and isolation from the atmosphere. When the volcanic activity resumed these soils were subsequently buried by several meter-thick lava layers flowing from land into water in a nearshore environment (Tarduno et al., 2002). The accumulation of lava flow with burial rates (~4 to ~25 metres/1000y) faster than subsidence rates (e.g., present-day subsidence rate ~ 2.0 mm/y and 2.5 mm/y) produced a relatively rapid burial with the decoupling of the soil from the atmosphere. During burial soils underwent a variety of conditions (e.g. heating, high pressure, and diagenesis effects), which might have caused differences from their still exposed typical tropical counterparts (red-brown Hawaiian oxisols, "ortox"; e.g., Soil Survey Staff, 1975).

3. Subsidence and decoupling from the ocean. The isolation from the Pacific Ocean probably occurred as a result of topographic (i.e., the upland location of soils), stratigraphic and tectonophysic factors (such as subsidence).

Soils at Site 1205 and Site 1206 are barren of calcareous nannoplankton and foraminifers (Holmes, 1995; Tarduno et al., 2002); in addition, they have a stratigraphic position, which is tens to hundred meters beneath marine deposits and basalt rock. These two lines of evidence would indicate that no direct contact between the soil and lagoonal/open sea waters occurred at the time of the two islands submergence. More likely those fossil soils were already buried and encapsulated in basalts prior to complete subsidence below sea level (~1 Ma to ~2 Ma) as it is observed for similar fossil soils cored beneath the Mauna Kea today. Finally, the Koko and Nintoku Seamounts subsided completely below sea level (~48 Ma to ~54 – 55 Ma) and the encapsulated soil beds likely maintained their decoupling from the ocean until when they were drilled.

4. Conclusions

There is further evidence supporting the use of the ODP fossil soils as model for Mars subsurface materials: 1. Their weathered nature and iron-rich composition. 2. Their extremely deep setting preserving basalt and palagonite/biosignatures for >48 Ma. 3. Bacterial isotope signature. 4. Genesis and composition similar to Recent/present-day Hawaiian oxisols that are well-established Mars surface compositional analogs (e.g., Bishop et al., 1993). The model above outlined can help drafting hypotheses for types of organics/biosignatures and potential physiological types of microbes preserved in the soil material. Furthermore, the soil samples underwent initial heating by overrunning lava flows that may have partially altered their original composition (e.g., organics and former microbial communities) throughout.

The existing elemental (e.g., Tarduno et al., 2002) and stable isotope data-set (Bonaccorsi in prep.) need to be compared with results obtained from standard microbiological techniques. This should be done in order to establish the potential for those Eocene soils to serve as a suitable analog for a possible near surface to deep biosphere on Mars.

5. Summary

Organic-poor deeply buried red paleosoils (late Paleocene-early Eocene) were cored during the ODP Leg 197 (North Pacific). The fossil soil model formation (i.e., surface formation, deep burial and isolation from the atmosphere and the ocean) is outlined here. We suggest a new model based upon literature and geochemical data which suggests considering these fossil soils as useful samples for astrobiology relevant studies, e.g., Mars soil analogues. One fundamental reason is preservation of organic traces in a deep earth system (no sun light, and reducing conditions), which remained isolated and decoupled from the ocean and the atmosphere for millions of years. This may be relevant to future studies of a possible deep biosphere on Mars where some preservation of subsurface to deeply buried past/present organics is possible.

6. References

Bishop, J.L., Pieters, C.M., and Burns, R.G. (1993) Reflectance and Mossbauer Spectroscopy of Ferrihydrite-Montmorillonite assemblages as Mars soil analog materials. Geochimica et Cosmochimica Acta **57**, 4583–4595.

Bonaccorsi, R. Lithological features and organic content in the sub-basement fossil soil from Nintoku (Site 1205) and Koko (Site 1206) Seamounts, In: J.A. Tarduno, R.A. Duncan and DW Scholl (eds.) Proc. ODP, Sci. Results 197 (in prep).

Bonaccorsi, R. Total Organic Carbon in Red Paleosoils and Basalts from ODP Leg 197 and their potential use as suitable models for Mars soil analogues, In: R. Norris (ed.) S-213 Bioastronomy 2002: Life Among the Stars (ASP) IAU Publications (in press). *http://www-odp.tamu.edu/sciops/staff/197/bonaccorsi.pdf.*

Bonaccorsi, R. (2002) Organic Matter and d13C Throughout a Sub-Basement Red Soil Unit in Hole 1206A Cored During Ocean Drilling Program Leg 197 (Koko Seamount): First Results, Eos Trans. AGU, 83(47), Fall Meet. Suppl., Abstract.

Holmes, M.A. (1995) Pedogenic alteration of basalts recovered during Leg. 144, In: J.A. Haggerty, I. Premoli-Silva, F. Rack, and M.K. McNutt (eds.) *Proc. of the Ocean Drilling Program*, Scientific Results, **144**, 381–398.

Kanavarioti, A, and Mancinelli, R.L. (1990) Could Organic Matter Have Been Preserved on Mars for 3.5 Billion Years? Icarus **84**, 196–202.

Mancinelli, R.L (2000) Accessing the Martian deep sub-surface to search for life. Planet. Space Sci., **48**, 1035–1043.

Soil Survey Staff (1975) Soil Taxonomy, Dept. of Agriculture. Handbook N. 436. Washington (U.S. Govt. Printing Office).

Tarduno, J.A., and the Leg 197 Science shipboard Party (2002) Proc. ODP, Init. Repts., 197 [Online]: *http://www-odp.tamu.edu/publications/197_IR/197ir.htm.*

SILICA-CARBONATE BIOMORPHS AND THE IMPLICATIONS FOR IDENTIFICATION OF MICROFOSSILS

ANNA M. CARNERUP[1], STEPHEN T. HYDE[1], ANN-KRISTIN LARSSON[1], ANDREW G. CHRISTY[1,2] and JUAN MANUEL GARCÍA-RUIZ[3]
[1]Department of Applied Mathematics, RSPhysSE, The Australian National University, Canberra ACT 0200 Australia. [2]Department of Geology, The Australian National University. [3]Laboratorio de Estudios Cristalograficos, Instituto Andaluz de Ciencias de la Tierra, CSIC—Universidad de Granada, E 18002 Granada Spain.

1. Introduction

The possibility of pseudofossils is a well-known obstacle to the identification of fossilized microorganisms (Cloud, 1973; Westall, 1999) and distinguishing abiotic from biotic origins is still a hotly debated topic (Dalton 2002; Schopf *et al.*, 2002; Brasier *et al.*, 2002). Here we report silica-carbonate aggregates, so-called 'biomorphs' that mimic—both morphologically and chemically-primitive microfossils.

2. The Formation of Biomorphs and Their Resemblance to Ancient Microfossils

A wide range of remarkable structures with curvilinear morphologies, reminiscent of simple biological organisms, can be produced (Figure 1).

The synthesis of these biomorphs involves very simple reaction conditions: an alkaline barium-rich silica solution that, through the absorption of carbon dioxide from the air, promotes precipitation of these complex materials within a day. They are self-assembled silica-carbonate composites that display a range of morphologies, dependent on pH, temperature and concentration of the reacting molecular species (Baird *et al.*, 1992; García-Ruiz and Moreno, 1997; García-Ruiz *et al.*, 2002). The reaction is geochemically plausible in the Archaean era (in alkaline, siliceous hydrothermal conditions) and biomorphs could have been naturally produced without biological intervention (García-Ruiz, 2000). Acid leaching produces hollow, siliceous biomorphic materials. Furthermore, these aggregates are traps for hydrophobic organic species (García-Ruiz *et al.*, 2002). Adsorption and thermal curing of small prebiotic molecules, such as phenol and formaldehyde, produces an organic skin covering the biomorph. This skin displays Raman spectral bands indicative of a kerogen-like phase (Figure 2), similar to those of the Warrawoona (micro)fossils (Schopf *et al.*, 2002; García-Ruiz *et al.*, 2003).

J. Seckbach et al. (eds.), Life in the Universe, 221–222.

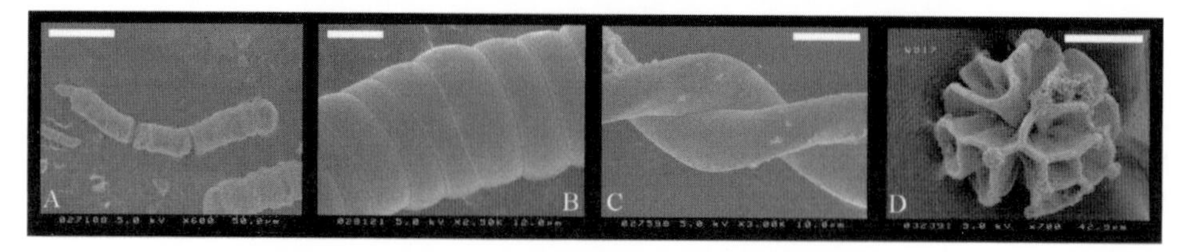

Figure 1. FESEM images of biomorphs (scale bars: A, D 50 μm; B, C 10 μm).

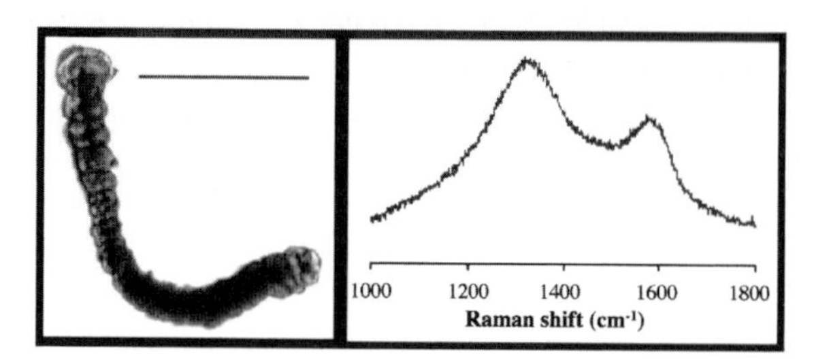

Figure 2. Biomorph with condensed hydrocarbons (scale bar 50 μm) and Raman spectra of carbonaceous coating (García-Ruiz *et al.*, 2003).

A purely inorganic route to the formation of carbonaceous microstructures with morphologies very similar to the most ancient microfossils has been demonstrated. These abiotic materials fulfill morphological and chemical criteria for biogenicity (García-Ruiz *et al.*, 2003). Therefore, the use of kerogen signatures and structural resemblance to modern bacteria cannot alone act as indicators of biogenicity. Abiotic formation of "microfossils" must be ruled out before biogenicity can be considered. The implications for the search and identification of the earliest of life forms are profound.

References

Baird, T., Braterman, P. S., Chen, P., García-Ruiz, J. M., Peacock, R. D. and Reid, A. (1992) Morphology of gel-grown barium carbonate aggregates—pH effect on control by a silicate-carbonate membrane. Material Research Bulletin **27**, 1031–1040.

Brasier, M. D., Owen, R. G., Jephcoat, A. P., Kleppe, A. K., van Kranendonk, M. J., Lindsay, J. F., Steele, A. and Grassineau, N. V. (2002) Questioning the evidence for Earth's oldest fossils. Nature **416**, 76–81.

Dalton, R. (2002) Sqaring up over ancient life. Nature **417**, 782–784.

Cloud, P. (1973) Pseudofossils: A plea for caution. Geology **1**, 123–127.

García-Ruiz, J. M. (2000) Geochemical scenarios for the precipitation of biomimetic inorganic carbonates. In: J. P. Grotzinger and N. P. James (eds.), *Carbonate sedimentation and diagenesis in the evolving Precambrian world.* SEPM special publication **67**, 75–89.

García-Ruiz, J. M. and Moreno, A. (1997) Growth behaviour of twisted ribbons of barium carbonate/silica self-assembled ceramics. Anales de Quimica Int. Ed **93**, 1–2.

García-Ruiz, J. M., Carnerup, A. M., Christy, A. G., Welham, N. J. and Hyde, S. T. (2002) Morphology: An ambiguous indicator of Biogenicity. Astrobiology **2**, 353–369.

García-Ruiz, J. M., Hyde, S. T., Carnerup, A. M., Christy, A. G., Van Kranendonk, M. J. and Welham, N. J. (2003) Science, *in press.*

Schopf, J. W., Kudryavtsev, A. B., Agresti, D. G., Wdowiak, T. J. and Czaja, A. D. (2002) Laser-raman imagery of Earth's earliest fossils. Nature **416**, 73–76.

Westall, F. (1999) The nature of fossil bacteria: A guide to the search for extraterrestrial life. Journal of geophysical research, **104** no E7, 16437–16451.

SOME STATISTICAL ASPECTS RELATED TO THE STUDY OF TREELINE IN PICO DE ORIZABA

L. CRUZ-KURI[1], C. P. MCKAY[2], and R. NAVARRO-GONZALEZ[3]
[1]Instituto de Ciencias Básicas. Universidad Veracruzana. Carr. Xalapa-Ver., Km. 3.5, Xalapa, Ver., Mexico, [2]NASA-Ames Research Center. Moffett Field, CA 94035, USA, and [3]Laboratorio de Química de Plasmas y Estudios Planetarios. Instituto de Ciencias Nucleares, UNAM, Circuito Exterior, Ciudad Universitaria, Apartado Postal 70-543, Mexico.

1. Introduction

In low latitude regions of the Earth we find surface temperature near 0° C in tropical alpine environments (Pérez-Chavez *et al.*, 2000). In these environments temperature and pressure rapidly decrease as the altitude increases and so organisms must adapt to them until they reach their physiological tolerance limits. A clear example of this is treeline at thermal gradients that represent an abrupt transition in life form dominance. In Pico de Orizaba, Mexico (19° N), there are such environments which can be used as plausible analogues for ancient and future life on Mars.

2. Methods

Several locations in the mountain were selected above, below and within the treeline of Pico de Orizaba (Cruz-Kuri *et al.*, 2001). In each station, we register relative humidity, air temperature and soil temperatures at various depths (up to 40 cm).

3. Results and Discussion

The meteorology in and outside treeline seems to be quite interesting. The temperature is modulated by a wave with a period of approximately twenty days; of course there is also the daily periodicity. This suggests performing more precise types of analyses for the corresponding time series. For each station and for every logger, cross-correlations of the corresponding time series were calculated. The time lags varied from −360 hours to +360 hours in units of 1 hour. Also, analyses of correlation were performed for the smoothed time series with daily averages, with minimum and with maximum temperatures; in this situation, the time lags were selected to vary from –28 to +28 days in units of 1 day. In addition, spectral analysis were performed for each series. From the resulting graphs it

J. Seckbach et al. (eds.), Life in the Universe, 223–224.

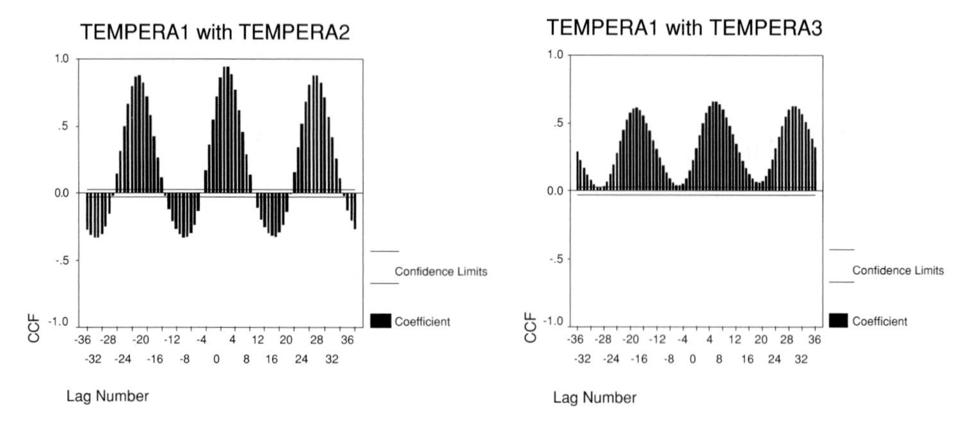

Figure 1. Cross correlations Temperature 1 (soil level) with Temperature 2 (10 cm depth) and with Temperature 3 (20 cm). Station NB, November 2002-March 2003. Time lags are from −36 to +36 hours.

became apparent that there are several patterns and that there is a 24-hour periodicity (which can be accounted for by the obvious natural phenomena) as well as a 7-day periodicity (for which no natural phenomena follows a seven-day cycle). A plausible explanation is that of human activity, is perturbating remote areas of the Earth. We have found some correlograms that even when they exhibit a daily periodicity, all the cross-correlations are positive; other ones for which the correlations are all negative; for others, some of the correlations are positive and some are negative. In the case of loggers which record soil temperatures at four depths, above patterns are related to heat conductions in the soil, for different types of soil as well as the air temperatures, the humidity and other climate variables. For some stations, the cross correlations were close to unity, thus indicating that the heat is transmitted very rapidly from one depth to the next ones. In other stations, the correlations are small. For others, the flow of heat has a reversal in direction as time goes on. Only one of the patterns is illustrated below.

4. Conclusions

The multivariate time series analysis makes apparent that the soil temperatures are driven by some other external climatic parameters. Understanding what type of relationships hold will greatly contribute to determine which factors are the most important ones in order to explain the positions of the treelines.

5. References

Cruz-Kuri, L., McKay, C. P. and Navarro-Gonzalez, R. (2001) Spatial and Temporary Patterns of Some Climate Parameters Around the Timberline of Pico de Orizaba, In: J. Chela-Flores, T. Owen and F. Raulin (eds.) *Astrobiology: First Steps in the Origin of Life in the Universe*, Kluwer Academic Publishers, pp. 293–301.

Pérez-Chavez, I., Navarro-Gonzalez, R., McKay, C. P. and Cruz-Kuri, L. (2000) Tropical Alpine Environments: A Plausible Analogue for Ancient and Future Life on Mars, In: J. Chela-Flores, G. A. Lemarchand and J. Oró (eds.) *Astrobiology: Origins from the Big-Bang to Civilisation*, Kluwer Academic Publishers, pp. 297–302.

IX. On the Question of Life on Mars and on the Early Earth

THE BEAGLE 2 LANDER AND THE SEARCH FOR TRACES OF LIFE ON MARS

ANDRE BRACK[1], COLIN T. PILLINGER[2], MARK R. SIMS[3]
[1]Centre de Biophysique Moléculaire, Rue Charles Sadron, 45071 Orléans cedex 2, France, [2]Planetary and Space Sciences Research Institute, The Open University, Walton Hall, Milton Keynes MK7 6AA, United Kingdom
[3]Space Research Centre, Department of Physics and Astronomy, University of Leicester, University Road, Leicester, LE1 7RH, UK

1. Introduction

The search for traces of life on Mars is presently focused on robotic *in situ* analyses (Westall et al., 2000) and the study of Martian meteorites (Brack and Pillinger, 1998). The early histories of Mars and Earth clearly show similarities. Geological observations from data collected by Martian orbiters suggest that liquid water was once stable on the surface of Mars. Mapping of Mars by Mariner 9, Viking 1 and 2, and by Mars Global Surveyor, revealed channels resembling dry river beds. Odyssey's gamma ray spectrometer instrument detected hydrogen (Boynton et al., 2002), which suggested the presence of water ice in the upper metre of soil, in a large region surrounding the planet's south pole, where ice is expected to be stable. The amount of hydrogen detected indicates 20 to 50 percent ice by mass in the lower layer beneath the top-most surface. The ice-rich layer is about 60 cm beneath the surface at latitude 60°S, and approaches 30 cm of the surface at latitude 75°S.

The Viking 1 and 2 lander missions were designed to address the question of extant life on Mars. The results were ambiguous, since although "positive" results were obtained, no organic carbon was found in the Martian soil, on the basis of measurements by gas chromatography-mass spectrometry. It was concluded that the most plausible explanation for these results was the presence, at the Martian surface, of highly reactive oxidants such as H_2O_2, which would have been photochemically produced in the atmosphere (Hartman and McKay, 1995). Direct photolytic processes can also be responsible for the dearth of organics at the martian surface (Stoker and Bullock, 1997). The Viking lander could not sample soils below 6 cm and therefore the depth of this apparently organic-free and oxidising layer is unknown. Bullock et al. (1994) have calculated that the depth of diffusion for H_2O_2 is less than 3 metres. The alpha proton X-ray spectrometer (APXS) on board the rover of the Mars Pathfinder mission measured the chemical composition of six soils and five rocks at the Ares Vallis landing site in 1997. The analysed rocks were partially covered by dust or a weathering rind similar in composition to the dust. Some rocks were found to be similar to terrestrial andesites, but it is not certain that these rocks are igneous. The texture of other rocks is difficult to interpret and might be sedimentary or metamorphic (Rieder et al., 1997).

J. Seckbach et al. (eds.), Life in the Universe, 227–231.

Beagle 2, the exobiology lander of ESA 2003 Mars Express mission, has been designed to search for evidence of life in subsurface and rock interior samples taking into account that the search for traces of life on Mars must necessarily be coupled with a geochemical exploration of the surface and near subsurface.

2. The Landing Site

Beagle 2 will land on Isidis Planitia, a large flat region which overlies the boundary between the ancient highlands and the northern plains. Isidis Planitia (11.6°N, 90.75°E) is a sedimentary basin, in fact, the third largest impact basin on Mars. It lies at ~10°N, which is the maximum latitude for a site to be warm enough for Beagle 2 to function properly. The number of rocks on the surface seems to be about right: not too much to threaten a safe landing, but enough for the sampling operations. The site is at a low enough elevation to allow the parachutes sufficient atmosphere to brake the lander descent.

3. Sample Collection

Protected samples will be acquired, by the rock corer, from the inside of any large rock which can be reached by the robotic arm. The 0.7 m robotic arm will also support all the instruments at once, a concentration referred to as Beagle 2 PAW, for Position Adjustable Workbench. The surface of the rocks will be ground (rock grinder) and the rocks will be investigated before and after grinding.

Another kind of protected sample will be acquired by the mole: from below the harsh, oxidising surface—if possible, from a region additionally protected by a boulder large enough not to have been disturbed since being emplaced. Developed by DLR Cologne, in conjunction with Transmash (Russia) and Technospazio (Italy), initially under the auspices of ESA TRP funding, the mole will provide an element of mobility to the stationary lander. The mole has the ability to crawl across the surface at the rate of 1 cm every six seconds, using a compressed spring mechanism to propel a drive mass. Samples are to be collected in the jaw-like mouth. It is expected that the mole will be able to crawl up to two metres away from the lander, including the burrowing phase; it is recovered by winding in the cable. The robotic arm aids the sample delivery from the mouth of the mole, into the opening to the miniaturised laboratory which is located inside the lander. The mole has a total weight of 900 g and power consumption of only a couple of watts. In addition to horizontal movement, the same process hammers the mole into the ground or under boulders a millimetre at a time in a vertical movement.

4. Cameras

The Beagle 2 lander carries three cameras:

- A stereo pair of cameras mounted on the robotic arm will provide a panoramic view of the scene around the landing site and monitor the activities during sampling.

- One of the cameras is equipped with a pop-up mirror which will provide the first wide angle picture of Mars soon after landing without having to lift the robotic arm from its stowage position.
- A third camera is part of a microscope, deployed by the robotic arm which will be used to examine fresh rock surfaces cleaned of weathered debris by a rock grinder. The microscope with 4 micron resolution will image at various wavelengths

5. Stepped Combustion and Gas Analysis Package

The solid sample (soil or rock) will be heated in steps of increasing temperature, and supplied with freshly generated oxygen at each increment. Any carbon compound present will burn to give carbon dioxide. The gas generated at each temperature will be analysed by the mass spectrometer. The instrument can distinguish between the two stable isotopes of the carbon and quantify the ratio. Other gases can be analysed by the same instrument, including methane. The design for the mass spectrometer is based on a 90° sector analyser and incorporates a magnet, weighing less than 1 kg, made from samarium-cobalt, together with an ion pump using the same material. It will operate in dual inlet mode, whereby samples and standards are sequentially compared, for high precision isotopic measurements; alternatively, for greatest sensitivity, the instrument can operate in static vacuum mode.

6. Spectrometers

The Mössbauer spectrometer will provide information about the oxidation state of iron in mineral samples. It will be used to determine the degree to which iron is in the oxidised form, Fe(III). The X-ray detector on Beagle 2 will provide elemental compositions from the energy spectrum of X-rays produced by bombarding samples with X-rays from ^{55}Fe and ^{109}Cd sources. A potential objective is measurement of the potassium content, for age dating purposes in conjunction with ^{40}Ar determination from the Gas Analysis Package. Quantifying major elements Mg, Al, Si, S, Ca, Ti, Cr, Mn and Fe and trace elements will help in identifying rock types and aid the interpretation of Mössbauer spectra.

7. Environmental Sensors

A suite of seven environmental sensors to provide measurements of the conditions on Mars with respect to its suitability for life and climatic parameters:

- An ultraviolet sensor to detect surface UV flux at 200–400nm.
- ESOS, a thin film able to detect an oxidising atmosphere.
- A radiation sensor to measure the total long term dose and dose rate of solar protons and high energy cosmic rays.

- Thermocouples which can detect air temperature variations to within ±0.05K.
- A pressure sensor which will measure both day and night air pressure to an accuracy of 0.1mBar.
- A wind gauge to monitor speed (to 0.1 m/s) and direction (to 5°)
- Dust impact detectors to register the momentum, direction and rate of martian dust.

8. Beagle 2 and Art

Beagle 2 is also the name of the music specially composed by the British rock band Blur for the project. Beagle 2 will be the call sign for the lander and will be beamed back from the surface of Mars to announce our safe arrival. The British artist Damien Hirst provided an image on the lander based on his famous spot paintings, which will serve as the colour calibration target for the on-board cameras. By using various iron oxides and other known minerals it will also be possible to obtain signals to standardise the Mössbauer and X-ray spectrometers.

9. The Beagle 2 People

The Beagle 2 Team comprises/compromised of C.T. Pillinger (Beagle 2 Consortium Leader), M.R. Sims (Beagle 2 Mission Manager), K. Arnold, C. Ashcroft, R. Asquith, C. Berry, M. Bonnar, A. Brack (Chairman of the Adjunct Science Team), G. Butcher, J. Clemmet, A. Coates, R. Cole, J. Dowson, W. Edwards, A. Ewbank, C. Farmer, G. Fraser, J. Gebbie, A. Griffiths, I. Hindle, A. Holland, H. Hamacher, S. Hurst, G. Johnson, J.L. Josset, B. Kirk, G. Klingelhöfer, H. Kochan, M. Leese, R. Marston, D. Moore, G. Morgan, N. Nelms, G. Paar, S. Peskett, N. Phillips, D. Pullan, L. Richter, T. Ransome, N. Roskilly, A. Senior, B. Shaughnessy, M. Sherring, R. Slade, A. Smolen, A. Spry, J. Standing, J.L.C. Stewart, R. Sulch, J. Sykes, J. Thatcher, N. Thomas, M. Towner, J. Underwood, L. Waugh, A. Wells, S. Whitehead, R. Wimmer, D. Wright, I.P. Wright, and J. Zarnecki with contributions from many others.

10. References

Boynton, W.V., Feldman, W.C., Squyres, S.W., Prettyman, T.H., Brückner, J., Evans, L.G., Reedy, R.C., Starr, R., Arnold, J.R., Drake, D.M., Englert, P.A.J., Metzger, A.E., Mitrofanov, I., Trombka, J.I., d'Uston, C., Wänke, H., Gasnault, O., Hamara, D.K., Janes, D.M., Marcialis, R.L., Maurice, S., Mikheeva, I., Taylor, G.J., Tokar, R. and Shinohara, C. (2002) Distribution of Hydrogen in the Near Surface of Mars: Evidence for Subsurface Ice Deposits, Science, **297**, pp. 81–85.

Brack, A. and Pillinger, C. (1998). Life on Mars: chemical arguments and clues from Martian meteorites. Extremophiles **2**, 313–319.

Bullock, M.A., Stoker, C.R., McKay, C.P. and Zent, A.P. (1994) A coupled soil-atmosphere model of H_2O_2 on Mars, Icarus, **107**, pp. 142–154.

Hartman, H. and McKay, C.P. (1995) Oxygenic photosynthesis and the oxidation state of Mars, Planet. Space Sci., **43**, pp. 123–128.

Rieder, R., Economou, T., Wänke, H., Turkevich, A., Crisp, J., Brückner, J., Dreibus, G. and McSween, Jr., H.Y. (1997) The chemical composition of Martian soil and rocks returned by the mobile alpha proton X-ray spectrometer: preliminary results from the X-ray mode, Science, **278**, pp. 1771–1774.

Stoker, C.R. and Bullock, M.A. (1997) Organic degradation under simulated Martian conditions, J. Geophys. Res., **102**, pp. 10881–10888.

Westall, F., Brack, A., Hofmann, B., Horneck, G., Kurat, G., Maxwell, J., Ori, G.G., Pillinger, C., Raulin, F., Thomas, N., Fitton, B., Clancy, P., Prieur, D. and Vassaux D. (2000) An ESA study for the search for life on Mars. Planet. Space Sci. **48**, pp.181–202.

MINIMAL UNIT OF TERRAFORMING AN ALTERNATIVE FOR REMODELLING MARS

HÉCTOR OMAR PENSADO DÍAZ
Instituto de Ciencias Avanzadas, A.C., Xalapa, Ver., México

Abstract. An alternative independent device for terraforming a planet is propossed. It will have an inner ecosystem protected from the outside which will interact with the surroundings and will began a microterraforming process. Such structure will be for Mars what cells are for plant life form, supporting the planet's oxygenation and global warming at a time, via the atmospheric densification.

1. Introduction

The main proposal for the process of terraforming Mars lead to a global warming of the planet by using several methods such as the one proposed by Lovelock & Allyby (1984) comprising a model for the addition of super green house gases, which was developed at the beginning by Lovelock, Allaby (1984) y Fogg (1989). Afterwards, Zubrin and McKay (1997) elaborated a model where they proposed the use of orbital mirrors that would reflect sun's rays directly into Mars South Pole, which has more frozen CO_2; they proposed as well the use of chemical ground processing factories in order to extract CO_2 and achieve the increase of the green house effect and densify the atmosphere. Finally, Marinova (2000) and Gestell (2001) propossed the addition of super green house gases by using perfluorocarbonate, microorganism and chemical factories.

2. The Minimal Unit of Terraforming

The current tendency for Mars terraforming is the global warming of the planet. However, this document suggests a model for terraforming Mars in sections with a device called the Minimal Unit of Terraforming (MUT) which implies an ecosystem as the main unit of nature, comprising a group of living organisms and their chemical and physical environment where they live, but applied to a remodelling and colonization on Mars. The MUT will be dome shaped structures built from ground stuck translucid plastic trends that will interact with the surroundings helped by gas exchange and biological ground prosecution. As a consequence, they will help to the rehabilitation of the surface and the creation of photolithographic and chemicalautotrophic organisms. They will take the CO_2 from the atmosphere via the photosynthesis by releasing the oxygen. Then, the MUT collaborate as well on the ground gas reduction process creating a

J. Seckbach et al. (eds.), Life in the Universe, 233–237.

difference in the temperature and releasing the gases out to the atmosphere for its densification.

Gaia principles by J. Lovelock and L. Margulis had been considered for the establishment of an environment inside the MUT since life means plasticity and it can be adapted and adapt the surroundings by bringing climate and biological balance. Life sown inside the MUT will not have an ecosystem like that of the Earth and Mars, but a neutral ecosystem which means the devices will be in thermodynamic lack of equilibrium with the Martian surface. Life will respond then to the dome weather conditions helped by life plasticity and would begin to adapt the inside according to its needs. Then, these structures would be located at latitude 20 degrees above and 20 degrees under the ecuatorian Martian region. The reason for its establishment on those latitudes is for its weather conditions, since the maximum temperature reaches 20° C and the minimum −140° C, moreover the atmospheric pressure fluctuates between the 7 and 10 milibars where the sun light rate is 43% higher than that received on Earth.

3. Minimal Unit of Terraforming Operational Model

The MUT should be capable of generate the water and energy need in order to give the inner environment could do its job of CO_2 receiving and increase its biomass. For that purpose the domes will have valves over their tops which will allow the access of martian air and the egress of photosynthetic gases. The valve functions will be the following: a) to let the entrance of martian air for the increasing of the MUT's inner pressure, achieving water existence; b) to let CO_2 enter the domes for photosynthetic purposes and for stabilization of partial pressures in the inside; c) exit of the resulting gases from photosynthesis such as oxygen and the degassing as steam.

The atmospheric pressure inside could stabilize in 60 milibars, since with that pressure; plants are in state of function properly. The resulting oxygen from the photosynthesis will be accumulating inside the MUT. After a while, the valves will activate releasing inner pressure so that the inner atmosphere escapes together with the generated oxygen. As a result, the amount of milibars will decrease in the inside and when the pressure reach 40 milibars an antenna will detect this decrease and will activate a device that will close the valves. With that, other valves will activate and they will let enter *sucked in* air from the outside and with that increase the inner pressure to the original 60 milibars. Martian air, containing 95 percent of CO_2 will be entering the MUT; however, this entrance should be regulated which means the partial pressures in the inside would have to be considered to avoid an acidosis in the plants (Hiscox, 1996) therefore, the death of the environment. Moreover, will have to be tolerant in high CO_2 levels. Hence, a gradual adaptation to a CO_2 concentrated environment should take place. The MUT will create a green house effect, which warmth will permit water existence, for which warmth generation and retention in the inside will be essential for the ecosystem growth. To achieve that aim, the dome should be a thermal structure that will avoid warmth liberation and cold access, protecting the inside from the Ultraviolet rays. The energy flow depends on the incidence of sun light, which in Mars are 500 watts/m2 warmth. This is important because it will determine the strength of the warming to permit the ground degassing process and the action of the photosynthetic organisms inside the MUT. The warmth concentration of the MUT will contribute to the release and storage of water via plant tissues, in which warmth will be extremely important in the activation of the inner hydrological cycle. It is known that Mars has water, but it is captured in the poles, in the underground and creating permafrost on the surface (Vázquez & Guerra, 1995); nevertheless, there is certain atmospheric moisture that may condensate and canalize into the dome for benefit of the inner ecosystem. A way of achieving this is through the zeolite, a mineral with important absorbent properties, which will take moisture from the martian atmosphere. Adam Bruckner (1995), propossed the use of a zeolite condenser, which could be part of the solution of water taking and its irrigation to the inside of the MUT. It can be obtained through photoelectric cells which will be constantly charging a battery system.

4. The Inner Ecosystem

The MUT could be considered as oxygen makers and processors of the martian surface. They will be balanced with the surroundings. The solar light that Mars receive is appropriate for the photosynthetic processes. However, the photosynthetic efficiency will depend on the amount of solar light on Mars, which is higher 43% than that of the Earth. The MUT will maintain the inner temperature, specially at night, protecting the plants from potential death by freezing. The ecosystem could be shaped by pioneer cosmopolitan, ground disintegrating and creating plants such as the bryophytes. These lead us to repeat an evolutive pattern that happened on Earth during the invassion of sea plants to the surface. Other possible organisms are plants from cold deserts such as the saxyfrague, and from high mountains such as the soldanella, which are water storing plants, the succulent and trees such as pines, specially the *pinus hartwegii lindl.* Other pioneer organisms are the gramineous, which are cosmopolitan as well and could be ideal for the formation of fertile ground. The pine species that could be part of the environment of the MUT are: *P. ayacahuite var. Veitchii Shaw, P. Johannis Robert, P. montezumae var. Lindleyi Loud, P. hartwegii Lindl,* and *white P. cooperi* which annual temperature ranges between −20 and 40 degrees C, and are located between 1500 to 4000 meters height. These pines have an electric charge that can express high resistance, favoring their ability of acclimatization. The ecosystems inside the MUT will gradually acclimatize to the planet's condition and they will be adapting Mars weather to their own conditions, with a feedback result. The consequence is a balanced point between the actual condition of the planet and that inside the MUT. Therefore, the MUT and their ecosystems in Mars will work according to the interaction Genotype + Environment has as a result phenotypical expressions that modify and form the planet's environment. The afore exposed ideas can be expressed as follows:

$$G + E_1 = {}^{E}P + ME_I = E_2 \tag{1}$$

Whereas G is the Genotype, E_1 is the Initial Environment, ${}^{E}P$ are the Phenotypical Expressions, the ME_I is the Initial Modelling of the surrounding or the resulting environment of the phenotypical expression and E_2 is the final resulting stabilized environment. However, the development and growth of the species inside the MUT will be in close relation with environmental factors and the available mineral nutrients. This is due to restriction factors according to the Liebig Law, (Odum, 1976) which refers to the minimum of nutrients an organism requires to live, which means that what is required inside the MUT is a *euri environment.* The martian ground (Zubrin, 1997) has enough nutrients for plant development, even in a higher amount than in Earth. Apart from nitrogen, which percentage is unknown and it is critical for the protein formation, could cause problem for its addition due to its low atmospheric concentration. Each unit will help with the atmospheric densification, therefore the addition of all the units that are spread over a martian given area will have to affect the current environment of the planet. In fact, a green house and a degassing effect will be generating in the inside of the unit, as suggested by Zubrin and McKay, but in the MUT the phenomenon is faster. Therefore, if a program comprising the installation of colonies and colony groups of MUT is accepted, simultaneously, a plant colonization and atmospheric processes, moreover each unit could be an experimental base which will feed with data the development, evolution, and improvement of plant colonies in Mars.

5. Conclusion

The MUT creates a thermodynamical disturbance of the environment in which they are exposed, generating an energy rich environment, supporting life development, which will gradually affect the martian environment, which will produce an effect on the genotypes of the various species developed in the MUT, creating a feedback system. With that system, Mars could be gradually terraformed, and its duration will depend on the number of MUT installed on the surface, because they will interact directly with the atmosphere and the ground and the consequence will be the reconversion generated by the interaction Genotype + Environment. However, not only the martian environment will be affected and remodelled, but the organisms of the environments, since an speciation process will look for its own firmness according to Liebig Law.

6. Acknowledgements

I want to thank very much Dr. Julian Chela-Flores for his important support for my participation in The Seventh Trieste Conference on Chemical Evolution and the Origin of Life. I want to thank as well, members of the Instituto de Ciencias Avanzadas A.C., Mr. Cirilo Santiago Cruz, Julieta Ivette Ramírez Enríquez and David Armando Morales Enríquez as well as all the people who supported me.

7. References

Allaby, M. and Lovelock, J. (1984) *The greening of Mars*, St Martin Press, New York.

Lovelock, J.E. (1985) Gaia. *Una nueva visión de la vida sobre la Tierra*, Ediciones Orbis, Madrid, España.

Hiscox, J.A. (1996) *Biology and the planetary engineering of Mars.* Department of Microbiology, University of Alabama, Birmingham, USA.

Odum, E. P. (1976) *Ecología*, C.E.C.S.A. México.

Odum, H.T., Odum, E.C., Brown, M.T., LaHart, D., Bersok, C., Sendzimir, J. (1988) *Environmental Systems and Public Policy*, Ecological Economics Program. University of Florida, USA.

Vázquez Abelardo, M. and Guerrero de Escalante, E.M. (1999) *La búsqueda de vida extraterrestre*, McGraw-Hill, España.
Williams, J. D. Coons, S.C., and Bruckner, A.P. (1995) Design of a Water Vapor Absorption Reactor for Martian in situ Resource Utilization, J. of the Brit. Interplanet. Soci. 48, 347–354.
Zubrin, R. and C.P. McKay (1993) Technological requirements for terraforming Mars. AIAA 93–2005. 1–14, 29th Joint Propulsion Conference and Exhibit.
Zubrin, R. (1997) *The case for Mars*, Touchstone, New York.

EARLY ARCHAEAN LIFE

FRANCES WESTALL
Centre de Biophysique Moléculaire, CNRS, Rue Charles Sadron, 45071 Orléans cedex 2, France

1. Introduction

After a number of decades of research on the evidence for life in the Early Archaean epoch (4.0–3.2 Ga), recent controversies have thrown the very concept of the existence of life on Earth in this time period into discussion (Brasier et al., 2001; Lindsay et al., 2003; Westall and Folk, 2003; van Zuilen et al., 2002, 2003; Lepland et al., 2002). In this contribution I will review the discussions, describe the procedures that should be followed in the search for fossil life, and briefly document the evidence for widespread and relatively abundant life in 3.5–3.3 Ga sediments.

1.1. THE OLDEST EVIDENCE FOR LIFE IN >3.7 GA ROCKS FROM GREENLAND

Greenland hosts the oldest sedimentary rocks on Earth (>3.7 Ga) at Isua and Akilia; they are, however, severely metamorphosed and deformed. The evidence for life at the time these rocks were formed is based on carbon isotope measurements (Shidlowski, 2001; Mojzsis et al., 1996; Rosing, 1999), and on observations of carbonaceous microfossils (Pflug and Jaeschke-Boyer, 1979; Robbins, 1987; Pflug, 1979, 2001). However, problems related to recent contamination of the rocks samples, as well as a metamorphic production of graphite, were highlighted by van Zuilen et al. (2002) and Westall and Folk (2003). One isotopic signal from rocks that are thought to have been deep-sea sediments (−18‰, Rosing, 1999) appears, however, to be original and is neither a product of contamination nor of later metasomatic origin.

1.2. LIFE IN 3.5–3.2 GA ROCKS FROM SOUTH AFRICAN AND AUSTRALIA: THE SCHOPF-BRASIER CONTROVERSY

Barberton Mountain Land (South Africa) and the Pilbara Craton (Australia) are the locations of the oldest, well-preserved sedimentary rocks on Earth. Relatively large, filamentous, carbonaceous structures were described from the silicified sediments as fossil cyanobacteria (relatively evolved, oxygen-producing photosynthetic bacteria), although a few rare, smaller filaments were described as filamentous bacteria (Schopf, 1993; Walsh, 1992). Recently, the interpretation of cyanobacteria has been challenged (Brasier et al., 2001) on the grounds that (1) the original sample actually came from the interior of a hydrothermal vein where

J. Seckbach et al. (eds.), Life in the Universe, 239–244.

cyanobacteria would not be expected to live, and (2) 3D-imaging showed that the "filaments" form part of much larger, highly irregular structures that do not resemble fossil microbes. Going one step further, these authors conclude that all evidence for life at this period is unfounded and that life probably did not exist at all (Lindsay et al., 2003). Is there any reliable evidence for life at this period and what procedures are needed to interpret it?

2. Early life in 3.5–3.3 Ga rocks from Barberton and the Pilbara

There is, in fact, abundant evidence for life in the Early Archaean cherts. However, a correct understanding of the nature of prokaryote life and biofilm formation, as well as appropriate techniques for detailed studies of structures on a micron scale (*i.e.* electron microscope techniques), are needed. In the first place, previous studies had concentrated on cyanobacteria-type microorganisms that are relatively large (generally an order of magnitude larger than other types of bacteria) and readily visible in thin sections. Most prokaryotes, however, are too small to be easily observed by light microscopy: it is difficult to distinguish a 0.5–1 μm coccoid from a mineralogical bacteriomorph of the same size in thin section.

2.1. METHODS OF IDENTIFICATION OF ANCIENT MICROFOSSILS

In order to avoid incorrect interpretations, the search for fossil life requires a wide-ranging, comprehensive approach. This means observations ranging from the macroscopic, (field context) to the microscopic scale (light microscope and electron microscopes), as well as chemical and isotopic analyses. There is no single biosignature that can be taken as evidence of *in situ* life in an ancient rock. Below I list the series of steps that I use in the search for fossil prokaryotes.

1. Macroscopic study of the field outcrop in order to understand the geological context (e.g. water-lain sediments or hydrothermal vein) (Fig. 1a). Larger-scale structures of potential microbial origin can be distinguished, such as tabular and domal stromatolites or thrombolites.

2. Macroscopic to hand lens investigation of selected samples and cut, fresh surfaces of the sample can provide further information relating to the sedimentological environment and potential macro-microbial constructions (Fig. 1b).

3. Thin section investigation provides information about the history of the rock and its mineralogical composition. Microfabric textures that can be important on the microbial scale can be studied. Larger-scale microbial structures such as biofilms and larger filaments can be identified in thin section.

4. Owing to the small size of the majority of prokaryotes, electron microscope investigations are necessary for more detailed study of the individual microbes, their colonies and their biofilms, keeping track, however, of the subsample location with respect to the macroscopic features. The SEM study entails delicate sample preparation, including HF etching. Artifacts can be produced in this process and great care needs to be taken to recognise and avoid them. Since a number of minerals, notably silica, produce structures that resemble certain microbial shapes, a rigorous

Figure 1. Volcaniclastic sediments deposited in shallow water conditions from a 3.46 Ga formation in the Pilbara: (a Field view, (b) cut surface (detail in (a)) showing truncated ripple marks and pumice fragments, suggestive of shallow water conditions. The arrow points to a microbialite horizon.

list of criteria need to be used in order to make a correct interpretation (see below) (Westall, 1999, 2003a; Westall et al., 2000). As noted above, recent contamination can also be a problem.

5. Analytical investigations to comprehend the chemistry of the potential microbes and their biofilms can be made through EDS point analysis or elemental mapping of selected structures, as well as microprobe studies. Further analysis of selected zones for carbon, nitrogen and sulphur isotopes can provide further important information, as well as trace element composition.

There are a number of characteristics relating to microorganisms that can be used in the identification of fossil prokaryotes. Whereas modern prokaryotes are identified on the basis of their (1) morphology, (2) colony-forming behaviour, (3) biochemistry, (4) genetic characteristics, of these, only the first two can be used in their entirety for identifying fossil prokaryotes. Rapid degradation of the organic molecules changes the biochemical and signature and genetic information is lost after some tens of thousands of years.

The morphological characteristics that can be used in identifying fossil prokaryotes include (i) size, (ii) shape, (iii) evidence for cell division, (iv) cell surface texture, which is related to (v) cell lysis, (vi) flexibility (for the long rod and filamentous forms) and (vii) cell differentiation (a particular characteristic of the cyanobacteria). Additional information can be provided by size/shape analysis of the microstructures.

The colony-forming characteristics include (i) the fact that bacteria are not solitary organisms and occur in colonies of a few to millions of individuals, (ii) the very frequent association of polysaccharide polymer (extracellular polymeric substances or EPS) associated with the organisms and the colonies, (iii) most colonies consist of more than one species forming a consortium, (iv) biofilm formation (biofilms provide a protected habitat in which the environmental conditions can be controlled by the microorganisms themselves), (v) the inclusion of gas pockets and gas escape structures in thick biofilms, and (vi) the inclusion of precipitated or trapped minerals in the biofilm.

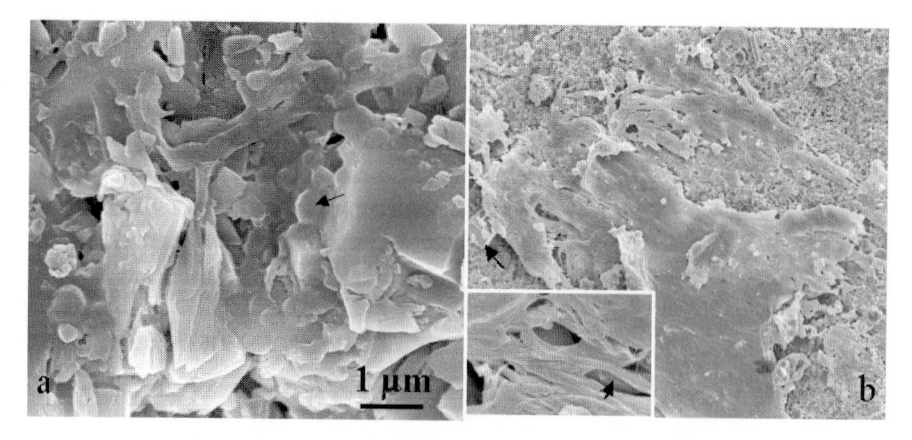

Figure 2. Microbial mats and fossil microorganisms from >3.4 Ga formations in the Pilbara (a) and Barberton (b). (a) Lysed coccoids (e.g. arrow) and degraded EPS in a biofilm formed under shallow water conditions. (b) Thick, robust, subaerial biofilm formed of fine filaments (arrow in inset) embedded by thick EPS. Note evaporite mineral associated with the biofilm (arrow in main photograph).

2.2. MICROFOSSILS AND MICROBIAL MATS FROM 3.5–3.3 GA

On the basis of the criteria outlined above, I have identified microbial mats formed in different environments by a variety of microorganisms in 3.5–3.3 Ga-old cherts from Barberton (South Africa) and the Pilbara (Australia) (Westall et al., 2001, 2002; Westall, 2003a,b) (*N.B.* Walsh (1992) and Walsh and Lowe (1999) have also previously described microbial filaments and mats from Barberton). Fine, laminar mats, formed by a consortia of small (0.5 μm) and larger (0.8 μm) coccoids and small filaments (0.25 μm diameter and up to tens of μm in length), associated with a fine film of EPS, occur at the surfaces of volcaniclastic sediments deposited in shallow water conditions (Fig. 2a). A laterally-extensive (over several meters), mini-stromatolite/thrombolite association occurs at the surface of shallow water sediments in one location (Fig. 1b). It consists of a consortium of small (0.5 μm) and larger (0.8 μm) coccoids forming a partly columnar to partly clotted fabric. Microbial mats also occur in partly subaerial environments, such as the littoral (beach) environment (Fig. 2b). Since these mats are exposed to the air, they exhibit a more robust appearance, being far thicker than their subaqueous counterparts. They have a dessicated appearance and evaporite minerals occur embedded in them, indicating subaerial exposure in an evaporitic environment. These mats are formed by small filamentous organisms (0.25 μm diameter and up to tens of μm in length). Other mats formed in the same environment consist of vibroid-shaped organisms (2–3.8 μm in length and 1 μm in width) embedded in EPS. Although all the mats and biofilms observed were formed in either a shallow water or littoral environment, there is no morphological evidence for organisms having the characteristics of cyanobacteria.

A preliminary study of a number of rock samples from different stratigraphic horizons and different locations within the two Early Archaean greenstone belts indicates that biofilm or microbial mat formation at sediment surfaces was relatively common in shallow water to subaerial environments (see also Walsh and Lowe, 1999).

3. Conclusions

Despite the problems associated with the identification of life in the Early Archaean period, it is possible to identify ancient fossil life in the cherts from Barberton and the Pilbara, if ALL information ranging from the macroscopic down to the microscopic level, as well as the biogeochemical analyses, is taken into account, and if care is taken concerning the possible presence of recent contamination and/or the production of artifacts during preparations for SEM observation.

Using these methods, I have identified microbial mats on the surfaces of volcaniclastic sediments deposited in shallow water to subaerial environments. These mats were formed by coccoidal, filamentous and vibroid-shaped organisms, often associated in consortia consisting of more than one type of organism. Life was common in the Early Archaean.

4. References

Brasier, M.D., Green, O.R., Jephcoat, A.P., Kleppe, A.K., van Kranendonk, M., Lindsay, J.F., Steele, A. and Grassineau, N. (2002) Questioning the evidence for Earth's oldest fossils. Nature, **416**, 76–81.

Lepland, A., Arrhenius, G. and Cornell, D. (2002) Apatite in early Archean Isua supracrustal rocks, southern West Greenland: its origin, association with graphite and potential as a biomarker. Precambrian Res., **118**, 221–241.

Lindsay J.F., Brasier M.D., McLoughlin N., Green O.R., Fogel M., McNamara K.M., Steele A. and Mertzman, S. A. (2003) Abiotic earth - establishing a baseline for earliest life, data from the Archaean of western Australia. LPSC **XXXIV**, # 1137.

Mojzsis, S.J., Arrhenius, G., McKeegan, K.D., Harrison, T.M., Nutman, A.P. and Friend, C.R.L. (1996) Evidence for life on Earth before 3800 millio-n years ago. Nature, **384**, 55–59.

Pflug, H.D. (1979) Archean fossil finds resembling yeasts. Geol. Palaeontol., **13**, 1–8.

Pflug, H.D., (2001) Earliest organic evolution, Essay to the memory of Bartholomew Nagy. Precamb. Res., **106**,79–92.

Pflug, H.D. and Jaeschke-Boyer, H. (1979) Combined structural and chemical analysis of 3,800-Myr-old microfossils. Nature, **280**, 483–486.

Robbins, E.I. (1987) *Appelella ferrifera*, a possible new iron-coated microfossil in the Isua Iron-Formation, Southwestern Greenland. In: P.W.U. Appel and G.L. LaBerge (eds.) *Precambrian Iron Formations*, Theophrastes, Athens, pp. 141–154.

Robbins, E.I. and Iberall, A.S. (1991) Mineral remains of early life on Earth? On Mars? Geomicrobiol. J., **9**, 51–66.

Rosing M.T. (1999) 13C depleted carbon microparticles in >3700 Ma seafloor sedimentary rocks from West Greenland. Science, **283**, 674–676.

Schidlowski, M. (2001) Carbon isotopes as biogeochemical recorders of life over 3.8 Ga of Earth history: evolution of a concept. Precambrian Res., **106**, 117–134.

Schopf, J.W. (1993) Microfossils of the Early Archean Apex Chert: new evidence of the antiquity of life. *Science*, 260, 640–646.

Van Zuilen, M., Lepland, A. and Arrhenius, G. 2002. Reassessing the evidence for the earliest traces of life. *Nature*, 418, 627–630.

Van Zuilen, M.A., Lepland, A., Teranes, J., Finarelli, J., Wahlen, M. and Arrhenius, G. (2003) Graphite and carbonates in the 3.8 Ga old Isua Supracrustal Belt, southern West Greenland. Precambrian Res., **126**, 331–348.

Walsh, M.M. (1992) Microfossils and possible microfossils from the Early Archean Onverwacht Group, Barberton Mountain Land, South Africa. Precambrian Res., **54**, 271–293.

Walsh, M.M. and Lowe, D.R. (1999) Modes of accumulation of carbonaceous matter in the early Archaean: A petrographic and geochemical study of carbonaceous cherts from the Swaziland Supergroup. In: D. R. Lowe and G.R. Byerly (eds.) *Geologic evolution of the Barberton greenstone belt, South Africa*, Geol. Soc. Am Spec. Paper, **329**, 115–132.

Westall, F. (1999) The nature of fossil bacteria. J. Geophys. Res., **104**: 16,437–16,451.

Westall, F. (2003a) The geological context for the origin of life and the mineral signatures of fossil life. In: H. Martin, M. Gardaud, G. Reisse, and B. Barbier (eds.) *The Early Earth and the Origin of Life*, Springer, Berlin, in press.

Westall, F. (2003b) Stephen Jay Gould, les procaryotes et leur évolution dans le contexte géologique. Palevol, in press.

Westall, F., and Folk, R.L. (2003) Exogenous carbonaceous microstructures in Early Archaean cherts and BIFs from the Isua greenstone belt: Implications for the search for life in ancient rocks. Precambrian Res., **126**, 313–330.

Westall, F., De Wit, M.J., Dann, J., Van Der Gaast., S., De Ronde., C. and Gerneke., D. (2001) Early Archaean fossil bacteria and biofilms in hydrothermally-influenced, shallow water sediments, Barberton greenstone belt, South Africa. Precambrian Res., **106**, 93–116.

Westall, F., Steele, A., Toporski, J. Walsh, M., Allen, C., Guidry, S., Gibson, E., Mckay, D. and Chafetz, H. (2000) Polymeric substances and biofilms as biomarkers in terrestrial materials: Implications for extraterrestrial materials. J. Geophys. Res., **105**, 24,511–24,527.

Westall, F., Brack, A., Barbier, B., Bertrand, M. and Chabin, A. (2002) Early Earth and early life: an extreme environment and extremophiles - application to the search for life on Mars. Proceedings of the Second European Workshop on Exo/Astrobiology Graz, Austria, 16–19 September 2002 **(ESA SP-518)**, pp. 131–136.

EXTRATERRESTRIAL IMPACTS ON EARTH AND EXTINCTION OF LIFE IN THE HIMALAYA

V. C. TEWARI
Wadia Institute of Himalayan Geology, Dehradun-248001, Uttaranchal, India and The Abdus Salam International Centre for Theoretical Physics 34136, Trieste, Italy

Abstract. The comets, meteorites and asteroids have collided with the Earth throughout geological history. The mass extinction at Permian-Triassic boundary and Cretaceous-Tertiary boundary is strongly supported by the extraterrestrial asteroidal impact theory in the Indian Himalayan sequences well exposed at Spiti in Western Himalaya and Um Sohryngkew section in Meghalaya, northeastern Himalaya. The carbon isotopic and palaeobiological events suggest extraterrestrial impacts at P/T and K/T boundaries all over the world. The early evolution of life, its diversification, carbon isotope chemostratigraphy, amino stratigraphy and extinction events have been discussed from the Indian Himalaya.

1. Introduction

The end of the Cretaceous period is named as K/T boundary and this boundary is marked by a catastrophic impact in the Yucatan Peninsula, Mexico. The Chicxulub impact crater is a possible trigger for the mass extinction that occured at K/T bondary about 65 million years ago in which the dinosaurs perished. The high concentration of platinum group elements particularly iridium at the K/T boundary in the Gubbio section of Italy strongly supports an extraterrestrial impact (Alvarez *et al.*, 1980). The platinum group elements have a rich concentration in extraterrestrial materials like meteorite, asteroid or comet and a very low concentration in the terrestrial volcanic eruptions for example Deccan basalt in India (Bhandari, 1991). The K/T boundary section at Um Sohryngkew River section (Figure 1) in Meghalaya, eastern lesser Himalaya, India contains a strong narrow peak of iridium (12.1 ng g). The iridium is enriched by a factor of about 10 in the broad band and by a factor of about 500 in the sharp peak above the Cretaceous Shales (Ir = 0.02 ng g, Bhandari, 1991). The K/T boundary sections at Padriciano, Trieste in North Italy and Slovenia has been established on the basis of palaentological extinction, sedimentological facies changes, carbon isotopic variations, iridium anomaly and micro tektites of extraterrestrial origin (Pugliese and Drobne 1995; Drobne *et al.*, 1996; Hansen 1995; Ogorelec *et al.*, 1995; Gregoric *et al.*, 1998 and Pugliese and Tewari, 2003). The iridium anomaly in Meghalaya of India is identical to that found globally and is extraterr-estrial. The Permo-Triassic boundary represents the major mass extinction on Earth. During the Permian period Siberia collided with the other landmasses and the largest super continent Pangea

J. Seckbach et al. (eds.), Life in the Universe, 245–248.

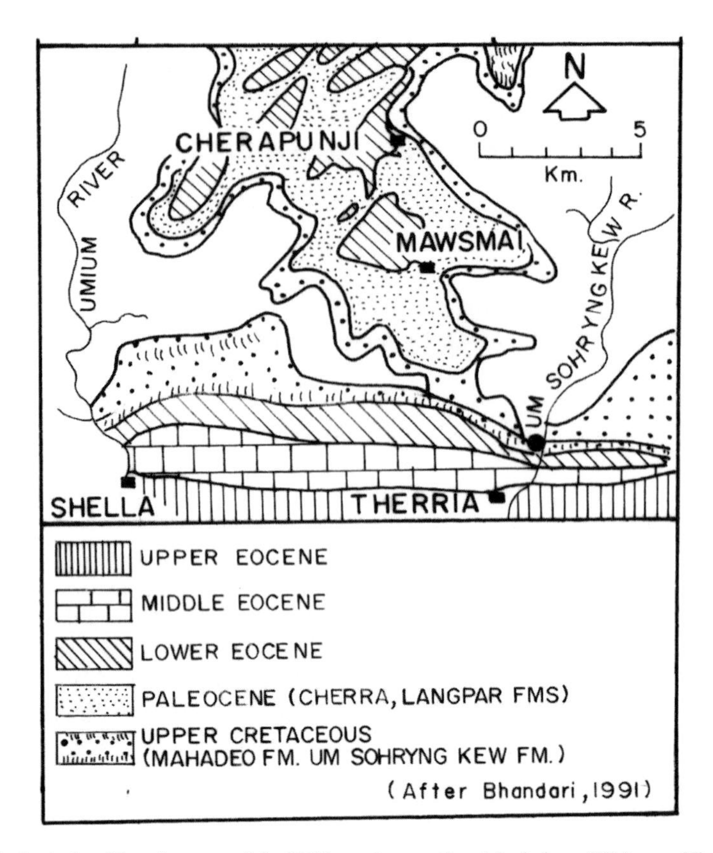

Figure 1. Geological and location map of the K/T boundary section, Meghalaya, NE Lesser Himalaya, India.

was created. The southern continent Gondwana started to drift away to the north creating new microcontinents. The Permo-Triassic boundary is characterized by Siberian volcanism, palaeoclimatic changes, oceanic anoxia events and extraterrestrial impacts. The recent carbon isotopic excursions from the P/T boundary sections of the Spiti valley in Tethys Himalaya of India has supported the impact hypothesis (Ghosh *et al.* 2002). In the present paper the major extinction events of life since the origin of life on Earth has been discussed.

2. Origins of Life from Amino Acids in Meteorites to Stromatolites

According to the recent research (Oro *et al.* 1971; Tewari, 2001a, 2002 a,b) comets and meteorites may have been a source of organic compounds on the early Earth. The Murchison meteorite found in Australia contains amino acids glycine, alanine, valine, proline, aspartic acid and glutamic acid (Oro *et al.* 1971). The discovery of amino acids phenylalanine, tyrosine and tryptophan by Laser Raman Spectroscopy from 4.5 billion year old Didwana-Rajod meteorite from Rajasthan, India (Tewari, 2002 a, b) confirms that life might have originated in space before being introduced to Earth by meteorites. The most convincing evidence for the extraterrestrial delivery of amino acids (alpha amino isobutyric acid and racemic isovaline) comes from the Cretaceous-Tertiary boundary section of the Stevens Klint in Denmark. The oldest sedimentary rocks on Earth are 3.9 billion years old Isua meta-Quartzite in Greenland but no convincing microfossils have been reported from these rocks. The oldest record of life on Earth, in the form of bacterial microfossils and stromatolites, is reported from the Apex chert of Western Australia (Schopf, 1993). The early evolution and diversification of life on Earth with special reference to the Himalaya has been discussed by Tewari (2001a, 2003 in press).

3. Discussion and Conclusion Major Extinction Events of Life on Earth

The stromatolites of Mesoproterozoic age (large Conophytons and other columnar forms) declined around 650 Ma. in Neoproterozoic period before the Ediacaran explosion of metazoans and metaphytes. The Ediacaran explosion of metazoan and metaphyte multicellular life took place after Neoproterozoic palaeoclimatic change from snow ball Earth to global warming (Tewari, 2001b). A major global decline of Mesoproterozoic stromatolites and planktonic acritarchs has been recorded on Earth related to Neoproterozoic/Varangian/Blainian glacial event. Explosive radiation of new acanthomorphic acritarchs, sponge spicules, multicellular brown sheet algae Vendotaenids and Ediacaran metazoans were recorded from Australia, Krol Group of the Uttaranchal Lesser Himalaya, India, China, Siberia, Canada and Namibia in South Africa (Tewari, 2001, 2003). During the Vendian 650 Ma. ago the dominant Precambrian flora and fauna perished in the first great extinction and has been correlated with a large glaciaton event. The Cambrian period is marked as an important turning point in the history of evolution of life on Earth. This is the time when most of the major groups of animals first appeared in the fossil record. This event is called the Cambrian Explosion of life on Earth. During the Cambrian, the trilobite Olenellids and archaeocyathids (reef building organisms) perished. Palaeobiological records also suggest that climate during Permocarboniferous was cool in Gondwanaland. The Cretaceous period about 65 Ma. ago suffered an asteroid impact and caused global cooling on Earth and extinction of dinosaurs from the planet. The organic compounds such as amino acids of extraterrestrial origin have been found in Martian (SNC) and other meteorites landing on Earth from the asteroid belt between Mars and Jupiter strongly suggests that the meteorites braught life to the Earth and can cause its extinction. Astrobiology in future may give us important clues regarding extraterrestrial origin of life on Earth and other planets of the universe (Tewari, 1998).

4. Acknowledgements

The author is grateful to Professor Julian Chela Flores, Abdus Salam International Centre for Theoretical Physics, Trieste, Italy for encouragement. Professor Nevio Pugliese of Dipartimento di Scienze Geologiche Ambientali e Marine, Trieste, Italy is thanked for kindly reviewing the article and valuable suggestions. Professor Joseph Seckbach, Hebrew University, Jerusalam, Israel is thanked for improving the article. The Directors of the Wadia Institute of Himalayan Geology, Dehradun, India and A.S. International Centre for Theoretical Physics, Trieste, Italy are thanked for providing facilities.

5. References

Alvarez, L.W., Alvaerz, W.F., Asaro, F. and Michel, H.V. (1980) Extraterrestrial cause for the Cretaceous-Tertiary extinction. Science, **208**, 1095–1100.

Bhandari, N. (1991) Collisons with Earth over geologic times and their consequences to the terrestrial Environment. Current Sci. **61**, 97–103.

Brazazatti, T., Caffau, M., Cozzi, F., Drobne, K. and Pugliese, N. (1996) Padriciano Section (Karstof Trieste, Italy). In: K. Drobne and B. Goriean (eds.) International Workshop POSTOJNA 96 *The role of impact processes in the geological and biological evolutionof Planet Earth.* pp. 189–198.

Drobne, K., Ogorelec, B., Dolenec, T., Marton, E. and Pugliese N. (1996) Cretaceous-Tertiary boundaryon the carbonate platform of the NW part of the Adriatic Plate. In: N. Bardet and E. Buffetaut (eds.) *Séance spec. Soc. Geol. France La Limite Cretace-Tertiare: aspects biologiques et* geologiques, Resume.

Ghosh, P., Bhattacharya, S.K., Shukla, A.D., Shukla, P.N., Bhandari, N., Parthasarathy, G. and Kunwar, A.C. (2002) Negative, D., 13 C excursion at the Permo-Triassic boundary in the Tethys. sea, *Curr. Science.* **83**, 498–502.

Hansen, H.J., Drobne, K., Gwozdz, R. (1995) The K/T boundary in Slovenia: dating by magnetic susceptibility, stratigraphy and ridium anomaly in a debris flow In: A. Montari and C. Coccioni (eds.) *ESF* 4th *International Workshop, 84–85*, Anacona.

Ogorelec, B., Dolence T., Cucchi, F., Giacomich, R., Drobne, K. and Pugliese, N. (1995) Sedimentological and geochemical characteristics of carbonate rocks from the K/T boundary to Lower Eocene in the Karst area (NW Adriatic Platform) *1st Croatian Geological Congress* Opatija, Zbornik radova Proceed., **2**, 415–421.

Oro, J., Gilbert, J., Lich tenstein, H., Wikstorm, S. An Flory, D.A. (1971) Amino acids, aliphatic and aromatic hydrocarbons in the Murchison meteorite. *Nature*, **230**, 105–106.

Pugliese, N. and Drobne, K. (1995) Palaeontological and palaeoenvironmental events at the K/T boundary in the Karst area. In: Montari, A. and Coccioni, R (eds.) *ESF 4th International Workshop.* 137–140.

Pugliese, N. and Tewari, V.C. (2003) Peritidal sedimentary depositional facies and preliminary carbon Isotope data from K/T boundary carbonates at Padriciano, Trieste, Italy. In: *32nd International Geological Congress, Italy to be held in Florence* from August 20–28, 2004 (abstract).

Schopf, J.W. (1993) Microfossils of the Early Archaean Apex Chert New evidence of the Antiquity of life, *Science*, **260**, 640–646.

Tewari, V.C. (1998) Earliest microbes on Earth and possible occurrence of stromatolites on Mars. In: J. Chela Flores and R. Raulin (eds.) *Exobiology Matter, Energy and Information in the Origin and Evolution of Life in the Universe.* Kluwer Academic Publishers, Netherlands, pp. 261–265.

Tewari, V.C. (2001a) Origins of life in the universe and earliest prokaryotic microorganisms on Earth. In: J. Chela Flores, T. Owen and F. Raulin (eds.) *First Steps in the Origin of Life in the Universe*, Kluwer Publishers, the Netherlands. pp. 251–254.

Tewari, V.C. (2001b) Neoproterozoic glaciation in the Uttaranchal Lesser Himalaya and the global Palaeoclimate change. In *National Symposium on Role of Earth Scientists in Integrated Development and Related Societal Issues.* GSI Spl. Publ., **65**, 49–55.

Tewari, V.C. (2002a) Discovery of amino acids from Didwana-Rajod meteorite and its implication on Origin of Life. *Journal of Geological Society of India*, **60**, 107–110.

Tewari, V.C. (2002b) Might life on Earth come from space. *Nature News India*, 2002, 11.

Tewari, V.C. (2003) Proterozoic diversity in Microbial life of the Himalaya. In: J. Seckbach (ed.) *Origins Genesis, Evolution and Biodiversity of Microbial Life in the Universe.* Kluwer Academic Publishers (in press).

PALAEOBIOLOGY AND BIOSEDIMENTOLOGY OF THE STROMATOLITIC BUXA DOLOMITE, RANJIT WINDOW, SIKKIM, NE LESSER HIMALAYA, INDIA

V. C. TEWARI
Wadia Institute of Himalayan Geology, Dehradun-248001, Uttaranchal, India and The Abdus Salam International Centre for Theoretical Physics 34136, Trieste, Italy

Abstract. The Mesoproterozoic (Riphean) stromatolite taxa are recorded from the Buxa Dolomite of the Ranjit Window, Sikkim Lesser Himalaya, India. The Riphean characteristic taxa are *Omachtenia, Colonnella columnaris, Kussiella kussiensis, Conophyton cylindricus, C. garganicum, Rahaella elontgata Tewari, Jacutophyton, Baicalia nova, Tungussia, Jurusania, Inzeria, Gymnosolen, Minjaria, Stratifera*, and *Gongylina.* The Neoproterozoic-Terminal Proterozoic (Vendian) stromatolite assemblage *Paniscollenia, Aldania, Tungussia, Linella, Colleniella, Linocollenia, Boxonia, linked Conophyton, Conistratifera, microstromatolites, Stratifera, Irregularia, Nucleella*, digitate stromatolites and oncolites are well developed in the Buxa Dolomite, Ranjit Window, Sikkim and its equivalents (Menga Limestone, Dedza Limestone and Chillipam Limestone) in the adjoining Arunachal and Bhutan Lesser Himalaya. The Mesoproterozoic to Terminal Proterozoic stromatolite diversification has been recorded for the first time from the Buxa Dolomite of the Sikkim Lesser Himalaya, India. The palaeobiological and biosedimentological significance of the stromatolites in the Buxa Dolomite has been discussed.

1. Introduction

The Buxa Group in the Northeastern Lesser Himalaya, India is well represented by dolomites, limestones, cherty stromatolitic-oolitic-intraclastic dolomite, calcareous quartzite and black carbonaceous shales in Arunachal and Sikkim areas (Acharyya, 1974; Tewari, 2001, 2002, 2003). The stromatolitic dolomite sequence is 800 m. thick in the Ranjit Window section of the Sikkim Lesser Himalaya.

The distribution of stromatolite assemblages, morphological variations and the palaeoenvironment of deposition has been established. The detailed study of the stromatolite morphology, microstructure, microfabrics and associated microbiota in the stromatolitic and bedded cherts suggests a Lower Riphean to Terminal Proterozoic age for the Buxa Dolomite in the Sikkim Lesser Himalaya. Tewari (2003) has done integrated sedimentological, palaeobiological, carbon and oxygen isotopic and Laser Raman Spectroscopic studies of the Buxa Dolomite from Arunachal Lesser Himalaya and has suggested a Neoproterozoic to Pre Cambrian—Cambrian boundary transition in the eastern lesser Himalaya.

J. Seckbach et al. (eds.), Life in the Universe, 249–250.

2. Palaeobiology and Biosedimentology of the Buxa Dolomite

Palaebiological remains discovered from the petrographic thin sections of the black cherts associated with the stromatolitic dolomites of the Buxa Group in Ranjit Window, Sikkim are organic walled microfossils (***Leiosphaeridia, Obruchevella, Myxococcoides, Siphonophycus, Eomycetopsis, Micrhystridium,*** and ***Acanthomorphic acritarchs***). The sedimentological studies of the stromatolites and sedimentary structures suggest that Buxa Dolomite was deposited in intertidal to subtidal, sandy intertidal and lagoonal environment The carbon and oxygen isotope analysis of the Buxa Dolomite from the Ranjit Window shows that carbon isotope ratios ($\delta^{13}C$) vary from -1.4 to $+1.0$ (PDB) and oxygen isotope values ($\delta^{18}O$) range from 18.9 to 23.9 (SMOW). The isotopic data also supports a shallow marine depositional environment. The geochemical analysis of stromatolitic carbonates has shown 16 to 22% MgO and 14 to 31% CaO. Laser Raman Spectra has shown the shift in wave number at 1100 cm and confirms the presence of amino acids (biomolecules).

3. Acknowledgements

The author is grateful to Professor Julian Chela Flores, ICTP, Trieste, Italy and Dr. F. Westall, CBM, CNRS, Orleans, France for discussions. I am indebted to Prof. Joseph Seckbach, Hebrew University, Jerusalam, Israel for reviewing the article and valuable suggestions. The Directors of the Wadia Institute of Himalayan Geology, Dehradun, India and Abdus Salam International Centre for Theoretical Physics, Trieste, Italy are thanked for providing facilities. The financial assistance from the D.S.T. for the project Palaeobiology and Biosedimentology of the Buxa Dolomite, NE Lesser Himalaya (No.SR/S4/ES-O) is thankfully acknowledged.

4. References

Acharyya, S.K. (1974) Stratigraphy and sedimentation of the Buxa Group, Eastern Himalaya. *Himalayan Geology* **4**, 102–116.

Tewari, V.C. (2001) Discovery and sedimentology of microstromatolites from Menga Limestone (Neoproterozoic/Vendian), Upper Subansiri District, Arunachal Pradesh, NE Himalaya, India. *Current Science* **80**, 1440–1444.

Tewari, V.C. (2002) Lesser Himalayan stratigraphy, sedimentation and correlation from Uttaranchal to Arunachal. *Aspects of Geology and Environment of the Himalaya.* Gyanodaya Prakashan, Nainital. pp. 63–88.

Tewari, V.C. (2003) Sedimentology, palaeobiology and stable isotope chemostratigraphy of the Terminal Proterozoic Buxa Dolomite, Arunachal Pradesh, NE Lesser Himalaya. *Journal of Himalayan Geology* **25, (1)**, 1–18.

X. Searching for Extraterrestrial Life, Europa, Titan and Extrasolar Planets

SEARCHING FOR EXTRATERRESTRIAL LIFE

TOBIAS OWEN
Institute for Astronomy, University of Hawaii
2680 Woodlawn Drive, Honolulu, Hawaii 96822 USA

1. Introduction

The search for life outside the Earth is currently stuck in the same debate that once surrounded the quest for other planetary systems. Excellent scientists support a spectrum of views ranging from a strong sense that life is such an improbable state of matter that life on Earth is probably all the life there is, through grudging admission that there may be other planets inhabited by colonies of bacteria, to the idea that millions of technically advanced civilizations are contemplating this same question throughout the Galaxy (Goldsmith and Owen, 2002).

By chance we happen to live in the era when, for the first time, we can transform this fascinating speculation into an experimental science: we can actually go out and look for signs of extraterrestrial life.

What are those signs? Life on Earth may not be the only life in the universe, but it is the only life we know. Hence we must start with a consideration of life as we know it, attempting to generalize from its most basic characteristics to find constraints that environments must satisfy to be habitable and then to identify the observable effects of living systems on those environments. As we are in no position to travel to extra-solar planets, the effects we seek are essentially confined to changes in atmospheric composition that we can detect spectroscopically from our remote vantage point. For example, the fact that our atmosphere is 21% oxygen would be a sure indication of life on Earth to an alien spectroscopist on the other side of the Galaxy (Owen 1980). Without the green plants to keep producing it, our present supply of oxygen would disappear in several million years, just 0.1% of geological time.

Looking into the fundamental chemistry of life, we are struck by its dependence on carbon as the main compound-forming element and water as the essential solvent. All life depends on carbon, and everywhere on Earth we find water there is life. The only places on our planet that are sterile are marked by the absence of this essential substance. A quick examination of the properties of carbon and water suggests that life's dependence on them is not the product of random chance.

Both carbon and water are extremely common in the universe. Water is one of the best solvents we know; its high heat capacity and latent heat of vaporization make it a good temperature regulator; it is an excellent greenhouse gas, ice floats so life can "winter over,"

J. Seckbach et al. (eds.), Life in the Universe, 253–256.

etc., etc. Carbon is the most versatile of all the elements in forming the complex molecules that will be essential for any form of life to store and transmit both energy and information. It moves easily between the most oxidized form, CO_2 and the most reduced form, CH_4, and both of these end members are excellent greenhouse gases, etc., etc.

We conclude that life on other worlds is highly likely to use carbon and water, while nitrogen's cosmic abundance suggests it will also be a component of other living systems just as it is on Earth. We are not being overly restrictive with this conclusion because we don't expect life elsewhere to use the same compounds of C, H, N, and O as we find on Earth. We don't expect the same amino acids, proteins, DNA, RNA, etc. It's not going to look like us and could well be inedible and smell terrible, but life elsewhere will probably rely on carbon as its main structural element, and water as its solvent, and incorporate nitrogen in its compounds. So where will we find it?

2. "Had We But World Enough and Time . . ."

Andrew Marvell was talking about love when he wrote that line, but it applies as well to life. We need a planet to provide the habitat for life, and we need the global climate on that planet to be reasonably stable for at least 4.6 billion years if we hope for intelligent life. We also need a long-lasting source of energy—thermal, chemical, or best of all, starlight.

These requirements and desiderata translate to the need for an Earth-like planet in a nearly circular orbit at the appropriate distance from a Sun-like star. The deeper requirement is to allow liquid water to exist on the planet's surface, which translates to a range of distances that define an annulus around the star (actually the space between two concentric spheres) known as the star's habitable zone. Stars slowly get hotter with time, so the habitable zone gradually moves outward. Furthermore, the temperature of a planet will also depend on the composition of its atmosphere: how much of a greenhouse effect it can produce. Thus the outer boundary of the habitable zone is not sharply defined.

In our own solar system, Venus (at 0.72 AU) is outside the habitable zone. It is too close to the Sun for water to be stable on its surface. Instead, a runaway greenhouse effect caused any original oceans to boil, and the resulting overheated atmosphere allowed water vapor to reach altitudes where it was easily dissociated by ultraviolet photons from the Sun. The hydrogen escaped, leading to a 150 times enrichment of D/H on the planet, which has been observed (Donahue et al., 1982; de Bergh et al., 1991).

On the other hand, Mars (at 1.5 AU) is still within the zone. The problem for Mars is not that it's too far from the Sun; instead, it is too small to sustain the thick atmosphere that would provide the necessary greenhouse effect to keep it warm. Asteroidal bombardment would remove more than 100 times the present mass of the atmosphere, unhindered by the planet's weak gravitational field (Melosh and Vickery 1989). An Earth-size planet in the orbit of Mars could be habitable.

Even the present Mars may have harbored the origin of life early in its history during an episode when the atmosphere was thicker and liquid water ran across the surface and pooled in its impact craters (Owen, 1997). Just how warm and wet Mars was in that early time and how long the periods of temperate climate lasted are still hotly debated (Squyres and Kasting, 1994; Forget and Pierrehumbert, 1997; Owen and Bar-Nun, 2000). At this stage in our ignorance we can even hold out the hope that life evolved to forms that survived in warm, wet regions underground on Mars, just as life has done on Earth (Onstott et al., 1999). The discovery that liquid water may still erupt from time to time onto the Martian surface (Malin and Edgett, 2000) adds to this hope.

A less likely but still intriguing target is offered by Jupiter's icy satellite Europa. This moon is warmed from the inside by the dissipation of tidal energy from Jupiter, the same engine that drives the astonishing volcanic activity on the inner satellite Io. Here the thought is that beneath the icy crust of Europa there may be a warm ocean of water, and in that ocean there could be life (Gaidos et al., 1999).

3. Finding Life on Brave New Worlds

What about all those other worlds out there of whose existence we now know? The sad fact is that all of these super-giant planets are themselves totally unsuited to be abodes of life. Not only that, most of them have migrated through the habitable zones of their systems to their present positions, thereby wiping out any Earth-like planets that might have existed. The other giants occupy eccentric orbits that will again cause them to wreak havoc on the types of planets we seek, or prevent them from forming in the first place. As of 1 October 2003, we know of only one suitable system. It has a giant planet in a nearly circular orbit at a distance of about 3 AU from its star, with no other giants between it and the star. However, we could only hope to detect such a system in the last year or so, when the accumulated observations would have covered enough of the giant planet's orbit to make identification certain. Thus we may hope for additional discoveries of systems whose basic architecture is like our own in the very near future.

This is reassuring, but we still won't have the certainty that these systems have Earth-size planets in their habitable zones. They have room for such planets, but are the planets there? This information will come from the Kepler and Eddington missions, which will start returning data on transits of terrestrial planets in 2007. The next step will be the use of interferometers in space to separate the light from these planets from the glare of their stars, allowing us to do spectroscopy to see what gases their atmospheres contain.

What gases do we seek? We have seen that plentiful oxygen is a sure sign of life. The blue-green bacteria created the first global abundance of oxygen on Earth some 2.5 billion years ago, and their descendants are still releasing oxygen today. Similar organisms, as well as trees and grasses, could be thriving on Earth-like planets throughout the Galaxy. Other bacteria on our own planet also give themselves away by the gases they produce, as the 2 ppm of methane in our atmosphere testifies. This perspective defines our task: we are looking for gases that would not exist in planetary atmospheres if living organisms were not present to produce them.

How can we be absolutely certain that the discovery of a disequilibrium gas in a planetary atmosphere is really a sign of life? In fact, there are other ways these gases could be generated, but we can eliminate these false positives with proper care.

Consider oxygen. Huge amounts of this gas will be produced by the runaway greenhouse phenomenon that desiccated Venus. This will be a temporary effect, however, as the oxygen liberated by the photolytic destruction of water vapor in a planet's upper atmosphere will soon combine with crustal rocks. By measuring the planet's temperature from its IR spectrum, there would be no doubt about the source of its atmospheric oxygen.

In the case of methane and other reduced gases, the key parameters will be the size of the planet, its distance from its star, and the age of the star. Any inner planet in any system throughout the Galaxy will ineluctably convert to a CO_2-dominated atmosphere if life fails to develop. We see this clearly with Mars and Venus in our own system. So if we find a significant amount of methane in the atmosphere of a distant Earth-sized planet that is in the habitable zone of a star over one billion years old, we can be quite certain that we are looking at an inhabited world. The inhabitants may simply be bacteria, but they would nevertheless demonstrate that the transformation of nonliving to living matter was not a unique event in the Galaxy.

4. The Future

All of these possibilities, and surely others as well, will receive detailed scrutiny and elaboration as we begin to undertake the first direct searches for evidence of life on other worlds like ours. The good news is that we have indeed reached the stage in the development of our own civilization where spacecraft and radio telescopes are taking the place of speculation in our quest to find life elsewhere in the universe. Is living matter a miracle or a commonplace phenomenon? By the year 2020 we should surely know.

5. References

de Bergh, C., Bézard, B., Owen, T., Crisp, D., Maillard, J.-P. and Lutz, B.L. (1991) Deuterium in Venus: Observations from Earth. Science **251**, 547–549.

Donahue, T.M., Hoffman, J.H., Hodges, R.R., Jr. and Watson, A.J. (1982) Venus Was Wet: A Measurement of the Ratio of D/H. Science **216**, 630–633.

Forget, F. and Pierrehumbert, R.T. (1997) Warming Early Mars with CO_2 Clouds that Scatter Infrared Radiation. Science **278**, 1273–1276.

Gaidos, E.J., Nealson, K.H. and Kirshvink, J.L. (1999) Life in Ice-covered Oceans. Science **284**, 1631–1633.

Goldsmith, D. and Owen, T. (2002) *The Search for Life in the Universe*, 3e, University Science Books, Sausalito, California.

Malin, M.C. and Edgett, K.S. (2000) Evidence for Recent Groundwater Seepage and Surface Runoff on Mars. Science **288**, 2330–2335.

Melosh, H.J. and Vickery, A.M. (1989) Impact Erosion of the Primordial Atmosphere of Mars. Nature **338**, 487–489.

Onstott, T.C., Phelps, T.J., Kieft, T., Colwell, F.S., Balkwill, D.L., Fredrickson, J.K. and Brockmann, F.J. (1999) A Global Perspective on the Microbial Abundance and Activity in the Deep Subsurface. In: J. Seckbach (ed.): *Enigmatic Microorganisms and Life in Extreme Environments.* Kluwer Academic Publishers, Dordrecht, The Netherlands, pp. 487–500.

Owen, T. (1980) The Search for Early Forms of Life in Other Planetary Systems: Future Possibilities Afforded by Spectroscopic Techniques. In: M. Papagiannis (ed.) *Strategies for the Search for Life in the Universe.* D. Reidel, Dordrecht, The Netherlands, pp. 177–185.

Owen, T. (1997) Mars: Was There an Ancient Eden? In: C. Cosmovici, S. Bowyer and D. Wertheimer (eds.) *Astronomical and Biochemical Origins and the Search for Life in the Universe.* Editrice Compositori, Bologna, Italy, pp. 203–218.

Owen, T. and Bar-Nun, A. (2000) Volatile Contributions from Icy Planetesimals. In: R. M. Canup and K. Righter (eds.) *Origin of the Earth and Moon.* University of Arizona Press, Tucson, pp. 459–471.

Squyres, S. W. and Kasting, J. F. (1994) Early Mars: How Warm and How Wet? Science **265**, 744–749.

SEARCH FOR BACTERIAL WASTE AS A POSSIBLE SIGNATURE OF LIFE ON EUROPA

ARANYA B. BHATTACHERJEE[1]* and JULIAN CHELA-FLORES[2]
*[1]INFM, Dipartimento di Fisica E. Fermi, Universita di Pisa, Via Buonarroti 2, I-56127, Pisa, Italy, *Permanent Institute: Department of Physics, A.R.S.D College, University of Delhi, Dhaula Kuan, New Delhi-110021, India. and [2]The Abdus Salam International Centre for Theoretical Physics, Trieste, Italy and Instituto de Estudios Avanzados, Caracas 1015A, Venezuela.*

1. Introduction: The Presence of Bacterial Waste and its Consequence

Of particular interest to the scientific community is the possible existence of extraterrestrial biological activity due to the presence of liquid water under the icy surface. This search is motivated by analogy with anaerobic life found in abundance in under sea volcanic vents on Earth (McCollom 1999; Pappalardo *et al.*, 1999) and the dry valley lakes of Antarctica. If Europa does indeed have a liquid water ocean beneath the outer ice crust as a result of interior volcanic heating, then it is possible that hydrothermal vents located on the seafloor may provide the necessary conditions for simple ecosystems to exist. The water ejected from the hydrothermal vents is typically rich in sulfur and other minerals. Bacteria present in the water extract all nutrients directly from the sulfur via chemosynthesis, making sunlight and oxygen unnecessary. Geochemical models have been proposed to explore the possibility that lithoautotropic methanogenesis ($CO_2 + 4H_2 = CH_4 + 2H_2O$) could be a source of metabolically useful chemical energy for the production of biomass at putative Europan hydrothermal systems (McCollom, 1999; Delitsky and Lane, 1997). In the absence of oxygen, anaerobic decomposition takes place in these hydrothermal vents. As a result of putrefactive breakdown of organic material (proteins), some elements are produced, such as hydrogen sulfide, methane, ammonia, and mercaptans, which are thiols/thio alcohols (RS-H, R-paraffinic, aromatic or cyclopraffine group). The sulfur in mercaptans found in bacteria ultimately derives from sulfate ($-SO_4^{2-}$), which is reduced in the cell. In bacteria that utilize sulfate as a source of sulfur, several steps in the reduction process eventually lead to hydrogen sulfide (H_2S) which is a direct precursor of the amino acid cysteine which is a thiol! The original source of sulphur on the Europan surface may be either: (a) ions implanted from the Jovian plasma, or alternatively, (b) much of the sulphurous material may be endogenic. The first possibility (a) has the difficulty that implantation would be expected to produce a more uniform surface distribution (Carlson *et al.*, 1999). On the other hand, sulphurous material on Europa's surface may have been formed internally and over geologic time it could have been emplaced onto the surface either geologically, or as we

J. Seckbach et al. (eds.), Life in the Universe, 257–260.

argue in this paper by the accumulated effect of biogenic processing over geologic time. Interestingly, evidence has been provided for the presence of mercaptans on the surface of Europa (McCord *et al.*, 1998), using reflectance spectra returned by the Galileo near infrared mapping spectrometer (NIMS) experiment. They found absorption in the 3.88 μm, attributed to S-H bond of mercaptans. A major scientific question to be answered about the possible existence of life is: If biological process such as methanogenesis and putrefaction are at work then how do they affect observable or measurable quantities? What would be the best way to detect these organisms? In order to answer such questions, lander missions would be needed since it seems unlikely that remote sensing techniques such as NIMS alone would be sufficient. An ideal approach to detect the presence of life would be to drop penetrating probes and an *in situ* vehicle, which would carry a chemical/physical laboratory (CPL). The *in situ* CPL will perform chemical and physical analysis of ice and ocean water to obtain information on the chemical constituents of the ice. Instruments required for these analyses would include UV-spectrometer, high frequency ultrasonic analyzers, chemical analyzer to look directly for bacterial metabolic wastes or indirect effect of these waste products on the icy surface or the water beneath. Interestingly, the use of fermentative products (organics and carbon dioxide) as metabolic signatures on Europa has been suggested (Prieur, 2002).

2. What Kind of Solute-Solvent Interactions We Should Expect on Europa's Surface?

The presence of various bacterial excreta is expected to perturb the normal water lattice as $(H_2O)_{liquid}$ _ $(H_2O)_{quasi\text{-}lattice}$. Release of these excreta over a prolonged period of time would certainly change the various physical properties of the icy layer and that of the water that lies beneath. The perturbations would be reflected in the various physical properties measured such as the viscosity, ultrasonic velocity, ultrasonic absorption coefficient, UV absorption spectra etc. UV absorption spectra are quite ideal for monitoring the formation/breaking of hydrogen bonds. An enthalpy change of 5–7 kcal/mole is expected during the formation/breaking of hydrogen bonds. The most likely hydrogen bond that may be formed is between the hydrogen of iodole chromophore and water molecule. The iodole chromophore may act as a proton donor in the hydrogen bonding. Low molecular weight and high functionality of water plays a significant role in biological processes. Water can be assumed to be a two-state system (Pethrick, 1982). Each molecule is joined to its nearest neighbors with four links. On the surface of these clusters exist molecules joined by less than the maximum number and may be considered to be in the process of either forming or breaking the cluster. The total can be divided into two regions, an open low density arrangement where the molecules are extensively hydrogen bonded and a high density region where the molecules are non-bonded. The equilibrium may be described as: $(H_2O)_{liquid}$ (bound)_n (H_2O) (free monomeric). Solute molecules (bacterial waste) which have the ability to form hydrogen bonds will attack the low density region and may act as acceptors and compete with the protons for the lone pair of electrons. The net result of such interaction is the disruption of the low density region and the equilibrium shifts towards the non-bonded form with a consequent increase in the density. Such a transition will change the freezing point of water and can be measured using high frequency ultrasonic transducers. Increase in solute concentration (bacterial waste) over a prolonged period of time is expected to produce an enthalpy change of 1–2 kcal/mole. This will be reflected as a red shift in some standard UV spectral lines. This happens because of the "Franck-Condon's "strain and the "Polarization shift" (Chaudhury *et al.*, 1994). The importance of experiments to detect bacterial wastes including mercaptans, that NIMS has demonstrated to be present on the Europan surface must be realized because non-living sources of mercaptans are difficult to anticipate.

3. Discussion and Conclusions

The detection and characterization of any bacterial excreta is an integral part of our search for evidence of life on Europa. In order to achieve this goal, it will be necessary to embark on an extensive survey of the surface of Europa using lander missions. Clearly an ideal approach to obtaining information on the composition of the European "ocean" is to sample the surface material which is thought to have originated at depth and become distributed on the surface by eruption like processes.

Any point on the surface of Europa which exhibits properties typical of contamination by bacterial waste should serve as a window to the underlying ocean. Before embarking upon any ambitious project, it would be helpful if we could test this proposal in the laboratory by mimicking the surface of Europa. This could be done by allowing for bacteria to grow for a prolonged period in an artificial pond and then freezing the water. Subsequently one can measure the various physical and chemical properties on the surface of this pond. This will help to understand what we should expect on the surface of Europa. However, in this context we should point out that such an experiment is probably mimicked by nature in the dry valley lakes of southern Victoria Land of Antarctica (Doran *et al.*, 1994; Parker *et al.*, 1982). These lakes lie in an ice-free area of just under 5000 km^2. This area contains more than 20 permanent lakes, which are warm environments containing liquid water that may reach room temperature under their iced surface of 4–6 m, while the outside temperature remains well below zero degrees Celsius. In the dry valley lakes there are abundant microorganisms underneath their iced surface. The estimated annual S removal is 104 kg in Lake Chad [cf., Table 5 in (Parker *et al.*, 1982)]. At any given time an important characteristic of the anaerobic prostrate cyanobacterial mats of, for instance Lakes Fryxell and Hoare, is their appearance as black mats, coarse and with a distinct H_2S odor (Doran *et al.*, 1994). Thus, at any given moment endogenic sulphur and other elements are to be found on the iced surface of, for instance, Lake Chad. To the best of our knowledge, there is no information on the presence of mercaptans in the dry valley lakes of Antarctica. Further investigations are required to search for mercaptans (or other bacterial wastes) in these dry valley lakes. There is, therefore, a suggestive analogy between the chemical composition of the surface of the dry valley lakes with the presence of sulfur compounds (such as mercaptans) on the iced surface of Europa.

4. Acknowledgements

A. Bhattacherjee acknowledges support by the Abdus Salam International Centre for Theoretical Physics, Trieste, Italy under the ICTP-TRIL fellowship scheme.

5. References

Carlson, R. W., Johnson, R. E. and Anderson, M. S. (1999) Sulfuric acid on Europa and the radiolytic sulfur cycle, *Science*, Vol. **286**, pp. 97–99.

Chaudhury, K., Bhattacherjee, A., Bajaj, M.M. and Jain, D.C. (1994) An analysis of 5-(p-Hydroxyphenyl)-5-phenylhydantoin induced perturbations in the 200–400 nm region, *Revue Roumaine de Chimie*, Vol. **39**, pp. 1091–1098.

Chela-Flores, J. (1998a) Europa: A potential source of parallel evolution for microorganisms, In: *Instruments, Methods and Missions for Astrobiology.* The International Society for Optical Engineering, Bellingham, Washington USA (R.B. Hoover, ed.), *Proc. SPIE*, Vol. **3441**, pp. 55–66; http://www.ictp.trieste.it/~chelaf/ss4.html

Chela-Flores, J. (1998b) A search for extraterrestrial eukaryotes, Physical and Biochemical Aspects of Exobiology, O*rigins Life Evol. Biosphere*, Vol. **28**, 583–596; http://www.ictp.trieste.it/~chelaf/searching_for_extraterr.html

Chela-Flores, J. (2003). Testing Evolutionary Convergence on Europa. International Journal of Astrobiology (Cambridge University Press), in press. http://www.ictp.trieste.it/~chelaf/ss13.html

Delitsky Mona L. and Lane, Arthur L. (1997) Chemical schemes for surface modification of icy satellites: A road map, *Jour. Geochem. Res.*, Vol. **102**, No. E7, pp. 16, 385.

Delitsky, L. and Lane, A. L. (1998) Ice chemistry in Galilean satellites *Jour. Geophys. Res.* Vol. **103**, pp. 31,391–31,403.

Doran, P.T., Wharton, Jr., R.A. and Berry, Lyons, W. (1994) Paleolimnology of the McMurdo Dry Valleys, Antarctica , *J. Paleolimnology* **10**, pp. 85–114.

Greenberg, R. (2002), Tides and the biosphere of Europa, *American Scientist*, Vol. **90**, pp. 48–55.

Horvath, J. Carsey, F., Cutts, J. Jones, J. Johnson, E. Landry, B., Lane, L., Lynch, G., Chela-Flores, J., Jeng, T-W. and Bradley, A. (1997) Searching for ice and ocean biogenic activity on Europa and Earth, Instruments, Methods and Missions for Investigation of Extraterrestrial Microorganisms, (R.B. Hoover, ed.), *SPIE*, Vol. **3111**, pp. 490–500; http://www.ictp.trieste.it/~chelaf/searching_for_ice.html

McCollom, T. M. (1999), Methanogenesis as a potential source of chemical energy for primary biomass production by autotrophic organisms in hydrothermal systems on Europa, *Jour. Geochem. Res.* Vol. **104**, No. E12, pp. 30,729–30,742.

McCord, T.B., Hansen, G.B., Clark, R.N., Martin, P.D., Hibbitts, C.A., Fanale, F.P., Granahan, J.C., Segura, NM., Matson, D.L., Johnson, T.V., Carlson, R.W., Smythe, W.D., Danielson, G.E., and the NIMS Team (1998) Non-water-ice constituents in the surface material of the icy Galilean satellites from the Galileo near-infrared mapping spectrometer investigation, *Jour. Geophys. Res.* Vol. **103**, No. E4, pp. 8603–8626.

Pappalardo, Robert T. James W. Head and Ronald Greeley (1999) The hidden ocean of Europa, *Scientific American*, pp. 34–43.

Parker, B.C., Simmons, Jr., G.M., Wharton, Jr., R.A. Seaburg, K.G. and Love, F. Gordon (1982) Removal of organic and inorganic matter from Antarctic lakes by aerial escape of bluegreen algal mats, *J. Phycol.* Vol. **18**, pp. 72–78.

Prieur, D. (2002) Life detection on Europa: Metabolic signatures, Europa Focus Group Workshop 3, Arizona, USA, p. 41.

Pethrick, R.A. (1982) Molecular Interactions Vol. 3, In: H. Ratajczak and W.J. Orville (eds.) Thomas, John Wiley and Sons, New York.

SULFATE VOLUMES AND THE FITNESS OF SUPCRT92 FOR CALCULATING DEEP OCEAN CHEMISTRY
The Situation in Europa's Ocean

STEVEN VANCE[1], EVERETT SHOCK[2], and TILMAN SPOHN[3]
[1]Box 351310, University of Washington, Seattle, WA 98195, USA, [2]Arizona State University Department of Chemistry, AZ, USA and [3]Westfälische Wilhelms [3]Universität, Institut für Planetologie, Münster, Germany

1. Introduction

Jupiter's three innermost moons experience tidal forcing due to the 4:2:1 orbital resonance they share. In Io, the closest, the energy dissipated by flexure of the moon's mantle is enough to make it the most volcanically active body in the solar system. In Europa, dissipation may be vary from a tenth to only a hundredth that in Io, but in theory this is still sufficient to melt much of the 170 km of ice (Anderson et al 1997) covering its surface and provide a moderate heat source (Hussman 2003). In fact, surface geology and planetological properties measured by the Galileo probe support the idea that Europa has an ocean perhaps as deep as 160 km, with a primary salt composition of Mg and Na sulfates (Greenberg et al 2000). With the aim of including pressure effects in future simulations of composition and dynamics in the ocean, we find predictions of supcrt92 software consistent with experimentally obtained molar volumes around 50°C and up to 2000 atm, roughly the pressure at the base of Europa's ocean. Understanding hydrothermal activity in Europa's crust, in the past and possibly in the present, is crucial to understanding dynamics and chemistry of the ocean as a whole, and for evaluting any biological activity that may have occured during the moon's history. In this paper we explore the limits of theoretical exploration of Europa's ocean, highlighting areas for further research.

2. Evidence for a Europan Ocean

Voyager in the 1970s and Galileo in the 1990s both scanned the surface of Europa, revealing an icy moon marked by a scarcity of craters and an array of cracking features. These suggest a young surface—less than 50 million years old—altered by continuous tidal flexing (Greenberg et al 2000). Gravity data from Galileo set the H_2O thickness in the range of 100–200 km (Anderson et al 1997). The possibility that this is consistent with a convecting ice subsurface is eliminated, at least in part, by the magnetic field signature. Induced by Jupiter's field, Europa's field appears to origininate near its surface, within the H_2O covering, for which the most likely explanation is some amount of ion-containing liquid (Kievelson

J. Seckbach et al. (eds.), Life in the Universe, 261–264.

1999). The ice covering's thickness is further limited by arcuate cracking features, which can be explained by a changing direction of crustal stress only if the ice is no more than a few kilometers thick (Hoppa 2001). The ice is probably at least a few km thick however, since moderately sized craters exhibit a central upwelling feature, excavated from solid material under the impact site (Turtle and Pierazzo 2001). Along surface cracks, and also in so-called chaos regions where melt-through events appear to have occured, a non-icy component is visible. This is thought to have resulted from deposition of dissolved constituents during sublimation of liquid. Surface near-infrared spectra for the non-icy component are consistent with sodium and magnesium sulfate, and carbonate salts, an indicator of ocean composition (McCord et al 2000).

3. Compositional Modeling

Some researchers have tried to constrain the composition of Europa's ocean based on surface data (Kargel et al 2000, Spaun and Head 2001, Zolotov and Shock 2001). Assuming the moon formed from representative chondrite meteorites—those representative of bulk composition of Jupiter's region in the early solar nebula—the authors take all volatile content as the proto-ocean and allow this to proceed to chemical equilibrium, precipitating brine salt as it does so. These qualitative models can hint at pH, nutrient flux, and geochemistry in Europa's ocean, but are not accurate enough to factor into dynamical calculations. A large problem, one that affects many domains of chemical research, is the lack of pressure data for aqueous solutions. Chemical models fit volumentric and other behavior to known values measured in the laboratory. Where these are not available the authors must extrapolate. Because pressure effects are secondary in significance to temperature effects, and also more difficult to obtain, they have been neglected as subjects for research. The physics of pressurized solutions may be crucial to the chemistry of Europa's ocean, however, as pressures in the base of Europa's ocean are much higher than those at average depth in Earth's ocean (2000 vs 500 atm). The situation is changing as new techniques for finding equations of state are developed (Abramson et al 1999). As will be shown in a future paper, we find supcrt92 (Shock et al 1992) accurate to 5% up to 2000 atm for Na_2SO_4 at 325–375 K. For much of Europa's ocean a model suited to lower temperatures is more appropriate. For this reason, Kargel et al (2000) and Spaun and Head (2001) used the FREZCHEM package (Mironenko 1997). Though FREZCHEM contains no parameterization for pressure, its extrapolated databases have proven accurate for applications at atmospheric pressure (1 atm) (Marion 1999). The author of FREZCHEM is currently working on a pressure-corrected version (personal communication). Supcrt may still be appropriate for modelling chemistry of Europa's ocean if one considers the possibility of hydrothermal activity in the crust.

4. Hydrothermal Systems

If the ice covering Europa's ocean is very thin, tidal dissipation will occur almost exclusively in the mantle. One can envisage a situation in which the initially higher dissipation from greater orbital eccentricity leads to a thin ice shell. The situation could then be maintained if the crust is sufficiently plastic and if enough tidal dissipation occurs throughout Europa's history. In any case, seafloor spreading is not expected to occur on Europa. Indeed, the absence of an intrinsic magnetic field suggests core convection is not occuring there, so mantle convection is not driven by the same mechanism as it is in Earth's mantle. In the absence of as active a mantle, seafloor volcanism may still occur on Europa—and probably did occur in its early history—due to the exothermic conversion of peridotite rock to serpentine. On Earth, the Lost City hydrothermal system is estimated to be 34,000 years, old based on radiocarbon dating (Fruh-Green et al 2003). Estimates of heat provided by serpentinization are consistent with this age, making for the feasible existence of hydrothermal systems supported by this mechanism alone. The depth of flow through porous rock under the system remains to be determined, leaving undetermined parameters in the understanding of chemically driven hydrothermal systems. In the crust of Europa's ocean it may be that water is able to penetrate more deeply owing to the lesser gradient in pressure, so that a greater reservoir for chemical heat generation is available. If porous flow is localized to the azimuth of the Europan body facing Jupiter, slow rotation of the mantle body may have sustained localized Lost City analogues for a significant portion of the moon's history, possibly to the present. The specifics of serpentinization in Europa's crust will be discussed in a future paper.

5. Prospects for Life

Some estimates of "biopotential" have been made for Mars and Europa, planets of interest to astrobiologists (Jakosky and Shock 1998, Chyba 2000, Zolotov and Shock 2003). These have always looked at total energy input to estimate the overall planetary productivity that could be supported, with the apparent intention of emphasizing that planetary explorers should not expect to find abundant or complex life under the Europan ice shell. Gaidos et al (2002), Jakosky et al (1998), and Shock et al (2003) note that only the less efficient metabolic pathways of methanogenesis and sulfur reduction are available in the absence of oxygen. Chyba (2000) points to the interaction of Jupiter's strong radiation field with impurities in Europa's ice as an additional source of energy for biota, but he still seems to agree with other authors that only microbial life can exist in the ocean. However, the global input of energy is less important if much of that energy is localized to support a hydrothermal community, and if the center of hydrothermal activity moves slowly enough for the hosted community to migrate with it. As described in the previous section, chemical heat is a more likely source for Europan hydrothermal energy, but even this may be sparse if fresh peridotite is not periodically exposed to Mg-containing water. Doubtless, Europan hydrothermal systems resemble colder systems on Earth—like those under Antarctic ice—more than the extensive, hot systems at mid-ocean ridges. Unfortunately, little is known about such systems, since their exploration even on Earth has begun only in the last decade (Dählmann et al 2001).

6. Summary

We review evidence for an ocean on Europa, pointing out where chemical and physical data are lacking for simulation and comparison with Earth-systems. Comparing experimental volumes for Na_2SO_4 with those predicted by supcrt92, we find supcrt92 agrees within 10% around 50°C, up to a pressure of 2000 atm. This is sufficient for future modeling of dynamics and composition of Europa's ocean, and to begin examining the role of serpentinization in ocean processes.

7. References

Abramson, E.H., Brown, J.M. and Slutsky, L.J. (1999) Applications of Impulsive Stimulated Scattering in the Earth and Planetary Sciences. Annu. Rev. Phys. Chem. **50**, 279–313.

Anderson, J.D., Lau, E.L., Sjogren, W.L., Schubert, G. and Moore, W.B. (1997) Europa's Differentiated Internal Structure: Inferences from Two Galileo Encounters. Science, **276**, 1236–1239.

Chyba, C.F. (2000) Energy for Microbial Life on Europa. Nature, **403**, 381–382.

Dählmann, A., Wallmann, K., Sahling, H., Sarthou, G., Bohrmann, G., Petersen, S., Chin, C.S. and Klinkhammer, G.P. (2001) Hot Vents in an Ice-cold Ocean: Indications for Phase Separation At the Southernmost Area of Hydrothermal Activity, Bransfield Strait, Antarctica. Earth Plan. Lett. **193**, 381–394.

Früh-Green, G. L., Kelley, D.S., Bernasconi, S. M., Karson, J.A., Ludwig, K.A., Butterfield, D.A., Boschi, C. and Proskurowski, G. (2003) 30,000 Years of Hydrothermal Activity at the Lost City Vent Field. Science, **301**, 495–498.

Gaidos, E.J., Nealson, K.H., and Kirschvink, J.L. (1999) Life in Ice-Covered Oceans. Science, **284**, 1631–1633.

Hussman, H. (2003) Europa's Ocean and the Orbital Evolution of the Galilean Satellites. Dissertation Thesis, Institut für Planetologie, University of Münster.

Jakocsky, B.M., and Shock, E.L. (1998) The Biological Potential of Mars, the Early Earth, and Europa. J. Geophys. Res., **103**, 19359–19364.

Kargel, J.S., Kaye, J.Z., Head, J.W., Marion, G.M., Sassen, R., Crowley, J.K., Ballesteros, O.P., Grant, S.A., and Hogenboom, D. (2000) Europa's Crust and Ocean: Origin, Composition, and the Prospects for Life. Icarus **148**, 226–265.

Lowell, P. and Rona, P.A. (2002) Seafloor Hydrothermal Systems Driven by the Serpentinization of Peridotite. Geophysical Research Letters **29**, (11), 10.1029/2001GL014411.

Marion, G.M., Farren, R.E., and Komrowski, A.J. (1999) Alternative Pathways to Seawater Freezing. Cold Regions Ci. Technol., **29**, 259–266.

Mironenko, M.V., Grant, S.A., G.M. Marion, and Farren, R.E. (1997) FREZCHEM2: A Chemical Thermodynamic Model for Electrolyte Solutions at Subzero Temperatures, *CRREL Rep 97–5*, USA Cold Regions Res. And Eng. Lab., Hanover, N.H.

McCord, T.B., Hansen, G.B., Fanale, F.P., Carlson, R.W., Matson, D.L., Johnson, T.V., Smythe W.D., Crowley, J.K., Martin, P.D., Ocampo, A., Hibbits, C.A., Granahan, J.C., and the NIMS Team (1998). Salts on Europa's Surface Detected by Galileo's Near Infrared Mapping Spectrometer. Science **280**, 1242–1245.

Sohl, F., Spohn, T. Breuer, D., and Nagel, K. (2002) Implications from Galileo Observations on the Interior Structure and Chemistry of the Galilean Satellites. Icarus, **157**, 104–119.

Spaun, N.A. and Head, J.W. III (2001) A Model of Europa's Crustal Structure: Recent Galileo Results and Implications for an Ocean. J. Geophys. Res. **106**, 7,567–7,576, 2001.

Turtle, E.P. and Pierazzo, E. (2001) Thickness of a Europan Ice Shell from Impact Crater Simulations. Science **294**, 1326–1328.

Zolotov, M.Y. and Shock, E.L. (2001) Evidence for a Weakly Stratified Europan Ocean Sustained by Seafloor Heat Flux. J. Geophys. Res. **106**, 32815–32827.

Zolotov, M.Y. and Shock, E.L. (2003) Energy for Biologic Sulfate Reduction in a Hydrothermally Formed Ocean on Europa. J. Geophys Res, **108**, (E4), 5022, doi:10.1029/2002JE001966.

THE CASE FOR LIFE EXISTING OUTSIDE OF OUR BIOSPHERE

Techniques for identifying molecular structures

RICCARDO SIDNEY GATTA
The International Center for Genetic Engineering and Biotechnology
New Delhi, India

> There is no fundamental difference between a living organism and lifeless matter. The complex combination of manifestations and properties so characteristic of life must have arisen in the process of the evolution of matter.
>
> *A.I. Oparin*

In order to identify life outside of a terrestrial paradigm we examine the Jovian moon, Europa as a potential source for biological material. The selection of biological targets is based on an investigation of convergent evolution as a universal precept in eukaryote development (Chela-Flores, 1998). Analytical techniques have benefited from multi-disciplinary efforts that have led to the creation of the first functional microscopic-scale laboratories that can perform a concerted series of "hyphenated" functions. Devices contained within a specialised Europan lander (Naganuma and Uematsu, 1998) will have the ability to gather samples, screen for promising biosignatures, and present sufficient data to accurately determine the characteristics of the samples. Application of these "labs-on-a-chip" (Wang *et al.*, 2001) holds great promise in the search for life out of our own biosphere. The question of whether life exists elsewhere in the universe needs to be pertinent to the novel search paradigms afforded by missions to other potential planets. Data inferring presence of liquid water under ice cover on the Jovian moon Europa has made it a candidate for being a possible source of extra-terrestrial life (Reynolds *et al.*, 1987). In the event of a rare opportunity for the direct sampling of the Europan surface and sub-surface (Chyba and Philips, 2002), we should examine whether life could be present in some form that has not evolved on Earth. Target and search criteria should be addressed as to whether evolution is a universal prerogative, or simply a terrestrial artifact. A search based on assays suitable for determining whether convergent evolution is a universally valid paradigm should also be open to the possibility of identifying higher organisms (Chela-Flores, 1998). Evolutionary classification of life gained a solid basis when Carl Woese propounded a novel method (Woese, 1987), using contemporary technology, that identifies those characteristics of organisms that could indicate evolutionary progress. It is necessary to identify aspects that are conserved enough in order to identify, yet are variable enough to allow an evolutionary development to be observed. This would imply an "earth-centric" approach (Conrad and Nealson, 2001) to the search for life in the universe and presupposes evolutionary criteria to be universally valid. The model of convergent evolution raises pertinent questions (Doolittle, 1994: Sette *et al.*, 2003), as different organisms often develop similar traits, that can be explained in

J. Seckbach et al. (eds.), Life in the Universe, 265–267.

a variety of manners (Dauplais *et al.*, 1997). In order to evidence the level of similarity, a common method is to create a battery of analytical probes that are known to conjugate with a previously identified analyte, and that can signal the level of interaction in a quantifiable manner. Interactions that have been evaluated in the known biosphere can be extrapolated to an unknown, but believed to be isolated, biosphere (Caetano-Anolles, 2003). In the case of a non earth-centric search for life, the method for evaluation must be de-coupled from what is considered to be life in our biosphere or in any habitability zone (Sagan, 1964). Life as an unknown target can still be identified and quantified according to consideration of its inextricable properties (Cleland and Chyba, 2002). An inherently earth-centric point of view of any earth-bound observer can be deconstructed in order to identify basic characteristics necessary for alternative forms of life to exist. Basic elements that can present emergent properties, such as topology and chirality (Bonner, 1995) could be identified and interpreted, though it may be some time before we obtain the knowledge necessary to prove the biogenicity of a purported biogenic signature, or at least that it has not been formed abiotically. Observable energy disequilibria may be an indication of the existence of life, suggesting a form of energy transduction such as in metabolism/catabolism (Bhattacherjee and Chela-Flores, 2003), or for replication at either molecular or cellular levels. In order to obtain sufficient data, micro-scale (chip based) devices permit most automated multistep assays (Anderson *et al.*, 2000) derived from bench-top systems with advantages of speed, cost, portability, and reduced energy/solvent consumption. Studies of mechanosynthesis of molecular machine systems will enable the development and production of a wide range of micro-components (Drexler, 1994). Attention must be paid to any influence that the analytical techniques may have on the sample under analysis (Fukushi *et al.*, 2003), as positive results in the search for life in the universe will surely have far-reaching effects on life and society in our own biosphere.

References

Anderson RC, Su X, Bogdan GJ and Fenton J (2000). A miniature integrated device for automated multistep genetic assays. Nuc. Acids Res. 28(12):e60.

Bhattacherjee AB and Chela-Flores J (2004). Search for bacterial waste as a possible signature of life on Europa, in this volume.

Bonner WA (1995). Chirality and life. Orig Life Evol Biosph. 25(1–3):175–90.

Caetano-Anolles G and Caetano-Anolles D (2003). An evolutionarily structured universe of protein architecture. Genome Res. 13(7):1563–71.

Chela-Flores J (1998). A search for extraterrestrial eukaryotes: physical and paleontological aspects. Orig Life Evol Biosph. 28(4–6):583–96.

Chyba CF and Phillips CB (2002). Europa as an abode of life. Orig Life Evol Biosph. 32(1):47–68.

Cleland CE and Chyba CF (2002). Defining 'life'. Orig Life Evol Biosph. 32(4):387–93.

Conrad PG and Nealson KH (2001). A Non-Earthcentric Approach to Life Detection. Astrobio. 1(1):15–24.

Dauplais M *et al.*, (1997). On the Convergent Evolution of Animal Toxins. JBC. 272(7):4302–4309.

Doolittle RF (1994). Convergent evolution: the need to be explicit. Trends Biochem Sci. 19(1):15–8.

Drexler KE (1994). Molecular nanomachines. Annu Rev Biophys Biomol Struct. 23:377–405.

Fukushi D *et al.* (2003). Scanning Near-field Optical/Atomic Force Microscopy detection of fluorescence in situ hybridization signals beyond the optical limit. Exp Cell Res. 289(2):237–44.

Naganuma T and Uematsu H (1998). Dive Europa: a search-for-life initiative. Biol.Sci.Space.12(2):126–30.

Oparin AI (1968). Life, its nature, origin, and evolution. Moscow: Nauka. 173pp.

Reynolds RT, McKay CP and Kasting JF (1987). Europa, tidally heated oceans, and habitable zones around giant planets. Adv Space Res. 7(5):125–32.

Sagan C (1964). Exobiology: a critical review. Life Sci Space Res. 2:35–53.
Sette A *et al.* (2003). Class I molecules with similar peptide-binding specificities are the result of both common ancestry and convergent evolution. Immunogenetics. 54(12):830–41.
Wang J, Ibanez A, Chatrathi MP, Escarpa A (2001). Electrochemical enzyme immunoassays on microchip platforms. Anal Chem. 73(21):5323–7.
Woese CR (1987). Bacterial evolution. Microbiol Rev. 51(2):221–71.

APPLICATION OF MOLECULAR BIOLOGY TECHNIQUES TO ASTROBIOLOGY

RICCARDO SIDNEY GATTA[1] and JULIAN CHELA-FLORES[2]
[1]The International Centre for Genetic Engineering and Biotechnology, New Delhi, India, and [2]The Abdus ICTP Strada Costiera 11; 34136 Trieste, Italy, and IDEA, Caracas, Venezuela

Abstract. The opportunity for direct examination of the Europan surface and sub-surface calls for a systematic and deductive approach to experimental design. To avoid the limitations of our inherent Earth-centric definition of life (Nealson *et al.*, 2002), we would be forced to examine a wide range of potential bio-signatures to guide more specific biological experiments (Chela-Flores, 2003). It is also important to look for recurring features that are important from the evolutionary history of our own biosphere (Zakon, 2003). Of the many candidate molecules, the structurally heterologous superfamily of voltage-gated cation channels is an evolutionary sensitive group of molecular structures, the single varieties of which can be easily distinguished. Implementation of the analytical aspects of this experiment would require remote control of miniaturized robotic systems. These mechanisms are under constant evolution since their uses are strongly tied to commercial, scientific and military interests. One paradigm for feasibility studies could come from data inferring the reprocessing of ice covering a Europan ocean. Reprocessing could be inducing life forms extant in the liquid water subsurface towards the ice covering, as it has already been demonstrated at the frozen surfaces of Antarctic lakes (Bhattacherjee and Chela-Flores, 2004), and as it has been suggested by geophysical analysis of the Galileo images of the icy surface of Europa (Greenberg *et al.*, 2002). The proposed series of experiments can be carried out *in situ* either within a submersible in the ocean beneath the ice layer (carried and launched from a cryobot), or even on the surface ice itself. Results from a preliminary examination of the environment would be used to determine the conditions necessary for sampling and pre-processing of any material of possible biological origin. Many techniques are currently available for identifying targets according to their molecular structure and their chemical-physical characteristics:

- novel sampling and isolation methods,
- specific antibodies or diabodies engineered as molecular probes,
- micro-arrays based on site-specific immobilization of complementary molecules,
- microscopy and micro-sensors for visualization/digital sampling of positive results.

New challenges will arise from the novel settings and will have to be addressed, singly, well in advance of any preliminary exercises. Moreover, a myriad of practical applications could be developed by addressing pertinent, emerging questions relating to:

- stability of sensitive organic material over a large period of time, and extreme conditions,

J. Seckbach et al. (eds.), Life in the Universe, 269–273.

- maintenance of biological activity within silica-, or hydro-micropatterned biogels,
- multiplexing in a microfluidic (lab-on-a-chip) environment,
- miniaturization of analytical devices such as microscopes and their power sources.

1. Introduction

Identification of biomolecules is a common process that is approached in a concise and systematic manner. When applied to the search for life in the universe many potential targets and recognition techniques can be envisioned. These can be refined by seeking with evolutionary criteria in an environment where water is available (DesMarais *et al.*, 2002). The problem of rational target selection is to single out known bio-signatures and evaluate their placement on an evolutionary timescale. Selection of a target based on the central dogma of molecular biology (Crick, 1958), which states that the flow of genetic information is from DNA, through RNA, to protein, is open to the idea that any of these key molecules could be, or could have once been (Woese, 2001), sources of hereditary data in some place in the universe (Crick, 1966 and 1970). The examination of samples of unknown composition yields results that are tied not only to the nature of the material under scrutiny but also to the conditions under which they are examined (Vreeland *et al.*, 2002). The delicate pleiotropic nature of protein-protein interactions, for instance, can be permanently affected by minimal variations that would alter important conformational interactions. There are methods for the successful recovery of biological material from problematical sources (Vreeland and Rosenzweig, 2002) since insufficient levels of sterility, contamination from other sources, or less than optimal reaction conditions always lead to unreliable results (Nicastro *et al.*, 2002).

2. Molecular Techniques

Structural information for the identification of life could be sought in membrane composition (channels, peptidoglycans, lipids, chirality) or in the form of genetic information (DNA, RNA . . .). The lack of available information about non-terrestrial macromolecules, however, makes it difficult to seek life through molecular probing of these components, though all can be analysed with specific assays: sugars (DeAngelis, 2002), proteins (Zakon, 2002) and regulatory machinery of gene expression (Conant and Wagner, 2003). Molecular subtyping methods can seek differences in control of fatty acid (Tornabene *et al.*, 1980), protein or nucleic acid (Woese and Fox, 1977) biosynthesis. Many evolutionarily conserved biomolecules could serve the purpose of attempting to ground a universal tree of life. Families of proteins where there are conserved structural elements, or domain-specific features (Marck and Grosjean, 2002), are thought to lead to ancient origins or even to a last universal common ancestor (LUCA) (Ouzounis and Kyrpides, 1996). The link between structure and function is apparent in the conservation throughout evolution of families of proteins that perform essential tasks (Ruta *et al.*, 2003). Ion channels belong to a large family of related genes that regulate vital functions. The simplest channels are found in all kingdoms of life. Ion channels consist of assemblies of subunit components (Hille, 2001) and thus share aspects of their membrane topologies (Miller, 2000). However, diverse ion specificities (Jeziorski *et al.*, 2000) and methods of functional regulation (Doyle *et al.*, 1998) make ion channels ideal targets for probes seeking differentiation by evolutionary criteria (Harte and Ouzounis, 2002) with, for example, semisynthetic libraries (Braunagal, 2003) that lend themselves to rational design and chemical synthesis. In order to assure the correct interpretation of image data obtained *in loco*, it is important to sample data at a sufficiently high resolution so that any further enhancement does not alter the acquired data. Light microscopy can utilize illumination sources that can be easily varied, such as with filtered short wavelength radiation that can cause fluorescent substances to produce emission spectra (Kain *et al.*, 1995). Green fluorescent protein (GPF) has been used as a specific reporter gene for ion channel expression (Marshall *et al.*, 1995). Confocal (laser scanning) microscopy (CLSM) offers a further advantage of being able to increase spatial resolution. Multi-photon microscopy uses short pulses of low energy, infra-red, light to excite a restricted cross-section within the sample without the need for confocal apertures. This increases the photostability of fluorescent molecules (Geddes *et al.*, 2003), though a recently developed alternative to organic molecules for immunocytochemical imaging is quantum dot technology. The higher quantum yield and stability of quantum dots could solve the major problem associated with a large amount of parallel assays and with the long latency period between assay set-up and performance (Tokumasu and Dvorak, 2003). They can be combined into highly specific bioconjugates for studying genes or proteins in applications that are not envisioned with traditional organic dyes and fluorescent proteins (Medintz *et al.*, 2003). Nuclear magnetic resonance (NMR) can be used to detect 3 dimensional (3D) placements of unmarked individual atoms (Doughtery *et al.*, 2000), even in extremely low magnetic fields (McDermott *et al.*, 2002). Analytical systems such as scanning probe microscopes already enable direct visualization and manipulation of individual macromolecules (Malayan and Balachandran, 2001): a key interest in process miniaturisation (PIM) in the field of molecular diagnostics (Brush, 1999). A new generation of scientific instruments is in development that can be adapted to the context of planetary exploration. Novel power sources (Fennimore *et al.*, 2003) and remote robotic control would guarantee the completion of a mission even in unforeseen circumstances. Microfabricated devices have already been adapted for transcript expression profiles of genes related to ion channels, with the ability to identify changes down to channel subunit level (LeBouter *et al.*, 2003). Biochips are used that permit electrophoretic separations and highly specialized applications of molecular biology. Though several different matrices and protocols are available for microarrays, storage periods remain very limited (Angenendt *et al.*, 2002). Biomimetic systems apply novel production methods and materials for creating surface moieties similar to those using proteins or nucleic acids that are made by bio-systems. The transport of fluids through nanoscopic conduits (Drexler, 1994) will allow single molecules of DNA to be analysed. It has been a long established goal to guide complex sequences of actions in simple nanoscale systems in order to create more intricate patterns (Deamer and Branton, 2002). Alternative biogels are being derived from a marine sponge, *Tethya aurantia*, that produces a protein group call silicateins responsible for biosilicate formation under benign conditions (Cha *et al.*, 1999). Molecular-scale channels are essentially entirely interfaced with no bulk fluid; thus, a complete understanding and control of interfacial chemistry on the nanometer scale to obtain a stable microfluidic network cannot be underestimated (Kim *et al.*, 2001). The challenge of finding life in the universe is driving research to develop more efficient scientific instruments that will not fail to benefit applications in every field, and in every biosphere that may be found to exist.

3. References

Angenendt P, Glokler J, Murphy D, Lehrach H and Cahill DJ (2002). Toward optimized antibody microarrays: a comparison of current microarray support materials. Anal Biochem. 309(2):253–60.

Bhattacherjee AB and Chela-Flores J (2004). Search for bacterial waste as a possible signature of life on Europa, in this volume.

Braunagel M (2003). Construction of semisynthetic antibody libraries. Methods Mol Biol. 207:123–32.

Brush M (1999). Automated Laboratories. The Scientist. 13(4):22.

Cha JN, *et al.* (1999). Silicatein filaments and subunits from a marine sponge direct the polymerization of silica and silicones in vitro. Proc.Natl.Acad.Sci.USA. 96(2):361–5.

Chela-Flores J (2003). Evolution of intelligent behaviour: Is it just a question of time? In this volume.

Conant GC and Wagner A (2003). Convergent evolution of gene circuits. Nat.Genet.34(3):264–6.

Crick FHC (1958). The biological replication of macromolecules. Symp.Soc.Exp.Biol. 12,138–63.

Crick CF (1966). The genetic code is probably universal. Nature 212(5069):1397.

Crick FHC (1970). Central Dogma of Molecular Biology. Nature. 227(258):561–563.

Deamer DW and Branton D (2002). Nanopore Analysis. Acc.Chem.Res. 35(10):817–25.

DeAngelis PL (2002). Evolution of glycosaminoglycans. Anat Rec. 268(3):317–26.

Des Marais DJ, *et al.* (2002). Remote sensing of planetary properties and biosignatures on extrasolar terrestrial planets. Astrobiology. 2 (2):153–81.

Dougherty WM, *et al.* (2000).The Bloch equations in high-gradient magnetic resonance force microscopy: theory and experiment. J.Magn.Reson. 143(1):106–19.

Doyle DA, *et al.* (1998). The structure of the K^+ channel. Science. 280(5360):69–76.

Drexler KE (1994). Molecular nanomachines: physical principles and implementation strategies. Annu Rev.Biophys.Biomol.Struct. 23:377–405.

Fennimore A.M., *et al.* (2003).Rotational actuators based on carbon nanotubes.Nature.424(6947):408–410.

Geddes CD, Gryczynski I, Malicka J, Gryczynski Z and Lakowicz JR (2003). Metal-enhanced fluorescence: potential applications in HTS. Comb Chem High Throughput Screen. 6(2):109–17.

Greenberg R, Geissler P, Hoppa G, and Tuffs BR (2002). Tidal-tectonic processes and their implications for the character of Europa's icy crust, Rev. Geophys. 40(2):1034–1038.

Harte R and Ouzounis CA (2002). Genome-wide detection and family clustering of ion channels. FEBS Lett. 514(2–3):129–34.

Hille B (2001). Ion channels of excitable membranes, 3rd ed. Sinauer, Sunderland, Mass. USA.

Jeziorski MC, Greenberg RM and Anderson PA (2000). The molecular biology of invertebrate voltage-gated Ca^{+2} channels. J Exp Biol. 203 Pt 5:841–56.

Kain SR, Adams M, Kondepudi A, Yang TT, Ward WW and Kitts P (1995). Green fluorescent protein as a reporter of gene expression and protein localization. Biotechniques. 19(4):650–5.

Kim YD, Park CB and Clark DS (2001). Stable sol-gel microstructured and microfluidic networks for protein patterning. Biotechnol Bioeng. 73(5):331–7.

Le Bouter S, *et al.* (2003). Microarray analysis reveals complex remodelling of cardiac ion channel expression with altered thyroid status. Circ.Res. 92(2):234–42.

Malyan B and Balachandran W (2001). Sub-micron sized biological particle manipulation and characterisation. J. Electrostat. 51–52:15–19.

Marck C and Grosjean H (2002). RNomics: analysis of tRNA genes from 50 genomes of Eukarya, Archaea, and Bacteria reveals anticodon-sparing strategies and domain-specific features. RNA.8(10):1189–232.

Marshall J, Molloy R, Moss GW, Howe JR and Hughes TE (1995). The jellyfish green fluorescent protein: a new tool for studying ion channel expression and function. Neuron. 14(2):211–5.

McDermott R, Trabesinger AH, Muck M, Hahn EL, Pines A and Clarke J. (2002). Liquid-state NMR and scalar couplings in microtesla magnetic fields. Science. 295(5563):2247–9.

Medintz IL, Clapp AR, Mattoussi H, Goldman ER, Fisher B, Mauro JM. (2003). Self-assembled nanoscale biosensors based on quantum dot FRET donors. Nat Mater. 2(9):630–8.

Miller C (2000). An overview of the K^+ channel family. Genome Biology. 1(4):reviews0004.1–0004.5

Nealson KH, Tsapin A, Storrie-Lombardi M (2002). Searching for life in the Universe: unconventional methods for an unconventional problem. Int. Microbiol. 5(4):223–30.

Nicastro AJ, Vreeland RH, and Rosenzweig WD (2002). Limits imposed by ionizing radiation on the long-term survival of trapped bacterial spores: beta radiation. Int J Radiat Biol. 78(10):891–901.

Ouzounis C and Kyrpides N (1996). The emergence of major cellular processes in evolution. FEBS Lett. 390(2):119–23.

Ruta V, Jiang Y, Lee A, Chen J and MacKinnon R (2003). Functional analysis of an archaebacterial voltage-dependent K^+ channel. Nature. 422(6928): 180–5.

Tokumasu F and Dvorak J (2003). Development and application of quantum dots for immunocytochemistry of human erythrocytes. J. Microsc. 211(Pt 3):256–61.

Tornabene TG, Lloyd RE, Holzer G and Oro J (1980). Lipids as a principle for the identification of archaebacteria. Life Sci Space Res. 18:109–21.

Vreeland RH and Rosenzweig WD (2002). The question of uniqueness of ancient bacteria. J.Ind. Microbiol. Biotechnol. 28(1):32–41.

Vreeland RH, *et al.* (2002). Halosimplex carlsbadense gen. nov., sp. nov., a unique halophilic archaeon. Extremophiles. 6(6):445–52.

Woese CR (2001). Translation: in retrospect and prospect. RNA. 7(8):1055–67.

Woese CR and Fox GE (1977). The concept of cellular evolution. J Mol Evol. 10(1):1–6.

Zakon HH (2002). Convergent evolution on the molecular level. Brain Behav Evol. 59(5–6):250–61.

TITAN
Current Status and Expected Exobiological Return of the Cassini-Huygens Mission

FRANÇOIS RAULIN[1], JEAN-PIERRE LEBRETON[2] and TOBIAS OWEN[3]
[1]LISA, CNRS and Universités Paris 12 et Paris 7, Avenue du Général de Gaulle, 94010 Créteil, Cedex France, [2]ESA Research and Scientific Support Department, ESTEC/SCI-SB, 2200 AG Noordwijk, The Netherlands, and [3]Institute for Astronomy, University of Hawaii, 2680 Woodlawn Drive, Honolulu, HI 96822 USA

1. Introduction

Many space missions of exo/astrobiological importance have been launched since the beginning of planetary exploration with space probes more than 40 years ago. The most exobiologically oriented one was certainly the Viking mission to Mars, which became the first extraterrestrial planetary target to be searched for evidence of (extinct and extant) life. However, there is another category of extraterrestrial planetary bodies of prime interest for Exobiology: bodies where a complex organic chemistry is taking place. Titan, Saturn's largest satellite, with its thick nitrogen atmosphere, rich in organics in the gas and aerosol phases, and with many analogies to the early Earth, is probably, with the comets, one of the most exobiologically interesting bodies of this second kind.

The Cassini-Huygens mission to the Saturn system is particularly devoted to study in great detail Titan's exotic world, where we may find many data of paramount interest for our understanding of prebiotic chemical and physical processes. Cassini-Huygens, the most ambitious mission ever sent to the outer solar system, is designed to explore in detail the Saturnian system. A joint collaboration between NASA and ESA, it involves a wide cooperation among 18 countries from both the United States of America and Europe.

This paper presents the current status of the Cassini-Huygens mission. It briefly describes what we already know about Titan, especially about the organic chemistry which is going on in the different parts of the satellite and what we do not know, especially the many questions of exobiological importance which are still unresolved. It finally presents which of these questions Cassini-Huygens is designed to address, and it discusses more generally the exobiological implications of the potential scientific return of this very exciting planetary mission.

J. Seckbach et al. (eds.), Life in the Universe, 275–280.

2. The Cassini-Huygens Mission

The NASA-ESA Cassini-Huygens spacecraft was successfully launched at 4:43 a.m. EDT by a Titan IVB/Centaur rocket on October 15, 1997, from Cape Canaveral, Florida. The spacecraft, which consists of a Saturn orbiter (Cassini) and a Titan atmospheric probe (Huygens) is on a seven-year interplanetary trajectory toward Saturn, that relied on four gravity-assist manoeuvres at Venus (in 1998 and 1999), Earth (August 1999) and Jupiter (December 2000) to reach its final destination. It is targeted to reach Saturn in late June 2004, with a Saturn orbit insertion on 1st July 2004. Then, Cassini will embark on a 4-year tour of the Saturnian system. At the end of 2004, after two initial Titan encounters, it will release the Huygens probe on the third orbit around Saturn on 25th December 2004. Huygens will penetrate Titan's atmosphere and parachute down to the surface on January 14, 2005 (Matson et al., 2002, Lebreton and Matson, 2002, Kazeminejad et al., 2002; Russell, 2002; see also http://www.jpl.nasa.gov/cassini/ and http://sci.esa.int/huygens/).

The Cassini/Huygens mission is designed to explore the Saturnian system in great detail, including the giant planet, its atmosphere, magnetosphere and rings, and many of its moons, especially Titan. Titan's exploration is indeed one of the main objectives of the mission which includes multiple opportunities for close remote sensing and in-situ observations of Titan with the Cassini orbiter which is planned to complete 74 orbits around Saturn, that include 44 close Titan's fly-by's, during the four years of the nominal mission (mid-2004 to mid-2008). Huygens descent will last about 2.5 hours, but if it survives landing, it may still function for up to two/three hours on the surface. Many of the twelve instruments of the Cassini orbiter and all six instruments of the Huygens probe will study in-situ the many chemical and physical aspects of the atmosphere and the surface of Titan. They will provide much information of crucial importance for extending our knowledge of the complexity of Titan's organic chemistry.

3. What Do We Already Know About Titan?

Thanks to the Titan flybys by two Voyagers in the early 80's and many later observations from the ground or Earth orbiting telescopes, we already have an important amount of information on Titan. With a diameter of more than 5500 km, Titan, is by the size, the second largest moon of the solar system after Jupiter's Ganymede. It is the only satellite to have a dense atmosphere, composed mainly of dinitrogen and methane, with a small fraction of dihydrogen, and possibly argon, which has yet to be detected. This atmosphere is nearly five times denser than the Earth's atmosphere, with a surface temperature of 90–100 K and a surface pressure of 1.5 bar. As in the Earth's atmosphere, with water vapor and carbon dioxide on the one hand and clouds on the other hand, Titan's atmosphere also contains greenhouse gases, (condensable CH_4, equivalent to terrestrial H_2O; non-condensable H_2, equivalent to terrestrial CO_2), and anti-greenhouse compounds (aerosols). The thermal profile of the lower atmosphere of Titan is very similar to the terrestrial one—although the temperatures are much lower there—with a troposphere (90–70 K), a tropopause (70 K) and a stratosphere (70–175 K). Moreover, the models of the surface of Titan suggest that it is covered—at least partially—with lakes or seas of methane and ethane. Water ice has very recently been detected at the surface (Griffith et al., 2003), which opens many important exobiological perspectives, including the possibility of presence of liquid water on Titan's surface episodically due to large impacts and the subsequent occurrence of prebiotic chemistry in these liquid bodies.

Titan's environment is very rich in organic compounds, present in the three components of what one can call, again by analogy with our planet, the "geofluid" of Titan: air (gas atmosphere), aerosols (solid atmosphere) and surface (solid and liquid bodies). Indeed several hydrocarbons and nitriles have already been detected in the gas and condensed phases in Titan's atmosphere. These organics are the products of the photochemistry of methane coupled with that of dinitrogen. All the organic compounds which were detected in the atmosphere of Titan were also produced in laboratory simulation experiments (Coll et al., 1999), together with complex macromolecular products made of C, H and N atoms (called "tholins") and supposed to be the main constituents of Titan's aerosols (Coll et al., 1999; McKay et al., 2001, Ramirez et al., 2002; and refs. included). Many others organics, which have not yet been detected in Titan, are also formed during these laboratory experiments, which strongly suggests that they are also present in the atmosphere of Titan. This is strongly supported by the very recent detection of benzene in Titan's atmosphere through ISO observations (Coustenis et al., 2003; and refs included), the presence of which was expected from the results of laboratory simulation experiments. In the same way, the detection of water, in gas phase, in the atmosphere (Coustenis et al., 1998) although at very low concentration, together with the presence of CO and CO_2, allows us to consider the presence of oxygenated organic compounds. Recent laboratory experiments performed on model atmospheres of Titan including traces of CO, show that the main O-containing organic product is oxirane (Coll et al., 2003).

4. What We Do Not Know About Titan?

Many questions of exo/astrobiological importance concerning Titan remain unsolved

1) What is the Origin of the atmosphere and in particular of methane? Indeed the existence of methane in Titan's atmosphere is a major puzzle at present. This gas is destroyed so rapidly by photochemistry that the amount we see today will be gone in just a few million years. Methane must be continuously re-supplied to the atmosphere. The needed source could be a surface reservoir of methane through the lakes and seas mentioned above, or an external source (cometary impacts?) or a subsurface reservoir involving methane hydrates (Mousis et al., 2002, and refs. included) and cryovolcanism, or even biological sources (Fortes, 2000), involving methanogen micro-organisms (even if this hypothesis looks very unlikely).
2) In relation to this first question, are there liquid bodies on Titan's surface? Earth based observations confirm a non-homogeneous surface with dark and clear regions. The very recent radar data of Titan (Campbell et al., 2003) is fully compatible with the presence of several areas of liquid hydrocarbons on Titan's surface. But the information about the surface state and chemical composition is still very limited.
3) What is the chemical composition of the aerosols? What are their elemental and molecular compositions? Are they made of organic tholins, covered by volatile

material, as suggested by experimental and theoretical modelling (Raulin and Owen, 2002 ; and refs. included).

4) Is there lightning in the troposphere and are there organic processes in the troposphere and surface, as suggested by theoretical models?

5) Is there a subsurface water-ammonia ocean, as indicated by modeling of the internal structure of Titan (Grasset and Sotin, 1996; and refs. Included)? Is there an organic/prebiotic chemistry going on in this hypothetical ocean? Is there life present in this environment?

6) Is Titan's organic and almost prebiotic chemistry even more complex than we expect? Are O-bearing atoms involved in the surface organic chemistry, because of dissolved CO and precipitated CO_2, or through a multiphase organic chemistry involving water ice recently detected (Griffith et al., 2003), or involving temporary liquid water produced through hypothetical cryovolcanism or meteoritic or cometary impacts?

7) In direct relation to the previous questions, are there macromolecular materials abundant in Titan's environment? If yes what is their structure? Do they include oligomers, in particular of HCN and C_2H_2? Are there purine and pyrimidine bases present? Are there amino acids or their analogues? Are pseudo-polypeptides included in Titan's organic oligomers (Raulin and Owen, 2002)?

8) Is chirality present in Titan organic chemistry? Is there an enantiomeric excess in chiral molecules which may be present in the aerosols and/or on the surface or even in the subsurface? Is chirality also present in the macromolecular materials? The low temperature of Titan's environment should protect any enantiomeric excess against racemization, thus the study of chirality in Titan's chemistry is of prime importance and could yield crucial information on the origin of chirality in living systems.

5. What Do We Expect from Cassini-Huygens?

Several of the questions discussed in the above section will be approached in great detail thanks to the Cassini-Huygens mission (Wilson, 1997; Matson et al., 2002).

Several instruments of the orbiter and five of the six instruments of the probe will provide data of exobiological interest. The optical remote sensing instruments of the orbiter (especially CIRS and UVIS) will determine the chemical composition of different zones of Titan's atmosphere and should be able, in particular, to detect many organics, including new species and allow the determination of their vertical concentration profile. The Cassini Radar will be able to map Titan's surface through the haze layers, and to determine the presence and distribution of liquid bodies. VIMS will also provide information on and mapping of the chemical composition of the surface.

On the Huygens probe (Russell, 2002), the GC-MS instrument, a gas chromatograph with three GC capillary columns, coupled to a quadrupole mass spectrometer, (Niemann et al., 2002) will perform a detailed chemical analysis of the atmosphere—including molecular and isotopic analysis—during the 2.5 hours of descent of the probe, and perhaps of the surface, after landing. The ACP experiment will collect the atmospheric aerosols, heat them at different temperatures including at high temperature to pyrolyse the refractory organic materials (tholins) and transfer the produced gases to the GC-MS instrument for molecular analysis (Israel et al., 2002). This will provide the first direct *in situ* molecular and elemental analysis of Titan's hazes. HASI will determine, in particular, the pressure and temperature vertical profiles. DISR will measure the atmosphere radiation budget, determine the cloud structure and take images of the surface. DISR should provide evidence before impact whether Huygens is approaching a liquid surface. SSP and the rest of the payload that will remain operational after impact, will provide information on the physical state and chemical composition of the surface.

6. Conclusions

Several of the conditions necessary for planetary chemistry to evolve toward complex organic and prebiotic systems are present on Titan (Raulin & Owen, 2002, and refs. included): a dense and midly reducing atmosphere, a variety of efficient energy sources, a complex atmospheric system with aerosols and the possible presence of surface liquid bodies.

As early as mid-2004, Cassini-Huygens, a great Human adventure with an unmanned planetary mission to explore the Saturn System, will provide many new data on Titan, essential for the field of exobiology. Its scientific return is expected to be at least one order of magnitude greater than that of the Voyager mission. As an example, the CIRS experiment has a sensitivity which is 10 to 100 times higher, a spectral resolution 10 times better and a spectral range much larger than IRIS, the Voyager IR spectrometer. Since 20 years after Titan's encounter by Voyager, IRIS data are still being analysed, one can easily forecast that it will take several decades to fully analyse and interpret all Cassini-Huygens data. The prospect of an extended mission after mid-2008 if the Orbiter is still performing well, promises to offer many new opportunities for Titan observations, taking advantage of the new knowledge acquired with Huygens and the first few years by the Orbiter.

Laboratory studies are still much needed to make available key data essential for the interpretation of the Cassini/Huygens data observations and for theoretically modelling Titan's environment. Much needed laboratory data includes kinetic data related to small polyynes and cyanopolynes (Vuitton et al., 2003), and the dielectric properties of the aerosol, rain, and surface material (Rodriguez et al., 2003).

The Cassini-Huygens mission will bring answers to many of the scientific questions concerning Titan. But several questions will still remain, in particular the problem of chirality and most of the questions related to the organic chemistry that is taking place on the surface and in the subsurface of Titan. Thus, the scientific planetological and exo/astrobiological community is already thinking about post Cassini-Huygens missions, able in particular to explore these aspects of the so mysterious world of Titan (Lorenz and Mitton, 2002).

7. Acknowledgments

This review work has been supported by grants from the French Space Agency (CNES: Centre National d'Etudes Spatiales), University Paris-Val de Marne, and the NASA.

8. References

Campbell, D.B., Black, G.J., Carter, L.M. and Ostro S.J. (2003) Radar Evidence for Liquid Surfaces on Titan, *Science*, **in press**.

Coll, P., Guillemin, J.-C., Gazeau, M.-C. and Raulin, F. (1999) Report and implications of the first observation of C_4N_2 in laboratory simulations of Titan's atmosphere, *Planet. Space Sci* **47** (12), 1433–1440.

Coll, P., Bernard, J.-M., Navarro-González, R. and Raulin, F. (2003) Oxirane: An exotic oxygenated organic compound in Titan? *Astrophys. J*, **in press**.

Coustenis, A., Salama, A., Lellouch, E., Encrenaz, Th., Bjoraker, G.L., Samuelson, R.E., De Graauw, Th., Feuchtgruber, F. and Kessler, M.F. (1998) Evidence for water vapor in Titan's atmosphere from ISO/SWS data, *Astron. Astrophys*. **336**, L85–L89.

Coustenis, A., Salama, A., Schulz, B., Ott, S., Lellouch, E, Encrenaz, Th, Gautier, D. and Feuchtgruber, H. (2003). Titan' s atmosphere from ISO mid-infrared spectroscopy, *Icarus*, **161**, 383–403.

Fortes, A.D. (2000) Exobiological implications of a possible ammonia-water ocean inside Titan, *Icarus* **146**, 444–452.

Grasset, Q. and Sotin, C. (1996) The cooling rate of a liquid shell in Titan's interior, *Icarus* **123**, 101–123.

Griffith, C., Owen, T., Geballe, T.R., Ravner, J. and Rannou, P. (2003) Evidence for the Exposure of Water Ice on Titan's Surface, *Science*, **300 (5619)**, 628–630.

Israel, G., Cabane, M., Brun, J.F., Niemann, H., Way, S., Riedler, W., Steller, M., Raulin, F. and Coscia, D. (2002) The Cassini-Huygens ACP experiment and exobiological implications, *Space Science Review*, **104** (1–4), 435–466.

Kazeminejad, B., Lebreton, J.-P., Matson, D. L., Spilker, L. and Raulin, F. (2002) The Cassini/Huygens Mission to Saturn and Titan and its relevance to Exo/Astrobiology, Proc. 2d European Workshop on Exo-/Astro-Biology, *ESA-SP* **518**, 261–268.

Lebreton, J.-P. and Matson, D. L. (2002) The Huygens probe: Science, Payload and Mission Overview, *Space Science Review*, **104**(1–4), 59–100.

Lorenz, R. and Mitton, J. (2002) *Lifting Titan's Veil*, Cambridge University Press, Cambridge, U.K.

Matson, D., Spilker, L.J. and Lebreton, J.-P. (2002) The Cassini/Huygens mission to the Saturnian system , *Space Science Review*, **104**(1–4), 1–58.

McKay, C.P., Coustenis, A., Samuelson, R.E., Lemmon, M.T., Lorenz, R.D. , Cabane, M., Rannou, P. and Drossard, P. (2001) Physical properties of the organic aerosols and clouds on Titan, *Planet. Space Sci.* 49, 79–99.

Mousis, O., Gautier, D. and Bockelée-Morvan, D. (2002) An evolutionary turbulent Model of Saturn's subnebula: Implications for the origin of the atmosphere of Titan, *Icarus*, **156**, 162–175.

Niemann, H.B., Atreya, S.K., Bauer, S.J., Biemann, K., Block, B., Carignan, G.R., Donahue, T.M., Frost, R.L., Gautier, D., Haberman, J.A., Harpold, D., Hunten, D.M. D.M., Israel, G., Lunine, J.I., Mauersberger, K., Owen, T.C., Raulin, F., Richards, J.E. and Way, S.H. (2002) The Gas Chromatograph Mass Spectrometer for the Huygens Probe, *Space Science Review*, **104** (1–4), 551–590.

Ramirez, S.I., Coll, P., Da Silva, A., Navarro-Gonzalez, R., Lafait, A. and Raulin, F. (2002) Complex Refractive Index of Titan's Aerosol Analogues in the 200–900 nm domain, *Icarus*, **156**(2), 515–530.

Raulin, F. and Owen, T. (2002) Organic chemistry and exobiology on Titan, *Space Science Review*, **104** (1–4), 379–395.

Russell, C.T. (ed.) (2002) The Cassini-Huygens mission, Overview, Objectives and Huygens Instrumentarium, *Space Science Review*, **104** (1–4).

Rodriguez, S., Paillou, P., Dobrijevic, M., Ruffié G., Coll, P., Bernard, J.M., and Encrenaz, P. (2003), Impact of aerosols present in Titan's atmosphere on the CASSINI radar experiment, *Icarus*, **164**, 213–227.

Vuitton, Gee, C., Raulin, F., Benilan, Y., Crepin, C. and Gazeau, M.-C. (2003) Intrinsic lifetime of C4H2*: Implications for the photochemistry of C4H2 in Titan's atmosphere, *Planet. Space Sci.*, **in press**.

Wilson A. (Ed.), European Space Agency (1997) Huygens : Science, Payload and mission, *ESA SP-* **1177**.

CHEMICAL CHARACTERIZATION OF AEROSOLS IN SIMULATED PLANETARY ATMOSPHERES
Titan's Aerosol Analogues

SANDRA I. RAMIREZ[1], RAFAEL NAVARRO-GONZALEZ[2], PATRICE COLL[3] and FRANÇOIS RAULIN[3]
[1]Centro de Investigaciones Químicas, UAEM Av. Universidad # 1001 Col. Chamilpa, Cuernavaca, Morelos 62210 Mexico; [2]Laboratorio de Química de Plasmas, Instituto de Ciencias Nucleares, UNAM Circuito Exterior C. U. Apdo. Postal 70-543, D. F. 04510 Mexico; [3]Laboratoire Interuniversitaire des Systèmes Atmosphériques, CNRS and Universités Paris VII et XII, 61 avenue du Général de Gaulle 94010 Créteil cedex France

Abstract. The surface of Titan is hidden, in the visible light, by two aerosol layers. The properties of these layers have been studied through ground-based and spacecraft observations, by theoretical modeling, and by different experimental approaches. Tentative analogues of Titan's aerosols have been synthesized in laboratories to determine their physical, chemical, and optical properties. It was precisely a careful analysis of the optical properties of the laboratory solid aggregates that shows that those properties frequently need a correction factor to adequately match Titan's geometric albedo. Trying to find an explanation to this fact, there has been a continuous search on the physical and chemical properties of the synthesized solids in relation to their quality as authentic laboratory analogues. A few reports are available concerning the chemical composition of laboratory aerosols and such studies varied significantly in experimental variables. Hence, there is a need for systematic studies of the solid products synthesized during simulation experiments devoted to determine their structural features. A brief description of the initial steps of such a study focused in the characterization, by analytical instrumental techniques, of laboratory aerosol analogues synthesized from 1-hour laser-induced plasma irradiation of a Titan's canonical atmosphere is presented. Understanding the chemical process that originate them and approaching to their chemical characterization can certainly help to easily interpret their role in Titan's atmospheric dynamics.

1. Introduction

Laboratory simulation experiments of planetary bodies initiated with one of the most popular and well documented: Miller's replication of Early Earth conditions. This chemically oriented experiment was formulated to test the idea that organic compounds that served as the basis of Life in our planet formed from a strongly reduced atmosphere (Miller, 1953). There has passed 50 years since the report of this classical experiment and during this time a great number of modified versions of it has been carried out. It has also inspired the

J. Seckbach et al. (eds.), Life in the Universe, 281–285.

formulation of innovative laboratory simulations not only to recreate Earth's scenarios but to look into the varied environments found in the planets and moons of our Solar System some of which have strongly called Exo/Astrobiology attention. Objects where a prebiotic-like chemistry is going on are of the most interesting ones. The satellite Titan with its dense atmosphere, rich in molecular nitrogen with a noticeable fraction of methane, appears as a planetary-size chemical reactor where a complex organic chemistry occurs in gas and condensed phases (Coll *et al.*, 1999; Raulin *et al.*, 1999).

2. Experimental Conditions

Laboratory simulations of Titan's atmosphere environment have demonstrated that long-time irradiations of CH_4 and N_2 mixtures with different energetic sources yield sticky, dark-color condensed materials proposed as good candidates to mimic Titan's aerosols (Sagan and Thompson, 1984; Khare *et al.*, 1984; Scattergood *et al.*, 1989; Ramírez and Navarro-González, 2000; Thompson *et al.*, 1991). However, significant differences arise because their properties display a strong dependence on the utilized experimental conditions. With the purpose of arriving to a more accurate simulation system of Titan's conditions that yields better quality aerosol analogues a protocol has been initiated (Coll *et al.*, 1999; Ramírez *et al.*, 2002) and continues under development. A mixture of ultra high purity methane and nitrogen gases (1:9), that simulates Titan's atmosphere, was introduced at a pressure of 670 mbar at room temperature, into a round 1.09-liter Pyrex glass flask. The mixture was subjected to laser-induced plasma (LIP) irradiation using a Nd-YAG laser with a pulse width of 7 ns operating at 10 Hz by 60 minutes. A 1.06 μm beam was focused at the center of the flask using a 5 cm focal distance plano-convex lens with anti-reflection coating. The solid collected from the glass flask with HPLC-grade methanol (CH_3OH) was dried and then analyzed by infrared (FTIR) and proton nuclear magnetic resonance (^{1}H NMR) spectroscopy. Infrared spectra from a KBr film were recorded in transmission mode from 4000 to 500 cm^{-1} with a Nicolet FTIR Magna 560 spectrometer at 2 cm^{-1} resolution, while ^{1}H NMR spectrum was obtained from D_2O-DMSO soluble fractions of the recovered aerosol analogue in a 300 MHz Varian instrument.

3. Results and Discussion

The strongest absorption bands found in the infrared spectra (Fig. 1) correspond to $-CH{=}CH_2$- or aromatic =C-H groups (>3000; 1415 cm^{-1}); -CH_3 (2960–2870; 1440–1400 cm^{-1}); -CH_2- (2925–2850; 1350–1150 cm^{-1}); -C≡N (2260–2240 cm^{-1}) -C=C- (1670–1640 cm^{-1}). The more interesting bands are those that show the presence of the double bonds and of the cyano group. It is important to mention that the spectra obtained resemble very closely that reported by Coll *et al.* (1999) where simulations were performed using a glow discharge. The fact of finding close similarities between these spectra is remarkable because it could mean that we have been able to produce a solid material with similar infrared spectroscopic characteristics using different energy sources. This implies less dependence on experimental parameters. McDonald *et al.* (1994) reported the transmission FTIR spectrum of a polymer obtained from a gas mixing ratio similar to the one

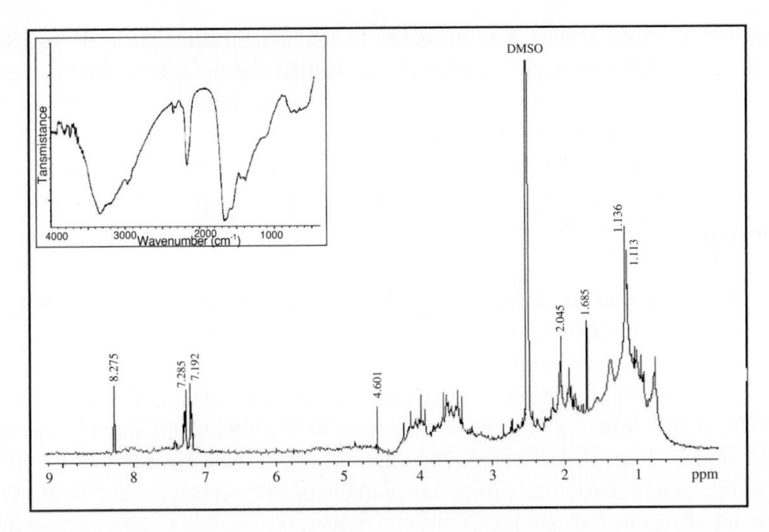

Figure 1. FTIR (inset) and ^{1}H NMR spectra of the recovered solid fraction from Titan's aerosol analogues synthesized after 1-hour LIP irradiation of a 1:9 methane in nitrogen gas mixture.

used in the present work, but irradiated by a Tesla coil. Even when the general profile of that spectrum does not resemble a lot our spectra, they reported the presence of saturated CH_3- and $-CH_2$- groups, as well as -C≡C-, C=C, -C≡N, N-H and tentatively carbonyl groups. The primary bands from the ^{1}H NMR analysis demonstrate that the aerosol analogue synthesized contains saturated and unsaturated aliphatic hydrocarbons due to the presence of $-CH_3$ (0.78 ppm), $-CH_2$- (1.11; 1.13 ppm); C=C-CH_3 or C≡C-CH_3 (1.68 ppm); -CH_2 =C-C (2.04 ppm); as well as aromatic hydrocarbons (7.19 and 7.28 ppm), and probably nitriles of the type Ph-CN (3.38 ppm) or –C=C-CN (3.45 ppm).

Limited work has been published so far about the chemical properties, and specifically the structural features of analogues for the Titan's atmospheric aerosols. The information collected by these two analytical techniques can help to proposed basic blocks expected to be found in the molecular "backbone" of the solids. It is not surprising to find the signatures of double and triple homo- and hetero-atomic bonds. It is believed that the origin of the solid phase begins with a linear polymer having the general structure $R_1C{=}NR_2$ arising from the interaction of gas-phase nitrile groups (McDonald *et al.*, 1994). It can also come from the condensation of residues of unsaturated hydrocarbons and a nitrile group or from two triple bond residues. The aromatic fragments detected can be formed from Diels-Alder type reactions, which are not difficult to occur since during the irradiation process the vessel contains a wide variety of reactive species: free radicals, excited molecules, and ionized species originated from the original gas mixture constituents as well as from the primarily produced molecules. It has been demonstrated that LIP discharges tend to produce mainly alkynes and benzene derivatives together with saturated and unsaturated hydrocarbons (Ramírez and Navarro-González, 2000). In a medium such as the one promoted during the proliferation of the plasma, the reactive character of the existent species is displayed and results in the recombination of all the primary species. How close the solids experimentally obtained, not only from the present work but from other experimental protocols, represent Titan's aerosols (*tholins*) is still in debate. However parallel to the argumentation proposed in that way, it must also exists argumentation aimed to understand the chemical processes that originate them and different approaches to elucidate their chemical structure. Studies in this manner will certainly help to easily interpret the role of the aerosols in Titan's atmospheric dynamics, and in terrestrial simulation experiments.

4. Conclusions

The initial steps of a study dedicated to determine the chemical characterization of Titan's atmospheric aerosol analogues has been initiated. Preliminary results show a lower dependence on experimental conditions and the presence of chemically interesting fragments of molecules. The study will continue with a detailed analysis of the production mechanisms that yields the solids from the gas-phase molecules to be able to approach to more accurate structural features of the aerosol analogues. Experimental investigations in this sense are needed and they are of particularly importance in the perspective of the exploration of Titan by the Cassini-Huygens (NASA-ESA) mission, which is expected to provide a tremendous amount of new observational data of Titan's environments, starting in 2004. An efficient retrieving of these data requires the availability of many laboratory data, concerning specifically those of Titan's aerosol analogues.

5. Acknowledgements

This work is being supported by grants from CONACYT (J40449-F) and PROMEP (103.5/03/1134). SIR is grateful to ICN-UNAM for permitting the use of their facilities for the synthesis of the aerosol analogues and to IQ-UNAM for providing the recorded spectra; she is also grateful to the ICTP and co-sponsors of the ***2003 Trieste Conference***, specially to Prof. J. Chela-Flores, for the afforded travel-grant. We appreciate the help provided by Prof. J. Seckbach in the formatting process of the manuscript.

6. References

Coll, P., Coscia, D., Smith, N., Gazeau, M.-C., Ramírez, S.I., Cernogora, G., Israël, G. and Raulin, F. (1999) Experimental laboratory simulation of Titan's atmosphere (aerosols and gas phase). Planet. Space Sci., **47**, 1331–1340.

Khare, B.N., Sagan, C., Arakawa, E.T., Suits, F., Callcott, T.A. and Williams, M.W. (1984) Optical Constants of Organic Tholins Produced in a Simulated Titanian Atmosphere: From Soft X-Ray to Microwave Frequencies. Icarus, **60**, 127–137.

McDonald, G.D., Thompson, W.R., Heinrich, M., Khare, B.N. and Sagan, C. (1994) Chemical investigation of Titan and Triton tholins. Icarus, **108**, 137–145.

Miller, S.L. (1953) A production of amino acids under possible primitive Earth conditions. Science, **117**, 528–529.

Ramírez, S.I., Coll, P., da Silva, A., Navarro-González, R., Lafait, J. and Raulin, F. (2002) Complex refractive index of Titan's aerosol analogues in the 200–900 nm domain. Icarus, **156**, 515–529.

Ramírez, S.I. and Navarro-González, R. (2000) Quantitative study of the effects of various energy sources on a Titan's simulated atmosphere. In: J. Chela-Flores, G. Lemarchand, and J. Oró (eds), *Astrobiology: Origins from the Big-Bang to Civilization*. Kluwer Academic Publisher, Kluwer Academic Publisher, 307–310.

Raulin, F., Coll, P., Smith, N., Bénilan, Y., Bruston, P. and Gazeau, M.C. (1999) New insights into Titan's Organic Chemistry in the Gas and Aerosol Phases. Adv. Space Res., **24**, 453–460.

Sagan, C. and Thompson, W.R. (1984) Production and Condensation of Organic Gases in the Atmosphere of Titan. Icarus, **59**, 133–161.

Scattergood, T.W., McKay, C.P., Borucki, W.J., Giver, L.P., Van Ghyseghem, H., Parris, J.E. and Miller, S.L. (1989) Production of Organic Compounds in Plasmas: a Comparison among Electric Sparks, Laser-Induced Plasmas, and UV light. Icarus, **81**, 413–428.

Thompson, W.R., Henry, T.J., Schwartz, J.M., Khare, B.N. and Sagan, C. (1991) Plasma Discharge in $N_2 + CH_4$ at Low Pressures: Experimental Results and Applications to Titan. Icarus, **90**, 57–73.

OBSERVATION, MODELING AND EXPERIMENTAL SIMULATION: UNDERSTANDING TITAN'S ATMOSPHERIC CHEMISTRY USING THESE THREE TOOLS

J.-M. BERNARD[1], P. COLL[1], C.D. PINTASSILGO[2], Y. BENILAN[1], A. JOLLY[1], G. CERNOGORA[3] and F. RAULIN[1]
[1]Laboratoire Interuniversitaire des Systèmes Atmosphériques, UMR CNRS 7583, Universités Paris 12 and Paris 7, CMC, 61 av. du Gal de Gaulle F-94010 Créteil cedex, France. [2]Departamento de Física, Faculdade de Engenharia, Universidade do Porto, 4200-465 Porto, Portugal and [3]Service d'Aéronomie, Université de Versailles Saint Quentin, 78035 Versailles Cedex, France

1. Introduction

Titan, the largest satellite of Saturn, has been studied as an exo/astrobiological object for several years. Its dense atmosphere is made of nitrogen with a few percent of methane. Since Miller's experiment, we know that reducing atmospheres are of great interest for prebiotic organic chemistry. Subjected to energetic particles bombardment (coming from solar radiations and Saturn's magnetosphere), both hydrocarbons, nitriles (like HCN, a precursor of amino-acids) and organic aerosols are produced inside the atmosphere in notable amounts.

For several years, at LISA, the atmospheric chemistry on Titan has been studied using laboratory simulation experiments. In order to mimic Titan's atmosphere, we initiate a glow discharge in a N_2/CH_4 (98/2) mixture at low temperature. Solid (tholins) and gaseous compounds including radicals and ions produced in the reactor are analyzed by different complementary analytical techniques.

During the past decade, many experimental simulations have been carried out in various laboratories. Experimental conditions are too different: temperature, pressure, composition of the initial gas mixture, energy sources, duration of experiment. The comparison between these experiments is then very difficult. The problem comes mainly from the lack of knowledge of the mechanisms leading to the formation of the products (gas and solid phases).

How to compare the reactor's chemistry and Titan's chemistry? In order to answer this question, we decided to deeply study the physical and chemical properties of the glow discharge. Pintassilgo *et al.* (1999) developed a kinetic model adapted to this type of experimental simulation in order to obtain the basic elementary mechanisms occurring in the discharge. We present here a coupling between the experimental simulation and this numerical model. This approach is used to calculate the electron energy distribution function

J. Seckbach et al. (eds.), Life in the Universe, 287–291.

(EEDF) and to determine the abundance of the species inside the plasma. This work allows us to have an insight into the fundamental mechanisms involved in these discharges and to go further on the study of the tholins' formation, as well as their composition.

2. Experimental Setup

The experimental device is described in Coll *et al.* (1999). An improvement of the reactor allows to determine:

- The species produced *in-situ* (in particular radicals and ions): an optical UV-VIS fiber with 200 μm core diameter is placed in front of a fused silica window at the end of the reactor in order to collect the UV-VIS radiations. An optical system composed of two lenses is installed between the end of the optical fiber and the entry of the spectrometer in order to improve the light collection adapting the angular aperture of the optical fiber to the one of the spectrometer. The analysis is achieved with a monochromator (THR Jobin-Yvon) of Czerny-Turner type with a focal length of 1.5 m, offering a resolving power of 20000. Currently, the 230–550 nm range can be studied due to the combination of the fiber and photomultiplier band pass respectively for the shorter and the longer wavelengths. The spectrometer is controlled by a LabVIEW (Natural Instrument) interface through an analogue output board (DDA06-Keithley) and a homemade electronics.
- The reduced electric field E/Ng, where E is the electric field and Ng is the total neutral gas density (linked to the gas temperature Tg through the ideal gas law Ng = p/KTg). E is determined by measuring the difference in floating potential between two electrostatic probes which are 15 cm apart and connected to a voltmeter through a high voltage probe. The gas temperature Tg is obtained from the rotational distribution of the ($\Delta v = 0$) transitions of the nitrogen second positive system $N_2(C^3\Pi_u) \rightarrow N_2(B^3\Pi_g)$ at 337.1 nm, assuming that the rotational temperature is in equilibrium with the kinetic gas temperature, under our discharge conditions. The rotational temperature is determined by a least square procedure simultaneously on the five ($\Delta v = 0$) transitions of the C-B transitions, from the comparison between experimental and calculated spectra. The uncertainty on these calculations is estimated to be less than 10 %.

3. Kinetic Model

In the plasma, the electrons are accelerated by the electric field in the discharge. As the plasma is weakly ionized, electrons collide mainly with neutral species. It was previously shown, for instance in a N2-H2 mixture (Loureiro and Ricard, 1993), that the energy of electrons is mainly controlled by inelastic collisions with the molecular nitrogen ground state $N_2(X_1\,\Sigma_g^+)$ (Allan, 1985). The plasma is not in thermal equilibrium: the electron temperature is much greater than the gas temperature. At pressures typically below 20 mbar, the mean temperature of the electrons in these discharges is in the range 1–5 eV (1eV∼11000 K), and the most energetic electrons have an energy of the order of 10 eV.

Theoretically, the reduced electric field E/Ng is determined from the requirement that, under steady-state conditions, the total ionization rate must compensate exactly the total loss rate due to ambipolar diffusion to the wall plus electron-ion recombination (Pintassilgo *et al.*, 1999). This formulation ensures that we obtain the quasineutral condition in the glow discharge, i.e., the ion density is equal to the electron density.

4. Results and Discussion

4.1. DETECTION OF PRODUCED SPECIES

Three kinds of species are studied: the radicals and ions which are produced inside the discharge and detected *in-situ* by the UV-visible spectrometer (Figure 1). By this last technique we have detected $N_2^{\dagger}$, CN, NH and CH; the solid compounds which are deposited on the walls of the reactor and analyzed in order to get their atomic composition (Bernard *et al.*, 2002); the produced gases which are collected at the end of the reactor and analyzed by IR spectrometry (Bernard *et al.*, in press) and GC-MS (Coll *et al.*, 1999).

4.2. REDUCED ELECTRIC FIELD E/Ng IN PURE NITROGEN COLD PLASMA

The aim of this work is to validate the kinetic model by experimental measurements. In Figure 2, a first comparison between theoretical and experimental E/Ng values is presented in pure nitrogen as a function of the value of the current. We can observe a good adequacy between the experimental and theoretical values. The next step will be to compare the data from a N_2/CH_4 mixture as a function of current and as a function of pressure.

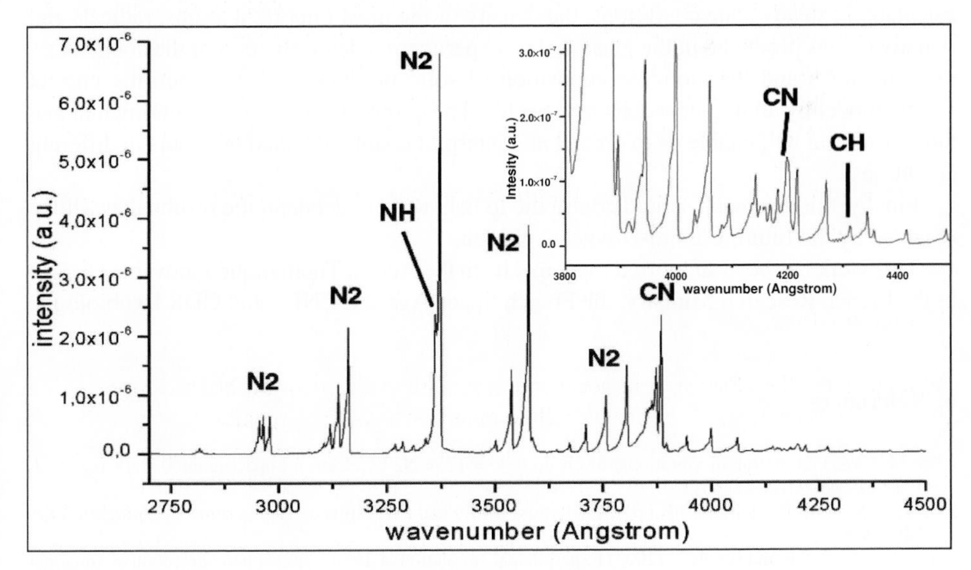

Figure 1. UV-visible spectrum of a N_2/CH_4 glow discharge (2% CH_4 in N_2). Pressure: 0.65 mbar; flow: 1200 sccm; current: 40 mA; spectral resolution: 0.5 nm.

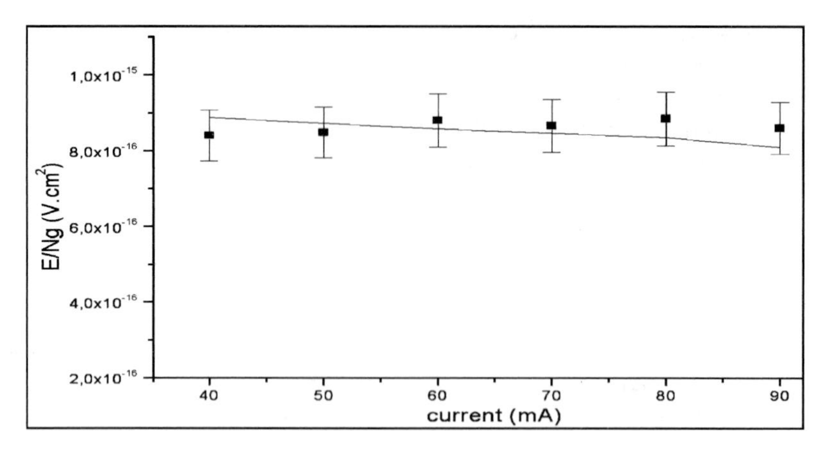

Figure 2. Reduced electric field (E/Ng) as a function of current in pure nitrogen. The dots are the experimental data and the curve represents the theoretical model.

5. Conclusion

The coupling of an experiment with a kinetic model of a N_2-CH_4 glow discharge helps to understand the physical and chemical processes in the reactor during the glow discharge. We have validated the physical properties developed by the kinetic model in pure nitrogen cold plasma.

The following step in the study of the discharge will be to determine the Electronic impact Energy Distribution Function (EEDF). Then, the rate coefficients for electronic reactions (ionization, dissociation, excitation) will be calculated using published values of reaction cross sections. As rate coefficients for neutral compounds are known, it will be possible to calculate, in steady state conditions, the density of the most important atoms, radicals and radiative states produced in the plasma. The experimental determination of the abundances of compounds and the atomic composition of solid products will constrain the current kinetic model by adding new detected species. The comparison between experimental and numerical data will enable us to predict and interpret results obtained by others in different conditions.

Finally, this approach should contribute to help us to understand the results that will be obtained by the future Cassini-Huygens mission.

This work has been supported by grants from the Action Thématique Innovante (A.T.I.) of the French Research Ministry, the French Space Agency, CNES and GDR Exobiologie.

6. References

Allan M. (1985) Excitation of vibrational levels up to v = 17 in N2 by electron impact in the 0–5 eV region. J. Phys. B: Mol. Phys. **18**, 4511.

Bernard J.-M., Coll P. and Raulin F. (2002) Variation of C/N and C/H ratios of Titan's aerosols analogues. ESA **SP-518**, 623–625.

Bernard J.-M., Coll P. and Raulin F. (2003) Experimental simulation of Titan's atmosphere: detection of ammonia and ethylene oxide. Planet. Space Sc, in press.

Coll P., Coscia D., Smith N. S., Gazeau M.-C., Ramirez S. I., Cernogora G., Israel G. and Raulin F. (1999) Experimental laboratory simulation of Titan's atmosphere: aerosols and gas phase. Planet. Space Sci. **47**, 1331–1340.

Loureiro J. and Ricard A. (1993) Electron and vibrational kinetics in a N2-H2 glow discharge with applications to surface process. J. Phys. D: Apll. Phys. **26**, 163.

Pintassilgo C. D., Loureiro J., Cernogora G. and Touzeau M. (1999) Methane decomposition and active nitrogen in a N_2-CH_4 glow discharge at low pressures. Plasma sources Sci. Technol. **8**, 463–478.

EXOBIOLOGY OF TITAN

MICHAEL SIMAKOV
Group of Exobiology, Institute of Cytology, RAS
Tikhoretsky Av., 4, St. Petersburg, 194064, Russia

Accretion models of the Saturnian satellite suggest that heating released during late stages of its formation was sufficient to create a warm, dense atmosphere with mass at least 30 times greater then the present value (Lunine and Stevemson, 1983) and large open ocean on its surface. Such juvenile Titan's ocean could exist during period of 10^8 years. As the great part of the primordial Titan's atmosphere could be supplied by comets during or after accretion, the composition of such atmosphere would have consisted of mostly H_2O, N_2, CO and CO_2, since the cometary carbon appears concentrated in the form of CO (ranging from a few to 45% relative to water), CO_2 (~15%) and heavy organic. The mass of volatile acquired by Titan from comets would be expected to be $\sim 10^{20}$–10^{22} g for CO and 10^{20}–10^{21} g for N (Griffith and Zahnle, 1995). So we can see that the Titan's primordial atmosphere could be warm, dense and consist of CO_2-(CO)-N_2.

The first stages of chemical evolution would have took place in these atmosphere and ocean under action of such energy sources as ultraviolet radiation, solar wind, galactic cosmic rays, magnetospheric plasma ion bombardment, electrical discharges and radiogenic heat. Recent attempts to establish a lower limit for the time required for emergence of life suggest that 10–100 million years was enough in case of Earth and the time of existence of the Titan's juvenile ocean was enough for arising of the first protoliving objects.

As the planet developed through time several energetic processes (irradiation, lightning, meteoritic and comet impacts) produced different forms of fixed nitrogen. All nitrogen could have been in the fixed form at the end of the planetary accretional period. Such scenario has been supposed by Mancinelli and McKay (1988) for the evolution of prebiotic nitrogen cycling on Earth, and the similar processes could be proposed in the case of the Saturn's satellite. Hence, in the absence of a recycling mechanism dissolved NO_2^- and NO_3^- would accumulate in the ocean.

During the phase of cooling, Titan's ocean was roofed over with icy crust. If life had originated by then, it could survive in some places up to the present (Fortes, 2000). On Earth microbial life exists in all locations where microbes can survive. In other case the variety of prebiotic processes can take place on Titan at present time. Many volatiles and inorganic salts were probably present in the primordial liquid layer and they must decrease the freezing temperature of the liquid at the stage of cooling. The compositions of the rich atmosphere, which is host to extensive organic photochemistry and internal liquid layer, must be very complex and Titan's putative ocean might harbor life or complex prebiotic structures.

J. Seckbach et al. (eds.), Life in the Universe, 293–296.

The most recent models of the Titan's interior lead to the conclusion that a substantial liquid layer exists today under relatively thin ice cover inside the satellite (Lunine and Stivenson, 1987; Grasset and Sotin, 1996; Grasset *et al.*, 2000). Lunin (1993) has shown that the underground ocean is the only structure that is consistent with all of the known constraints (chemical, tidal, ground-base radar and near-infrared observation) and Lorenz (2001) has found that internal oceans are mandated for the large icy satellites. Thermal evolution models also predict the existence of thick ($\sim$300 km) liquid layer with relatively thin ($\sim$80 km) ice cover (Grasset *et al.*, 2000). Spohn and Schubert (2003) have shown that even radiogenic heating in a chondritic core may suffice to keep a water ocean inside large icy satellites.

The present composition of the putative liquid layers of the ice satellites is probably very complex. Mass balance calculations modeled an extraction of elements into the aqueous phase from chondritic material show that Titan's extensive subsurface ocean likely also contains dissolved salts from endogenic materials resembling to carbonaceous chondrite rocks incorporated into the satellite during its formation and released at the time of planetary differentiation.

There are three sources of organic carbon for Titan's ocean: complex atmospheric chemistry (Clarke and Ferris, 1997), carbonaceous chondrites and cometary bodies, and seafloor hydrothermal systems. Since the light energy in the form of the solar radiation is not accessible in such conditions (the solar flux to Titan's surface is $\sim$1.1% from the Earth's one) the chemical energy has to be the main source which drives the life and other disequilibrium processes. So, the initial components, such as NO_3^-, SO_4^{2-}, CO_3^{2-} for the origin of lithoautotrophic processes could exist in the Titan's putative ocean from the earlier stages of the satellite's evolution and provide biologically useful electron donor-acceptor pairs in the upper layer where the temperature and pressure are not very hostile. Nitrate accumulated in the ocean at the first stage of atmosphere's evolution would have allowed the first protobiosystems to use it as the primary source of energy. We would like to propose the idea that the first protoliving systems in Titan's ocean could had internal energy source, namely, the chemical potential of an inorganic reaction – "Basic Reaction" (BR).

There are some candidates on the role of BR. Electron acceptors such as NO_3^-, SO_4^2, Fe^{3+}, Mn^{4+}, or CO_2 have to be coupled with the electron donors. Electron donors that may be important in such process include H_2, CO, CH_4, Fe^{2+}, Mn^{2+}, pyrite, sulfur compounds and organic material. Some of these molecules could be generated abiotically on the bottom of the internal ocean by the reaction of water with rocks of the silicate mantle and by the reaction of water with meteoritic components and others could be synthesized under the action of radiation.

Four energetic full operative biogeochemical cycles are possible inside Titan's ocean, namely nitrogen (N-cycle), sulfur (S-cycle), iron (Fe-cycle) and carbon (C-cycle) and all of them could be connected each with other (Simakov, 2003).

The basic reaction of nitrate reduction to dinitrogen is a more thermodynamically favorable in the row of different inorganic substrates. The all gaseous nitrogen in the contemporary Titan's atmosphere can be the product of this reaction (Simakov, 2000). Very interesting bacteria have been discovered recently which use ammonium as an inorganic electron donor for denitrification (Jetten *et al.*, 1999). This reaction has a very favorable energetic (-357 kJ/mol). Hydroxylamine (NH_2OH) and hydrazine (N_2H_4) are formed as intermediates and bicarbonate is the sole carbon source. This is the first case when hydrazine, a rocket fuel, is a free intermediate in any biological system. Both these components could be widespread in the Titan's environments and could be used microorganisms for energy transduction and the buildup of an electrochemical gradient. And we can hypothesize a start reaction as $N_2H_4 \Rightarrow N_2$ which can evolve through $NH_2OH \Rightarrow N_2H_4 \Rightarrow N_2$ to $NO_3^- \Rightarrow NH_2OH \Rightarrow N_2H_4 \Rightarrow N_2$ at the rout of microbial evolution.

Dissimilatory ferric iron-reducing and ferrous iron-oxidizing organisms also can form the basis for a closed ecosystem which gains energy through cyclic reduction and oxidation of iron minerals, sometime by NO_3^--dependent way. Fe(III) oxyhydroxides are readily reduced with H_2S, inorganic sulfides, elemental sulfur and various organic acids. On Earth microbial Fe(III) reduction is the major way of organic carbon oxidation in anaerobic environment. Microorganisms utilizing Fe(III) as an electron acceptor were discovered into mesobiotic marine and freshwater anoxic sediments and submerged soils. The denitrifying bacterium was isolated from the mud of Mariana Trench. It shows greater tolerance to low temperature and high hydrostatic pressure (50 MPa). Thermobiotic ecosystems also contain bacteria able to reduce Fe(III) with formate, lactate or molecular hydrogen. The production of reduced end products, e.g. Fe(II), FeS by Fe-reduction and H_2S with such processes could resupply the basic reaction with reagents. Ferrous iron is oxidized chemically by a number of inorganic compounds, most notably molecular oxygen, manganese oxide (MnO_2) and nitrate. So we can imagine a biogeochemical cycle for maintain of the primordial Titan's ecosystem. The putative life inside Titan does not depend on solar energy and photosynthesis for its primary energy supply and it is essentially independent of the surface circumstances. There could be microorganisms having a great similarity with the last common ancestor (LCA) on Earth.

Rich chemosynthetic ecosystems could be associated with methane clathrate areas on the icy bed. The cold methane vents induced by liquid methane could serve as a source for forming of the chemosynthetic ecosystems. These processes could involve a transfer of electrons from methane to sulfate or others electrons acceptors. The examples of such systems could be found around methane clathrate on the Earth's sea bed. Some organisms are capable of disproportioning of methanol, methylamines or methylsulphides to methane and carbon dioxide, oxidizing -CH_3 groups to CO_2 anaerobically. Microbial consortia based on anaerobic oxidation of methane coupled to sulfate reduction can support other microbial communities by generating of substantial biomass accumulation derived from methane.

Along with the upper layer of the internal water ocean when a temperature and pressure are suitable for living processes there are some additional appropriate sites for biological and/or prebiological activity at present day (Simakov, 2001): (1) water pockets and liquid veins inside icy layer; (2) the places of cryogenic volcanism; (3) macro-, mini- and micro-caves in the icy layer connecting with cryovolcanic processes; (4) the brine-filled cracks in icy crust caused by tidal forces; (5) liquid water pools on the surface originated from meteoritic strikes; (6) the sites of hydrothermal activity on the bottom of the ocean.

The environments mentioned above indicate that all conditions capable of supporting life are possible on Titan. All requirements needed for exobiology—liquid water which exists within long geological period, complex organic and inorganic chemistry and energy sources for support of biological processes are on Saturnian moon. On Earth life exists in all niches where water exists in liquid form for at least a portion of the year. Sub glacial life may be widespread among such planetary bodies as Jovanian and Saturnian satellites and satellites of others giant planets, detected in our Galaxy at last decade (Perryman, 2000).

The low temperature hypersaline brines have been proposed as habitat for microbial communities on Mars. The existence of rich atmosphere is the main difference from the Jupiter's moons. This atmosphere could supply the large quantity of different organic compounds to putative ocean. There are some possible mechanisms for extensive, intimate interaction of a liquid water ocean with the surface of the ice crust. Titan provides also insights regarding the geological and biological evolution of early Earth during ice-covered phase. There is a huge deficiency of carbon in the contemporary environment and this disappeared carbon could be contained as biomass and dissolved organic carbon in the putative ocean. Possible metabolic processes, such as nitrate/nitrite reduction, sulfate reduction and methanogenesis could be suggested for Titan. Nitrate and sulfate could be predominant forms of N and S in the ocean and nitrate and/or sulfate reduction would have been potential sources of energy for primitive life forms. Given the possibility that organic compounds may be widespread in the ocean from synthesis within hydrothermal systems, derived from atmospheric chemistry and delivered by comets and meteorites these putative nitrate and sulfate reducers may have been either heterotrophic or autotrophic. Furthermore, at the presence of substantial amount of methane the methanogenesis along with methanotrops also have been energetically favorable. Excreted products of the primary chemoautotrophic organisms could serve as a source for other types of microorganisms (heterotrophes) as it has been proposed for Europa.

References

Clarke, D.W. and Ferris, J.P. (1997) Chemical evolution on Titan: comparisons to the prebiotic Earth. Orig. Life Evol. Biosphere, **27**, 225–248.

Fortes, A.D. (2000) Exobiological implications of a possible ammonia-water ocean inside Titan. Icarus, **146**, 444–452.

Grasset, O. and Sotin, C. (1996) The cooling rate of a liquid shell in Titan's interior. Icarus, **123**, 101–112.

Grasset, O. *et al.* (2000) On the internal structure and dynamics of Titan. Planet. Space Sci., **48**, 617–636.

Griffith, A.C. and Zahnle, K., (1995) Influx of cometary volatiles to planetary moons: The atmospheres of 1000 possible Titans. J. Geophys. Res., **100**, 16907–16922.

Jetten, M.S.M. *et al.* (1999) The anaerobic oxidation of ammonium. FEMS Microbiol. Rev., **22**, 421–437.

Lorenz, R.D. (2001) *32nd Lunar Planet. Sci. Conf.* Abstract #1160.

Lunine, J.I. (1993) Does Titan have an ocean? A review of current understanding of Titan's surface. Rev. Geophys., **31**, 133–149.

Lunine, J., Stevenson, D. (1983) Formation of the Galilean satellites in a gaseous nebula. Icarus, **52**, 14–38.

Lunine, J.I. and Stevenson, D. (1987) Clathrate and ammonia hydrate at high pressure: Application to the origin of methane on Titan. Icarus, **70**, 61–77.

Mancinelli, R., McKay, C. (1988) The evolution of nitrogen cycling. Orig. Life Evol. Biosphere, **18**, 311–325.

Perryman, M.A.C. (2000) Extra-solar planets. Rep. Progr. Phys., **63**, 1209–1272.

Pizzik, A.J. and Sommer, S.E. (1981) Sedimentary iron monosulfides: Kinetics and mechanism of formation. Geochim. Cosmochim. Acta, **45**, 687–689.

Simakov, M.B. (2000) Dinitrogen as a possible biomarker for exobiology: The case of Titan. In: G. A. Lemarchand and K. J. Meech (eds.) *Bioastronomy'99: A new era in bioastronomy*, Sheridan Books, pp. 333–338.

Simakov, M.B. (2001) The possible sites for exobiological activity on Titan. In: *Proc. First European Workshop on Exo/Astro-Biology, Frascati, 21–23, May 2001*, pp. 211–214.

Simakov, M.B. (2003) Possible biogeochemical cycles on Titan. *In press*

Spohn, T. and Schubert, G. (2003) Oceans in the icy Galilean satellites of Jupiter? Icarus, **161**, 456–467.

XI. The Search for Extraterrestrial Intelligence (SETI)

SETI-ITALIA
2003 Present Activities and Future

S. MONTEBUGNOLI[1], J. MONARI[1], C. BORTOLOTTI[1], A. CATTANI[1], A. MACCAFERRI[1], M. POLONI[1], A. ORLATI[1], S. RIGHINI[2], S. POPPI[1], M. ROMA[1], M. TEODORANI[1], C. MACCONE[3], C.B. COSMOVICI[4], N. D'AMICO[5]
[1]*CNR- Istituto di Radioastronomia,—Via Fiorentina, 40060 Villafontana,-Bo-Italy;* [2]*Osservatorio Astronomico di Torino Italy;* [3]*Centro di Astrodinamica "C. Colombo", Via Martorelli 43, 10155 To Italy;* [4]*-CNR Istuto di Fisica dello Spazio Interpl., Via Fosso Del Cavaliere, 00133 Roma, Italy;* [5]*Osservatorio Astronomico di Cagliari, Via salita dei Pini, Cagliari, Italy.*

Abstract. Observation activities within the Seti program have started at the Medicina radiotelescope (near Bologna—Italy-) since March 1998. An ultra high frequency resolution Serendip IV spectrometer module, in a 4 million channels configuration, was used. Observations with the Serendip IV spectrometer connected in piggyback mode have been carried out so far at the 32-mt VLBI dish antenna in the microwaves astronomical bands. In order to increase the detection possibilities, a more flexible version of the data post processing procedure (SALVE II) and a quite new KLT (Karhunen Loeve Transform) for data detection have been introduced. The final tests of such a KLT have been carried out with an expandable fast Mercury Altivec multinode CPUs cluster. This is planned to operate in parallel to the already existing Serendip IV high-resolution spectrometer.

1. Introduction

There are at least 10^{22} stars in the universe. The nearest 10^{11} are organized into the Milky Way, a lens shaped sistem of stars, gas, dust and dark matter wich is about 100.000 ly in diameter. This is our galaxy: the "island" where we live in the Universe! A fundamental question has arisen, from the ancient greek phylosophers up to the present scientists: *are we the only intelligent inhabitants of this boundless island?*

One of the way to answer this question is to search for some manifestation of a distant technology as, for instance, radio signals. The Seti program is aimed to the search for radio signals coming from the outer space and generated by extraterrestrial technological civilizations.

The Italian Institute of Radioastronomy of the Istituto Nazionale di Astrofisica, has been directly involved in Seti activities since March 98 when a Serendip IV, Werthimer (1996), system, an ultra high frequency resolution FFT (Fast Fourier Transform) based spectrum analyzer, was installed at the Medicina 32 m dish VLBI antenna (Fig. 1).

J. Seckbach et al. (eds.), Life in the Universe, 299–302.

Figure 1. View of the Medicina 32 m dish.

Since it operates in parallel, Werthimer (2000), to the ongoing observing programsat the 32 m VLBI dish, dedicated time for this particular radio signals search is not required. This approach drammatically reduces the cost of the program. While the radioastronomers use the radiotelescope, the Serendip IV searches for a CW signal at the same frequency and position in the sky as programmed in the observing schedule. In this way, the Seti search exploits the 100% of the antenna working time without any additional charge and, at the same time, takes under control the RFIs (Radio Frequency Interferences).

2. Seti–Italia: Present Situation

We tested the system during summer 1998 when the european VLBI network was engaged (under NASA request) to check some Mars Global Surveyor orbit parameters: the system post processing software provided an alarm due to the frequency shifted (Doppler effect) radio carrier coming from the spacecraft.

So far, the total estimated observed time, since the installation of the system, is summed up in the following table.

We haven't received any evidence of ETI radio signals during this observation time. We detected only man made radio interference (RFIs). Unfortunately the RFIs situation in Italy is continuously getting worse and an increasing effort is devoted to the radio astronomical frequency bands protection.

At present the instruments potentially involved in this program are:

Single Dish (GHz)	VLBI (GHz)	Geodynamic (GHz)	Calibration & Tests	Maintenance	Other
1.4	1.6	2			
1.6	5	8			
5	8				
8	22				
22					
43					
510 days	400 days	100 days	210 days	150 days	≈40 days

1. ***The Northern Cross Radiotelescope*** 564 × 640 mt (Medicina -Bo-): a very large transit radiotelescope able to cover the whole northern hemisphere ($0^0 \div 90^0$) with 30.000 sqm of collecting area.
2. ***The 32 mt VLBI dish*** (Medicina -Bo-)
3. ***The 32 mt VLBI dish*** (Noto -Sr- equipped with a new active mirror surface.

- The *SRT 64 mt dish* (under contructions near Cagliari, Sardinia Island) equipped with an active surface mirror allowing to work up to 100 GHz with a very good efficiency. It will work from 300 MHz up to 100 GHz

A new release of SALVE (SALVE 2) *(salve in Italian means "hello")* for Serendip IV data post processing procedures, based on the Hough transform, Monari (2000), for pattern recognition, was installed and tested. This is a very useful utility, especially exploited during the very crucial dopplered line detection algorithms post processing phase.

3. New Developmente in Data Acquisition

A prototype (MEDALT-1) of a fast vector multinode (Motorola 7400) system was assembled to test a Karhunen Loeve Transform (KLT) fast algorithm. The prototype is equipped with a double 12 bit 41 Ms/s A/D converter (Pentec™) triggered through an ad hoc designed programmable Direct Digital Synthesiser (DDS) clock generator. Data are moved from the A/Ds board to the CPUs carrier board (each VME carrier board can house two piggy-back modules with 2 CPUs each) through a FPDP bus.

It is well known the advantage obtained by the application of the KLT, Dixon (1993), instead of the FFT, in searching for any kind of signal in a radio band. The FFT always uses *cos* and *sin* as base functions and, in addition, it considers the signals periodic as the length of the data acquisition window. In this way it works well with monochromatic signals but not so properly with more complex radio signals. The KLT extracts the base functions from the signal itself and, then, works properly with any kind of signals (of course for a monochromatic signal it finds *cos* and *sin* base functions as the FFT). For this reason the KLT is a very suitable key transform, Maccone (1994), able to "understand" whether or not a radio band, supplied by the radio telescope, contains any kind of modulated signals, Maccone (2003). If an extraterrestrial technological civilization exists and communicates on its planet with radio waves, we cannot know in advance which radio signals they are spreading out the space (unless they intentionally transmit a CW signal). In this situation the KLT is the more suitable transform to detect unknown radio signal. The fast approach to the KLT (eigenvalues and eigenvectors computation) we worked on, is based on the following steps:

- N points data acquisition
- N points autocorrelation vector computation
- Autocorrelation matrix computation
- Arnoldi/Lanczos factorisation
- Givens Rotation (Eigenvalues approximation)

Assuming that our radio band will not contain more than 40 signals (searched signal + RFIs), we compute only these first 40 eigenvalues (in practice the first more important eigenvalues) and not all the N eigenvalues. For instance, operating with N = 1024 points, 10 MHz input bandwidth and using only one processor, it takes about 1.5 sec to compute the first 40 eigenvalues.

4. Summary

SETI observations have started at the Medicina radio telescopes since March 98. No evidence of ETI signals came out from our SETI activity so far. It is anyway necessary to go ahead with SETI observations because with the same high-resolution back end, the man-made radio interference can be effectively taken under control. A fast approach to the KLT computation (used as a detection tool in SETI), has been tested with good results, anyway a more detailed investigations on this transform needs to be done.

5. References

Dixon, S. R., Klein C. A. (1993) On the detection of unknow signals. The Third decennial US-USSR Conference on Seti ASP Conference series, Vol 47, 1993 Pag. 129–140.

Maccone C. (1994) Telecommunications, KLT and Relativity, Volume 1, IPI Press, Colorado Springs, CO, 1994, ISBN #1-880930-04-8.

Maccone C. (2003) Innovative SETI by the KLT", Pešek Lecture 2003, paper # IAA.9.1.01 presented at the 2003 IAF Bremen International Astronautical Congress.

Monari J., Montebugnoli S., Ravaglia F., Cecchi M. (2000), SALVE, New Era in Bioastronomy ASP Conference Series, Vol. 213, 479–483.

Werthimer D., Boyer S., David N., Donnely C., Cobb J., Lampton M., Aireau S. (1996) The Berkeley SETI Program Serendip IV instrumentation, Proceedings of 5th International Conference on Bioastronomy IAU Colloquium, No 161, 683–688.

Werthimer D., Boyer S., Cobb J., Lebofsky M., Lampton M. (2000) The Serendip IV Arecibo Sky Survey, A New Era in Bioastronomy ASP Conference Series, Vol. 213, 479–483.

SETI ON THE MOON

CLAUDIO MACCONE
Alenia Spazio S.p.A., Strada Antica di Collegno, 253,
Torino (TO) 10146—Italy

1. Introduction

The "Lunar Farside Radio Lab" Study of the International Academy of Astronautics (IAA), started in 1998 by late French astronomer Jean Heidmann (1923–2000), underwent substantial extensions and revisions since its coordination was taken up by this author. These modifications can be summarized as follows:

1) The goal of the Study was enlarged so as to encompass the whole of radio astronomy, rather than just SETI.
2) It was stressed that, from the Lunar Farside, one can detect radio frequencies lower than 15 MHz (i.e. wavelenghts longer than 20 m) impossible to detect from the Earth because of the blocking effect of the Earth's ionosphere. By detecting these radio waves from the Farside of the Moon, new discoveries should be expected especially in the fields of Cosmology and Stellar Astrophysics.
3) Lunar Farside Crater Saha, initially selected by Heidmann to host a radiotelescope, was replaced by Lunar Farside Crater Daedalus, located just at the Earth's Antipode on the Moon Farside. In fact, Daedalus is much more shielded than "Saha", not only against the radiation emitted by any future spacecraft orbiting the Earth at distances higher than the geostationary orbit, but even against the radiation emitted by the future Space Stations located at the triangular Lagrangian points L4 and L5 of the Earth-Moon system, as proposed long ago by Jerry O'Neill of Princeton University.
4) Four different scenario were envisaged for the relevant space mission, dubbed RadioMoon:
   a) Cheapest and easiest of all, just a spacecraft orbiting the Moon in its equatorial plane and carrying a 3-meter inflatable antenna detecting radio signals from the Universe when in the Shadow of the Earth, and downloading the data when above the Nearside.
   b) More expensive but not-so-hard-to-make: just the same as a) but with two or more spacecrafts, so as to create and interferometric Array in orbit around the Moon.

J. Seckbach et al. (eds.), Life in the Universe, 303–306.

c) More expensive still and hard-to-make, landing a Phased Array inside Daedalus and keeping the link with the Earth by a relay satellite in circular orbit around the Moon.
d) Very expensive and difficult: the same as in b) but with the goal of creating of an Array of Phased Arrays inside Daedalus. Much robotic work would then be requested plus one or more relay satellites orbiting the Moon.

Finally, legal protection of Daedalus from radio-pollution was sought, initially by virtue of the IISL. This is a very important issue, that should be soon presented to the United Nations Committee for the Peaceful Use of Outer Space. The "Lunar Farside Radio Lab" Study of the IAA should be published in 2004. The current status and future prospects of this "Cosmic Study" of the IAA are explained and justified in this paper.

2. Terminal Longitude λ on the Moon Farside for Radio Waves Emitted by Telecommunication Satellites in Orbit Around the Earth

In this section we just mention, without mathematical proof, an important equation, vital to select any RFI-free Moon Farside Base. We want to compute the small angle α (see Figure 1) beyond the limb (the limb is the meridian having longitude 90° E on the Moon) where the radio waves coming from telecommunications satellites in circular orbit around the Earth still reach, i.e. they become tangent to the Moon's spherical body. The new angle $\lambda = \alpha + 90°$ we shall call "terminal longitude" of these radio waves. In practice, no radio wave from telecom satellites can hit the Moon surface at longitudes higher than this terminal longitude λ. Then, λ turns out to be given by the equation

$$\lambda\,(R) = \operatorname{atan}\left(\frac{R - R_{Moon}}{\sqrt{D^2_{Earth-Moon} - (R - R_{Moon})^2}}\right) + \frac{\pi}{2}.$$

Here the independent variable R can range only between 0 and the maximum value that does not make the above radical become negative, that is $0 \leq R \leq D_{Earth-Moon} + R_{Moon}$.

3. Selecting Crater Daedalus at 180° E

The Committee claims that the time will come when commercial wars among the big industrial trusts running the telecommunications business by satellites will lead them to

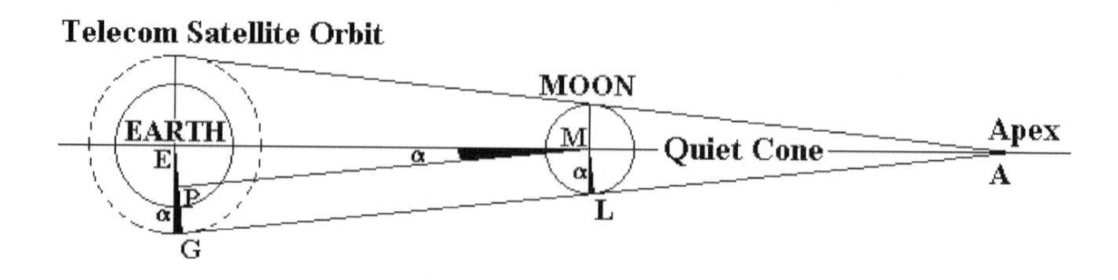

Figure 1. The simple geometry defining the "Terminal Longitude, λ" on the Farside of the Moon, where radio waves emitted by telecom satellites circling the Earth at a radius R are grazing the Moon surface.

grab more and more space around the Earth, pushing their satellites into orbits with apogee much higher than the geostationary one, with the result that ***crater Saha will be blinded as soon as a company decides to go higher than the geostationary orbit.*** The last remark is important for Bioastronomers. If we, the supporters of Bioastronomy, bet everything on a SETI and Bioastronomy Base located at Saha, then we may loose everything pretty soon! ***A "safer" crater must be selected further East along the Moon equator. How much further East?*** The answer if given by the above equation for λ.

Next we are now led to wonder: what is the Moon Farside terminal longitude corresponding to the distance of the nearest Lagrangian point, L1? The answer is given by the equation upon replacing $R = 323050$ km, and the result is $\lambda = 154.359°$. In words, this means the following basic, new result: ***the Moon Farside Sector in between 154.359 E and 154.359 W will never be blinded by RFI coming from satellites orbiting the Earth alone.***

In other words still, the ***limit*** of the blinded longitude as a function of the satellite's orbital radius around the Earth is 180° (E and W longitudes just coincide at this meridian, corresponding to the "change-of-date line" on Earth). But this is the ***antipode*** to Earth on the Moon surface, that is the point exactly opposite to the Earth direction on the other side of the Moon. And our mathematical theorem simply proves that the antipode is the most shielded point on the Moon surface from radio waves coming from the Earth. An intuitive and obvious result, really.

So, where are we going to locate our SETI Farside Moon base? Just take a map of the Moon Farside and look. One notices that the antipode's region (at the crossing of the central meridian and of the top parallel in the figure) is too a rugged region to establish a Moon base. Just about 5° South along the 180° meridian, however, one finds a large crater about 80 km in diameter, just like Saha. This crater is called Daedalus. So, ***the Committee proposes to establish the first RFI-free base on the Moon just inside crater Daedalus, the most shielded crater of all on the Moon from Earth-made radio pollution!***

4. The Committee's Vision of the Moon Farside for RFI-free Searches

Let us replace the value of $\lambda = 154.359°$ by the simpler value of $\lambda = 150°$. This matches also perfectly with the need for having the borders of the Pristine Sector making angles orthogonal to the directions of L4 and L5. The result is this vision of the Farside:

1) The Nearside of the Moon is left totally free to activities of all kinds: scientific, commercial and industrial.

2) The Farside of the Moon is divided into three thirds, namely three sectors covering 60° in longitude each, out of which:
   a) The Eastern Sector, in between 90° E and 150° E, can be used for installation of radio devices, but only under the control of the International Telecommunications Union (ITU-regime).
   b) The Central Sector, in between 150° E and 150° W, must be kept totally free from human exploitation, namely it is kept in its "pristine" radio environment totally free from man-made RFI. This Sector is where crater Daedalus is, a ~80 km crater located in between 177° E and 179° W and around 5° of latitude South. At the moment, the Committee is not aware of how high is the circular rim surrounding Daedalus.

c) The Western Sector, in between 90° W and 150° W, can be used for installation of radio devices, but only under the control of the International Telecommunications Union (ITU-regime). Also:

1) The Eastern Sector is exactly opposite to the direction of the Lagrangian point L4, and so the body of the Moon completely shields the Eastern Sector from RFI produced at L4. Thus, L4 is fully "colonizable".

2) The Western Sector is exactly opposite to the direction of the Lagrangian point L5, and so the body of the Moon completely shields the Western Sector from RFI produced at L5. Thus, L5 is fully "colonizable" in the Committee's vision, whereas it was not so in Heidmann's vision. In other words, the Committee's vision achieves the *full bilateral symmetry* of the vision itself around the plane passing through the Earth-Moon axis and orthogonal to the Moon's orbital plane.

3) Of course, L2 may not be utilized at all, as it faces crater Daedalus just at the latter's zenith. Any RFI-producing device located at L2 would flood the whole of the Farside, and must be ruled out. L2, however, is the only Lagrangian point to be kept free, out of the five located in the Earth-Moon system. Finally, L2 is not visible from the Earth since shielded by the Moon's body, what calls for "leaving L2 alone"!

5. References

Heidmann J. (1994), Saha Crater: a candidate for a SETI Lunar base, Academy Transactions Note, Acta Astronautica, **32**, 471–472.

Heidmann J. (2000) Sharing the Moon by Thirds: An Extended Saha Crater Proposal. Advances in Space Research. **26**, 371–375.

Heidmann J. (2000), Recent Progress on Lunar Farside Crater Saha Proposal, Acta Astronautica, **46**, 661–665.

Maccone C. (2001), Searching for Bioastronomical Signals from the Farside of the Moon, First European Workshop on Exo/Astrobiology, ESRIN, Frascati, Italy, 21–23 May 2001. **ESA SP-496**, 277–280.

Maccone C. (2001), The Lunar SETI Cosmic Study of IAA: Current Status and Perspectives, **IAA-01-IAA.9.1.05**, 52nd Int. Astronaut. Congress., 1–5 Oct. 2001.

Maccone C. (2002), Planetary Defense and RFI-Free Radioastronomy from the Farside of the Moon: A Unified Vision, Acta Astronautica, **50**, 185–199.

Maccone C. (2002), The Lunar Farside Radio Lab Study of IAA, **IAA-02-IAA.9.1.4**, World Space Congress – 2002, held in Houston, Texas, 10–19 October 2002.

Maccone C. (2003), The Quiet Cone Above the Farside of the Moon, Acta Astronautica, **53**, 65–70.

PROPOSING A UNITED NATIONS SECRETARY GENERAL SETI INTERNATIONAL ADVISORY BOARD

GIANDOMENICO PICCO[1], GIANCARLO GENTA[2], PIERO GALEOTTI[3] and DANILO NOVENTA[4]
[1]Advisor to the United Nations Secretary General, New York, USA;
[2]Department of Mechanics, Politecnico di Torino, Torino, Italy;
[3]Department of Physics, University of Torino, Torino, Italy; and
[4]Non Governmental Peace Strategies Project—Geneve, Switzerland

1. Introduction

Since the mid-1980s the SETI Committee of the International Academy of Astronautics started a study on the attitude researchers in the field should take when (and if) a serious candidate signal is detected. The results were published in a number of papers printed in a special Issue of Acta Astronautica (Tarter and Michaud, 1990) under the heading of *SETI Post Detection Protocols*. This research work was later at the base of a document with the formal title *Declaration of Principles Concerning Activities Following the Detection of Extraterrestrial Intelligence*.

At point 8 of this Declaration the possibility of broadcasting an answer is dealt with. This issue was first considered by Ney (Ney, 1985) and Goodman (Goodman 1990), who suggested the formalization of *ad hoc* International Agreements. In a subsequent paper Goldsmith (Goldsmith 1990) suggested that the International Astronautical Federation and the International Astronomical Union should give way to a Committee aimed to create an international agreement on how to formulate an *answer from planet Earth* to a confirmed detection of an alien message.

Michael Michaud, president of a subcommittee of the SETI Committee of the International Academy of Astronautics, suggested that future projects related to the issue of sending radio (or optical) signals towards the outer space should be discussed in an international forum. Such a document was approved by the Board of Trustees of the Academy, and become a formal Position Paper (Tarter and Michaud, 1990). It was also approved by the Board of Directors of the International Institute of Space Law. It was also suggested that this document should be submitted to the COPUOS. In 1996 John Billingham, chairman of the SETI Committee of the International Academy of Astronautics, published this document with the title of *Post-Detection SETI Protocol*. In various points of this document there is an explicit reference to the United Nations. In particolar, it is stated that a the Secretary General of the U.N. should be immediately informed of a confirmed detection.

J. Seckbach et al. (eds.), Life in the Universe, 307–309.

In the *Draft Declaration of Principles Concerning the Sending of Communications to Extraterrestrial Intelligence*, included in the Protocols it is stated that the problem od answering should be discussed by the COPUOS and the General Assembly of the UN. The UN should base its decisions on the exerience and knowledge of scientists and experts. The final decision on how to deal the matter of the answer should be entrusted to the Secretary General, who is the only authority who has the duty of *speaking for humankind.*

2. Proposal

The present proposal, formulated by the Italian SETI Study Center, is aimed to create an international advisory board aimed to assist the UN Secretay General in the task of representing humankind in the event of a contact with alien civilizations. It is articulated in the following points:

1. The preparation of an ad hoc study on SETI by an international panel of experts of various disciplines for the United Nations, addressed to the Secretary General. This study should cover all possible aspects (including scientific, philosophical, social, media, intelligence and security aspects) and be updated on a bi-annual basis.
2. The identification and formalization of the role of the United Nations Secretary General and his International Advisory Board in the functions of decision-making, management of communication with the international media system vis-à-vis the possible detection of artificial signals of proven extraterrestrial origin.
3. To create an effective, temporary board for the United Nations Secretary General, with information and situation support capabilities for SETI. It should act as the United Nations Secretary General's situation support staff in instances of extraterrestrial signal detection and, in particular, to assist him in the decision-making process upon successful detection of such signals.

The board will, in effect, develop, and make official, the contents of the SETI Post-Detection Protocols, expanding its activity to managing communication of any signal detection to the international media system. It should be composed of experts with internationally recognized credibility and competence in the following areas: Astrophysics and Cosmology, Bioastronomy, Radioastronomy, Astronautics and Space Exploration, Mathematics and Informatics, International Space Law, Social Sciences, International Relationships, Philosophy of Science, Religious Aspects, Mass Media, Intelligence and Military Aspects. The International Advisory Board should meet every two years to update and follow-up on current needs.

The funds required to create, maintain and support the activities of the international advisory board will not be provided by the United Nations but rather will be raised in the private sector, through various foundations, companies and associations with adequate international standing.

3. References

Goldsmith, D., (1990), Who Will Speak For Earth?, *Acta Astronautica*, Vol. 21, No. 2, 149–151.
Goodman, A.E., (1990), Diplomacy and the Search for Extraterrestrial Intelligence, *Acta Astronautica*, **21**, No. 2, 137–142.
Ney, P. (1985), An Extraterrestrial Contact Treaty?, JBIS, **38**, 521–522.
Tarter J.C. and Michaud M.A. (Eds.) (1990), SETI Post-Detection Protocol, *Acta Astronautica*, **21**, No. 2.

SOME ENGINEERING CONSIDERATIONS ON THE CONTROVERSIAL ISSUE OF HUMANOIDS

GIANCARLO GENTA
Department of Mechanics, Politecnico di Torino,
C. Duca degli Abruzzi 24, 10129, Torino, Italy

Abstract. Many papers have been published in the past on the issue of the possible existence of humanoid extraterrestrial intelligence (ETI) and the prevalent opinion is now that the humanoid form is rather an exception than a rule. The aim of the present paper is to consider an intelligent being as a sort of machine which has to perform a number of tasks, and to discuss whether the humanoid form is dictated by them. While not intending to supply answers but only to formulate some problems, it is suggested that, although there is no doubt that a close relationship between our layout and our essence of intelligent beings exists, this is not enough to support any claim that the humanoid form is prevalent.

1. Introduction

The *predominance of the humanoids*, i.e. whether extraterrestrial intelligent living beings (ETIs) may be similar, in a general sense, to humans is a very old issue: the first to take a firm stance on the subject was Galileo (Galilei, 1613), who stated that living beings may exist on the Moon or on other planets, but that their characteristics must be not only different from those of the beings on the Earth, but also from what our wildest imagination can produce. Nevertheless when the idea of ETIs became popular it was very often embodied in humanoid form: the aliens of early science fiction novels were humanoid, as well as almost all of those starring in science fiction moves and the popular image of aliens is that of humanoids. However, ETIs with different body shapes and even based on a completely different biology have been described (Pickover, 1999).

The very concept of 'humanoid' can refer either to an ETI with a body similar to that of humans, or a living being with any body shape but with an intelligence comparable with human intelligence both in quantity and quality. In humans intelligence goes together with consciousness and we take usually for granted that this is the case for all intelligent beings, but this is another point which can be debated.

Since most ETIs are assumed to be much older than us, by million (or billion) years, the very assumption that humanoid (in any sense) ETIs exist means that this is the ultimate form toward which evolution tends. There is no doubt that the human body and intelligence are very successful products of the evolutionary pressures on this planet. It can be argued that physical evolution may stop with the birth of intelligence and cultural evolution takes its place. A conscious species which attribute a large value to the life of individuals tends to

J. Seckbach et al. (eds.), Life in the Universe, 311–315.

oppose with increasing success natural selection as soon as science and technology advance: medicine, in its effort to grant a life as healthy and normal as possible to all individuals, independently from their fitness in a Darwinian sense, is a long battle against natural selection. But medicine and biology could have also another task: not to stop evolution, but to facilitate it producing favourable mutations. There is a general agreement to consider ethically unacceptable this last issue.

These considerations apply to all ETIs, who could behave in a different way facing this ethical choice. It then seems that we could find two types of ETIs: either species undergoing very little physical changes even in long times, and others which design their bodies using genetic engineering and eventually differentiate in many species to adapt themselves to different environments, different tasks or even just personal taste.

The aim of the present paper is to present some considerations on the human body as a peculiar machine designed (or better evolved) for a set of tasks and on its intelligence as the control program of such a machine. Note that the term evolution has been used to describe the advance of technology (Basalla, G. 1989).

2. Mobility and Manipulation

Mobility of a living being is strictly linked with how it gets food and energy. Autotrophic beings may not need any sort of mobility, while heterotrophic ones, and particularly predators (Endler 1993, Cockell and Lee, 2002), usually need to move to obtain food. Large animals either are supported on a solid surface, float in a fluid under the effect of hydrostatic (fluidostatic) forces or fly using aerodynamic (fluid-dynamic) forces (Azuma, 1989). Very small beings may use other supporting mechanisms, like surface tension (e.g. insects of the order Hemiptera), molecular interactions, etc. Since it is likely that an intelligent being has a minimum size larger than allowing to use these mechanisms, they will not be considered. Other solutions, like magnetic levitation or jets, are conceivable but are quite hypothetical and will not be considered.

On the Earth most animals moving on a solid surface use some sort of legs. Evolution is characterized by a gradual reduction of the number of the legs and with the terrestrial vertebrates their number reduced to four. A high number of legs, together with a low position of the center of mass allows the animal to remain easily in static equilibrium during all phases of walking. In general, the larger is the animal and the lower is the gravity of the planet, the easier is to remain in equilibrium on a small number of legs, in the sense that the response of the nervous system to avoid falling down may be less quick. From this point of view low gravity simplify all operations related to motion.

On the other side, a maximum walking speed exists for any animal; then a change of gait and the transition from walking to running or jumping (Cavagna et al., 1998, Minetti, 2001) is needed. This speed is larger for taller animals and higher gravity. Since a high speed is in general an important factor in natural selection, there is a strong incentive to shift from walking (a sequel of static equilibrium positions) to running (which includes positions in which static equilibrium is not guaranteed). Large animals, possibly with a smaller number of legs, may have then an advantage, and bipeds are a very good configuration for beings having an adequate control system.

For what we can infer from the case of the Earth, the rise of intelligence and consciousness is linked to the development of the nervous system. Neurogenesis seems to be linked with multicellularity, and as soon as the nervous system started to become complex also the process of encephalization started. To reach a complexity sufficient for intelligence and consciousness, the nervous system must have a minimum size.

The energy consumed by the human brain is quite large: it constitutes only 2% of the weight of the body, but it consumes 20% of energy. From an energetic viewpoint it seems that an intelligent species needs to be heterotrophic, possibly high in the food chain, since it needs a much energetic food which can be assimilated quickly. These considerations would indicate that ETIs should be, at least at the beginning, predators. Moreover, it is a common opinion that the difficulties experienced by a generalist animal for hunting had a positive effect on the development of intelligence (Leakey, 2000).

On the Earth, hominids started developing technology when still their consciousness and intelligence were far less advanced than those characterising *Homo Sapiens Sapiens*. If an intelligent being is a 'generalist' animal, this is mostly due to the fact that he is able to use objects that increase the potentialities of its body, working like prostheses: what other species accomplish by slowly modifying their body, intelligent life forms obtain in an incomparably shorter time by implementing purposely designed objects (Righetto, 2000). Technology is a consequence of the development of intelligence, but causes itself problems which require an increase of the size of the brain and then a development of intelligence. As a consequence, intelligent beings need to have a technology and some manipulatory organs, arms and hands (at least in the sense used in robotics).

ETIs can thus be expected to have some sort of locomotory and manipulatory organs; the latter being best derived from locomotion organs like legs. The humanoid layout with two arms and two legs seems to be optimal.

3. Symmetry

The general body plan of humanoids seems to be very suitable for an intelligent being, and the development of intelligence may well have been favoured by the challenges caused by the humanoid form. In particular, a biped layout with erect stance seems to be a logical consequence of an intelligent being with manipulatory appendages deriving from quadrupeds and carrying a heavy brain-case with many sensory organs at the front end. But such configuration is very demanding on the equilibrium organs and on the brain to obtain a sure and fast biped walk and run. Only with *Homo* (*Abilis* and then *Erectus*) these features could be obtained, while earlier hominids were much apelike (even early members of the *Homo* genus were more primitive from this viewpoint).

However, these considerations are not sufficient to state that it is likely that ETIs have a body form resembling humans. On the Earth all terrestrial and many sea animals have a bilateral symmetry, at least in the exterior body shape, with an even number of legs and the few ones that stand on three supports use a strong tail (in the plane of symmetry) for this purpose. Likewise they have an even number of organs for sight and hearing and, if they have a single mouth in the symmetry plane, it is because they have a single digestive tube. Yet the nostrils are two, but this seems to be completely arbitrary.

On our planet evolution developed also several living beings with radial symmetry of various order, like starfish (mostly of order five) or octopuses (order eight), but none of them has articulated limbs or adapted to live on land; the mainstream of evolution leading to intelligence took a different path.

The presence of a symmetry plane has some peculiar dynamic characteristics for a moving object (Genta, 2003) and robots with radial symmetry (mostly of order six) did not show particular advantages as walking machines (Genta and Amati, 2002), but this does not seem to rule out the possibility that in other biospheres intelligence may develop in animals with completely different symmetry characteristics.

No doubt that we can imagine (although with difficulty) ETIs with no symmetry at all. Radial symmetry is another possible choice, but other symmetries, much less familiar to us, are possible. In the mineral world, crystals exhibit many possibilities in this field; perhaps astrobiology will supply other examples. The fact that we cannot imagine them is immaterial: following Galileo it is what we should expect from extraterrestrials.

4. Conclusions

The humanoid layout has been shown to be an optimized response to the needs of an intelligent being living on the surface of a rocky planet, and our intelligence and our body evolved together and are closely matched. In spite of all this the belief that the humanoid layout is common among ETIs is quite naive. Evolution proceeds at random through small changes and cannot explore all the space of the configurations to search the optimal one to fill any ecological niche. It reaches near-optimum solutions which may be close to each other (convergent evolution) only if the genes producing them are potentially present in common ancestors.

Much then depends on the way life started. If the supporters of panspermia are right (and the author thinks it is unlikely), and all living beings derive from some sort of organised matter containing genetic instructions, similarities can be larger and still larger will be if what is carried trough space are directly viruses or spores. Finally there is the problem of symmetry. On our planet the common origin didn't prevent to experiment with different symmetries. The humanoid form might at best be the optimal solution for an intelligent being with bilateral symmetry, while different shapes can have been reached in places where different symmetries prevail.

5. References

Azuma, A. (1989), *The Biokinetics of Flying and Swimming*, Springer, Tokyo.
Basalla, G. (1989), *The Evolution of Technology*, Cambridge University Press.
Cavagna, G. A., Willems, P.A. and Heglund, N.C. (1998), Walking on Mars, *Nature*, Vol. 393, June, p. 636.
Cockell, C.S. and Lee, M. (2002), Interstellar Predation, *JBIS*, Vol. 55, pp. 8–20, 2002.
Coffey, E.J. (1985), The Improbability of Behavioural Convergence in Aliens—Behavioural Implications of Morphology, *JBIS*, Vol. 38, pp. 515–520.
Endler, J.A. (1993), Interactions between Predator and Prey, In: J.R. Krebs and N.B. Davies (eds.) *Behavioural Ecology: an Evolutionary Approach.*, Scientific Publications, Oxford. pp. 169–196.
Galilei, G. (1613), *Istoria e dimostrazioni intorno alle macchie solari e loro accidenti*, Rome.
Genta, G. (2003), *Motor Vehicle Dynamics*, World Scientific, Singapore.

Genta, G. and Amati, N. (2002), Non-Zoomorphic versus Zoomorphic Walking Machines and Robots: a Discussion, *European Journal of Mech. & Env. Engineering*, Vol. 47, n. 4, 2002, pp. 223–237.
Leakey R. (2000), *The Origin of Humankind*, Orion Books Ltd., London.
Minetti, A.E. (2001), Invariant Aspects of Human Locomotion in Different Gravitational Environments, *Acta Astronautica*, Vol. 39, No 3–10, pp. 191–198, 2001.
Pickover, C. (1999), *The Science of Aliens*, Basic Books, Boulder, Colorado.
Righetto, E. (2000), *La scimmia aggiunta*, Paravia, Torino.

XII. The Search for Evolution of Intelligent Behavior and Density of Life

THE NEW UNIVERSE, DESTINY OF LIFE, AND THE CULTURAL IMPLICATIONS

STEVEN J. DICK
U. S. Naval Observatory, 3450 Massachusetts Avenue, NW
Washington, DC 20392-5420 USA.

Abstract. Cosmic evolution embraces at least three vastly different possibilities for the destiny of life in the universe. The ultimate product of cosmic evolution may be only planets, stars and galaxies—a "physical universe" in which life is extremely rare. By contrast cosmic evolution, through biological evolution, may commonly result in life, mind and intelligence, an outcome that I term the "biological universe." Taking a long-term view not often discussed, cultural evolution may have already produced artificial intelligence, constituting a "postbiological universe." Astronomical, biological and cultural evolution are the three components of the Drake Equation, and each component must be taken seriously. Each of the three possible outcomes of cosmic evolution results in a different destiny for life. But the destiny of life is not predictable; where intelligence is involved, the philosophical problem of free will must also play a role.

1. Cosmic Evolution: Three Possible Outcomes

Before discussing the destiny of life in the new universe, we must clarify what we mean by "the new universe". A century ago, most astronomers argued that the universe was only 3,600 light years in extent, that the solar system was nearly at its center, and that humans were its ultimate purpose. The destiny of life was the destiny of humans, and that destiny was tied to religious and philosophical beliefs.

Our view of the universe today has immensely enlarged. We now know that the visible horizon of our universe is about 13.7 billion light years in extent, and full of galaxies. We now know about the expanding universe, the accelerating universe, Einsteinian space-time, inflationary cosmology, and dark energy. But no concept has been so radical as cosmic evolution. The universe a century ago was static. Ours today is evolving, and cosmic evolution is the guiding principle for all of astronomy. Thus, in contrast to a century ago, we can speculate on the destiny of life—on Earth and in the universe – based on what we now know about cosmic evolution.

The intellectual basis for this guiding principle of cosmic evolution had its roots in the 19th century when a combination of Laplace's nebular hypothesis and Darwinian evolution gave rise to the first tentative expressions of parts of this worldview. But cosmic biological evolution first had the potential to become a research program in the 1950s and 1960s when its cognitive elements had developed enough to become experimental and observational sciences. Harvard College Observatory Director Harlow Shapley was an early modern

J. Seckbach et al. (eds.), Life in the Universe, 319–326.

proponent of this concept, and already in 1958 spoke of it in now familiar terms, elaborating his belief in billions of planetary systems, where "life will emerge, persist and evolve" (Shapley, 1958).

Already as the Space Age began, the concept of cosmic evolution – the connected evolution of planets, stars, galaxies *and* life – provided the grand context within which the enterprise of exobiology was undertaken. The idea of cosmic evolution spread rapidly over the next 40 years, both as a guiding principle within the scientific community and as an image familiar to the general public. NASA enthusiastically embraced, elaborated and spread the concept of cosmic evolution from the Big Bang to intelligence as part of its SETI and exobiology programs in the 1970s and 1980s. When in 1997 NASA published its Origins program Roadmap, it described the goal of the program as "following the 15 billion year long chain of events from the birth of the universe at the Big Bang, through the formation of chemical elements, galaxies, stars, and planets, through the mixing of chemicals and energy that cradles life on Earth, to the earliest self-replicating organisms—and the profusion of life". With this proclamation of a new Origins program, cosmic evolution became the organizing principle for most of NASA's space science effort.

Today, the Big Question remains—how far does cosmic evolution commonly go? Does it end with the evolution of matter, or the evolution of life and intelligence? In this sense two astronomical worldviews hang in the balance in modern astronomy, just as they did four centuries ago when Galileo wrote his *Dialogue on the Two Chief World Systems*, and marshaled all the arguments for and against the geocentric and heliocentric theories. The two chief world systems today, I argue, are the physical universe, in which cosmic evolution commonly ends in planets, stars and galaxies, and the biological universe, in which cosmic evolution routinely results in life, mind and intelligence. We are on the brink today of being able to decide between these two worldviews. And that is why astrobiology is so important—it is the science that will decide which of these worldviews is true. I also argue that there may be a third worldview opened up by cosmic evolution—the postbiological universe based on cultural evolution. But it is not a worldview yet commonly discussed.

2. The Physical Universe

Almost all of the history of astronomy, from Stonehenge through much of the 20th century, deals with the people, the concepts, the techniques that gave rise to our knowledge of the physical universe. Babylonian and Greek models of planetary motion, medieval commentaries on Aristotle and Plato, the astonishing advances of Galileo, Kepler, Newton and their comrades in the Scientific Revolution, the details of planetary, stellar and galactic evolution—all these and more address the physical universe. The physical universe now boasts a whole bestiary of objects unknown a century ago—from blazars and quasars to pulsars and black holes. The quest for a biological universe should in no way obscure the fact that the physical universe is in itself truly amazing.

For millennia the destiny of life in the universe was synonymous with the destiny of life on Earth, and was tied to the geocentric system associated with Aristotle, the Earth at the center and the heavens above. This cosmological worldview provided the very reference frame for daily life, religious and intellectual. The heliocentric system changed that, thus the societal uproar following this daring new cosmological worldview. Since then the history of modern astronomy has been one of increasing decentralization of humanity. Even though we are no longer physically central, the destiny of life in a physical universe in which life is very rare remains closely tied to humanity's destiny. Ward and Brownlee (2003) have set forth the scenario of how Earth will eventually become uninhabitable. Long before that time, space travel will have taken us beyond the Earth, and of necessity it will have to take us beyond the solar system if we are to survive the death of the Sun. Then, the destiny of life is limited by the physical laws of the universe. As Chaisson (2001) has emphasized, we are now in the life era, but we will not always be. Adams and Laughlin (1999, 47–50) have laid out the fate of life on Earth in a cosmic context: in two billion years the Sun will have increased in brightness enough to induce a runaway greenhouse effect on our home planet. If we can escape to another star, however, our longevity rises considerably: the gas depletion rate indicates that star formation in galaxies will not halt for another 100 trillion years.

Long before we have to face the challenges of Earth's future, much less of the future of the universe, cultural evolution will have changed *homo sapiens* in fundamental ways. Indeed, some believe that cultural evolution will most likely have resulted in artificial intelligence within a few thousand years, and in the replacement of the species. Moravec (1988) and Kurzweil (1999) among others foresee this happening within a few hundred years. In science fiction terms, this physical universe, in which life and mind are rare but in which artificial intelligence is common, is the universe of Isaac Asimov's Foundation series in science fiction. No non-human biological life is found in Asimov's entire Foundation series; the galaxy is populated by humans and by intelligent robots intimately associated with humans. There are no independent extraterrestrials. What may happen on Earth in the next few thousand years may have already happened among extraterrestrials. I will return to this idea of a postbiological universe later.

If life is to be played out in this physical universe, the destiny of life is for humans, or their robotic ancestors, to populate the universe. In such a universe, where we are unique or very rare, stewardship of our pale blue dot takes on special significance.

3. The Biological Universe

The second possible outcome of cosmic evolution is the biological universe—the universe in which cosmic evolution commonly ends in life. Ideas about a possible biological universe date back to ancient Greece, in a history that is now well known (Dick, 1982; Crowe, 1986; Guthke, 1990; Dick, 1996). The Copernican revolution, which made the Earth a planet and the planets potential Earths, provided the theoretical underpinnings for extraterrestrial life. Much of the history of exobiology is an elaboration of this theme, attempting to show just how similar to the Earth the other planets really are. The idea of planets beyond the solar system also has a deep history stretching back at least to the 17th century with the vortex cosmology of René Descartes and the explicit depiction of planetary systems with Bernard le Bovier de Fontenelle's *Entretriens sur la pluralité des mondes* (1686). Theoretical ideas about the formation of planets, and empirical searches for them, are an elaboration of this theme.

Over the last 40 years science has probed in a substantial way this new worldview of the biological universe. In the 1950s and 1960s four cognitive elements—planetary science, the search for planetary systems, origin of life studies, and SETI—converged to give birth to the field of exobiology (Dick, 1996). After the Viking missions failure to find life on Mars, in the 1990s many events conspired to revitalize exobiology: the Mars rock, the Mars Global Surveyor observations of the gullies of Mars, the Mars Odyssey detection of water near the surface, the Galileo observations of Europa, circumstellar matter, extrasolar planets, life in extreme environments including hydrothermal vents, and complex interstellar organics. All these elements fed into NASA's new astrobiology program, which emerged from a deep organizational restructuring at NASA in 1995. Astrobiology now is a much more robust science than exobiology was 40 years ago. Astrobiology places life in the context of its planetary history, encompassing the search for planetary systems, the study of biosignatures, and the past, present and future of life. Astrobiology science added new techniques and concepts to exobiology's repertoire, raised multidisciplinary work to a new level, and was motivated by new and tantalizing evidence for life beyond Earth.

Astrobiology's image of a biological universe raises deep questions of the destiny of life and societal impact at many levels. When addressing societal impact, one must distinguish between a biological universe full of microbial life, and one full of intelligent life. Certainly the implications would be different. One is tempted to say that the impact of the discovery of microbial life would be limited to science, as biologists contemplated their first data for a universal theory of biology. But the reaction in 1996 when possible nanofossils were announced in the Mars rock clearly shows that the impact will be greater than that.

The idea of *intelligent* life in the universe has already generated a long history of discussion of potential impact, especially in the theological arena. In the Christian tradition, for example, discussion of the implications of extraterrestrial life for the doctrines of Redemption and Incarnation now have a 500 year history (Crowe, 1986; Dick 2000b). More recently, NASA itself has sponsored a number of societal impact studies, in accordance with the National Aeronautics and Space Act goal of identifying the impact of the space program on society. In conjunction with the launching of the NASA SETI program, in 1991–1992 NASA sponsored a systematic series of workshops on the cultural aspects of SETI (Billingham et al., 1999). And shortly after the astrobiology endeavor was launched, another group gathered to discuss broader concerns (Dick, 2000a; Harrison and Connell, 1999). Although the cultural impact of discovering primitive life continues to receive little attention, the impact of contact with intelligent life has been the subject of much speculation, and some serious study. One approach is that historical analogs form a useful basis for discussion, not in the form of the usually disastrous physical cultural contacts on Earth, but by studying the transmission of knowledge across cultures and the reception of scientific worldviews (Dick, 1995). The Foundation for the Future, which focuses on the question of the state of humanity a thousand years hence, sponsored a meeting on the implications of deciphering a message with high information content (Tough, 2000). And earlier this year the American Association for the Advancement of Science Program on the Dialogue between Science and Religion launched a series of workshops on the societal implications of astrobiology. Thus the societal impact of extraterrestrial life has been recognized from a variety of viewpoints, but much remains to be done.

From another angle both scientists and historians of science have seen the idea of a universe full of life as a kind of worldview similar in status to the Copernican and Darwinian worldviews. Whether seen as a worldview, as one of the landmark questions of human thought, or as an essential element in the question of humanity's place in nature, astrobiology has the potential to impact society no less than the other great revolutions of science. Indeed, one can see the UFO debate and alien theme in science fiction as attempts of popular culture to work out the new worldview.

The destiny of human life in a biological universe is quite different from that in a physical universe. Rather than populating a universe empty of life, the destiny of humanity is perhaps to interact with extraterrestrials, to join what has been called a "galactic club" whose goal is to enhance knowledge.

4. The Postbiological Universe

In the remainder of this paper I want to argue that there is another option aside from the physical and the biological universe, an option that thus far has not been taken seriously. But if we take seriously physical and biological cosmic evolution, we also need to take seriously cultural evolution as an integral part of cosmic evolution and the Drake Equation. For those of you familiar with the vast sweep of time in Olaf Stapledon's *Last and First Men*, you will know what I mean when I say that we need to think in Stapledonian terms. While astronomers are accustomed to thinking on cosmic time scales for physical processes, they are not accustomed to thinking on cosmic time scales for biology and culture. But cultural evolution now completely dominates biological evolution on Earth. Given the age of universe, and if intelligence is common, it may have evolved far beyond us. I have recently argued (Dick, 2003) that cultural evolution over thousands, millions or billions of years will likely result in a "postbiological universe" populated by artificial intelligence, with sweeping implications for SETI strategies and for our worldview. It also has implications for the destiny of life, indeed, artificial intelligence may be the destiny of life on Earth if it has already happened throughout the universe. We may see our future in the evolution of extraterrestrial civilizations. This is another motivation for searching.

The three scientific premises for a postbiological universe are 1) that the maximum age of ETI is several billion years; 2) L, the lifetime of a technological civilization is >100 years and probably much larger; and 3) in the long term cultural evolution will supersede biological evolution, and produced something far beyond biological intelligence.

It is widely agreed that the maximum age of ETI, if it exists, is billions of years. Recent results from the WMAP place the age of the universe at 13.7 billion years, with a 1% uncertainty, and confirm that the first stars formed at about 200 million years after the Big Bang. The oldest Sun-like stars probably formed within about a billion years, or 12.5 billion years ago. By that time enough heavy element generation and interstellar seeding had taken place for the first rocky planets to form. Then, if Earth's history is any guide, it may have taken another 5 billion years for intelligence to evolve. In a universe 13.7 billion years old, this means that the first intelligence could have evolved 7.5 billion years ago. Livio (1999) and Kardashev (1997), among others, have argued that extraterrestrial civilizations could be billions of years old, and it is commonly accepted among SETI practitioners.

L, the lifetime of a technological civilization, is notoriously uncertain. Even pessimists admit 10,000 years is not unlikely. But the key point is that the age of ETI does not have to be large for cultural evolution to do its work. Even at our low current value of L on Earth, biological evolution by natural selection is already being overtaken by cultural evolution, which is proceeding at a vastly faster pace than biological evolution (Dennett, 1996). Technological civilizations do not remain static; even the most conservative technological civilizations on Earth have not done so, and could not given the dynamics of technology and society. Unlike biological evolution, L need only be thousands of years for cultural evolution to have drastic effects on civilization.

The course of cultural evolution is unpredictable even on Earth, much less in the universe. Darwinian models of cultural evolution have been the subject of much recent study (Lalande & Brown, 2000), but they are fraught with problems and controversy, whether drawn from sociobiology, behavioral ecology, evolutionary psychology, gene-culture co-evolution or memetics.

While theoretical and empirical studies of cultural evolution hold hope for a science of cultural evolution, lacking a robust theory to at least guide our way, we are reduced at present to the extrapolation of current trends supplemented by only the most general evolutionary concepts. Several fields are most relevant, including genetic engineering, biotechnology, nanotechnology, and space travel. But one field—artificial intelligence—may dominate all other developments in the sense that other fields can be seen as subservient to intelligence. Biotechnology is a step on the road to AI, nanotechnology will help construct efficient AI and fulfill its goals, and space travel will spread AI. Genetic engineering may eventually provide another pathway toward increased intelligence, but it is limited by the structure of the human brain. In sorting priorities, I adopt what I term the central principle of cultural evolution, which I refer to as the Intelligence Principle: *the maintenance, improvement and perpetuation of knowledge and intelligence is the central driving force of cultural evolution, and that to the extent intelligence can be improved, it will be improved.* The Intelligence Principle implies that, given the opportunity to increase intelligence (and thereby knowledge), whether through biotechnology, genetic engineering or AI, any society would do so, or fail to do so at its own peril. I have elsewhere attempted to justify this principle (Dick, 2003), but the argument comes down to is this: culture may have many driving forces, but none can be so fundamental, or so strong, as intelligence itself.

The field of AI is a striking example of the Intelligence Principle of cultural evolution. Although there is much controversy over whether artificial intelligence can be constructed that is equivalent or superior to human intelligence—the so-called Strong AI argument—several AI experts have come to the conclusion that AI will eventually supersede human intelligence on Earth. Moravec (1988) spoke of "a world in which the human race has been swept away by the tide of cultural change, usurped by its own artificial progeny." Kurzweil (1999) also sees the takeover of biological intelligence by AI, not by hostility, but by willing humans who have their brains scanned uploaded to a computer, and live their lives as software running on machines. Tipler (1994), well known for his work on the anthropic principle and the Fermi paradox, concluded that machines may not take over, but will at least enhance our well-being. But the self-reproducing von Neumann machines that Tipler foresaw in his explanation of the Fermi paradox may well exist if his view of the Fermi paradox is wrong.

Thus, it is possible that L need not be millions of years for a postbiological universe scenario. It is possible that such a universe would exist if L exceeds a few hundred or a few thousand years, where L is defined as the lifetime of a technological civilization that has entered the electronic computer age (which on Earth approximately coincides with the usual definition of L as a radio communicative civilization.

The postbiological universe cannot mean a universe totally devoid of biological intelligence, since we are an obvious counterexample. Nor does it mean a universe devoid of lower life forms. Rather, the postbiological universe is one in which the majority of intelligent life has evolved beyond flesh and blood. The argument makes no more, and no fewer, assumptions about the probability of the evolution of intelligence or its abundance than standard SETI scenarios; it argues only that if such intelligence does arise, cultural evolution must be taken into account, and that this may result in a postbiological universe.

Thus it is possible that the destiny of life on Earth is artificial intelligence, and that other civilizations in the universe have already realized this destiny. How such postbiologicals—whether terrestrial or extraterrestrial—would use their knowledge and intelligence is a value question that is at present unanswerable. The likelihood of a postbiological universe and its implications should be systematically considered. Whether one relishes or opposes the idea of a universe dominated by machines, the transition to such a universe presents many moral dilemmas and raises with renewed urgency the ancient philosophical question of destiny and free will.

5. Summary

The new universe, driven by the astronomical, biological and cultural components of cosmic evolution, may result in any of the three outcomes described here: the physical universe, the biological universe, or the postbiological universe. Which of the three the universe has produced in reality we do not yet know—this is the challenge of astrobiology. But the outcome depends largely on what Davies (1998) has called the biofriendly universe. The question of the biofriendly universe brings us full circle to the physical universe with which we began, for the anthropic principle (more accurately termed the biocentric principle) postulates that a universe full of life is written into the very fabric of the physical universe, into its constants, laws and atomic structure. Our particular universe—the object of our contemplation and study over thousands of years—may indeed be a product of a highly evolved intelligence, albeit a natural intelligence. This brings us into the realm of theology, but not the usual supernatural and anthropocentric theology. It brings us to a cosmotheology (Dick, 2000), in which we need not enter the realm of the supernatural, any more than we require the fundamental Aristotelian dichotomy between the Earth and the Heavens that held sway for two millennia. It is a cosmos in which humanity is not central, yet where it is at home in the universe in which it plays its role. Whatever its long-term destiny, it is surely the destiny of humanity in the near future to follow the trail of scientific evidence wherever it may lead, even if it means abandoning old scientific, philosophical and theological ideas.

References

Adams, F. and Laughlin, G. (1999). *The Five Ages of the Universe*. The Free Press, New York.
Billingham, J. et al. (1999). *Social Implications of the Detection of an Extraterrestrial Civilization*. SETI Press, Mountain View, Ca.
Chaisson, E. (2001). *Cosmic Evolution: The Rise of Complexity in Nature*. Harvard University Press, Cambridge, Mass.

Crowe, M. J. (1986). *The Extraterrestrial Life Debate, 1750–1900: The Idea of a Plurality of Worlds from Kant to Lowell.* Cambridge University Press, Cambridge; Dover reprint, 1999.

Davies, P. (1998). *The Fifth Miracle: The Search for the Origin of Life.* Allen Lane, London.

Dick, S. J. (1982). *Plurality of Worlds: The Origins of the Extraterrestrial Life Debate from Democritus to Kant.* Cambridge University Press, Cambridge.

Dick, S. J. (1995). Consequences of Success in SETI: Lessons from the History of Science, In: G. Seth Shostak (ed.) *Progress in the Search for Extraterrestrial Life*PASP, San Francisco, 521–532.

Dick, S. J. (1996). *The Biological Universe: The Twentieth Century Extraterrestrial Life Debate and the Limits of Science.* Cambridge U. Press, Cambridge.

Dick, S. J. (2000a). Cultural Aspects of Astrobiology: A Preliminary Reconnaissance at the Turn of the Millennium," In: G. LeMarchand and K. Meech (eds.) *Bioastronomy 99: A New Era in Bioastronomy.* PASP: San Francisco. pp. 649–659.

Dick, S. J. (ed.) (2000b). *Many Worlds: The New Universe, Extraterrestrial Life and the Theological Implications,* Templeton Press, Philadelphia.

Dick, S. J. (2003). Cultural Evolution, the Postbiological Universe and SETI. *International Journal of Astrobiology,* 2, 65–74.

Guthke, K. S. (1990). *The Last Frontier: Imagining other Worlds from the Copernican Revolution to Modern Science Fiction.* Cornell University Press, Ithaca, N.Y.

Harrison, A. A., Connell, K. (1999). Workshop on the Societal Implications of Astrobiology, Full Report posted on the web at http://astrobiology.arc.nasa.gov/workshops/societal/

Kardashev, N. S. (1997). Cosmology and civilizations, *Astrophysics and Space Science,* **252**, 25–40.

Kurzweil, R. (1999). *The Age of Spiritual Machines: When Computers Exceed Human Intelligence.* Penguin Books, New York.

Lalande, K. N. and Brown, G. R. (2002). *Sense & Nonsense: Evolutionary Perspectives on Human Behaviour.* Oxford University Press, Oxford.

Livio, M. (1999a). How rare are extraterrestrial civilizations and when did they emerge? *Astrophysical Journal,* **511**: 429–431.

Moravec, H. (1988). *Mind Children: The Future of Robot and Human Intelligence* Harvard U. Press: Cambridge, Mass.

Shapley, H. (1958). *Of Stars and Men.* Beacon Press, Boston.

Shklovskii, J. & Sagan, C. (1966). *Intelligent Life in the Universe.* Holden-Day, San Francisco, pp. 360–361.

Shostak, S. (1998). *Sharing the Universe: Perspectives on Extraterrestrial Life.* Berkeley, Ca., 103–109.

Tipler, F. (1985). Extraterrestrial intelligent beings do not exist, In: E. Regis (ed.) *Extraterrestrials: Science and Alien Intelligence* (Cambridge, 1985) pp. 133-150.

Tipler, F. (1994). *The Physics of Immortality.* New York.

Tough, A. ed. (2000). *When SETI Succeeds: The Impact of High-Information Contact,* Foundation for the Future, Bellevue, Wash.

Ward, P. and Brownlee, D. (2000). *Rare Earth: Why Complex Life is Uncommon in the Universe.* Copernicus, New York.

Ward, P. and Brownlee, D. (2003). *The Life and Death of Planet Earth: How the New Science of Astrobiology Charts the Ultimate Fate of our World.* Henry Holt, New York.

EVOLUTION OF INTELLIGENT BEHAVIOR
Is it just a question of time?

JULIAN CHELA-FLORES
The Abdus Salam International Centre for Theoretical Physics. Strada Costiera 11; 34136 Trieste, Italy and Instituto de Estudios Avanzados, Caracas 1015A, Venezuela.

Abstract. Several issues have been raised regarding the nature of biology in a universal context: (1) life is a cosmic imperative (De Duve, 1995); (2) multicellular life is a rare phenomenon in the cosmos, although the existence of microbial life may still be widespread. This possibility has been referred as the "Rare-Earth" Hypothesis. (Ward and Brownlee, 2000). We shall develop a third possibility: (3) evolution of intelligent behavior is just a matter of time and preservation of steady planetary conditions, and hence ubiquitous in the universe (Chela-Flores, 2003a, b). Darwin's theory of evolution is assumed to be the only theory that can adequately account for the phenomena that we associate with life anywhere in the universe. This question is motivated by the problem of understanding the bases on which we can get significant insights into the question of the distribution of life in the universe. Such information would also have deep implications on the other frontier of astrobiology mentioned above, the destiny of life in the universe. We argue in favor of the inevitability of life by assuming that Darwinian evolution is a universal process (Dawkins, 1983) and that the role of contingency has to be seen in the context of evolutionary convergence, not only in biology, but also in other realms of science. We shall restrict our discussions to astrobiology. The four areas which define this new science are: the origin, evolution, distribution and destiny of life in the universe. It is undoubtedly the fourth one, which is most likely to encourage interdisciplinary dialogue (Aretxaga, 2004; Vicuña and Serani-Merlo, 2004).

1. Introduction

We approach the empirical question of how to test the earliest stages of biological evolution in our own solar system, including Europa. For this purpose we may benefit from the results of the Galileo Mission to the Jovian system: Some of the Galileo results suggest that Europa has an inner core, a rocky mantle and a surface layer, mainly of liquid water. Impact craters also suggest that there is an ice-covered inner ocean, since they are shallower than would be expected on a solid (silicate) surface, such as that of the Moon. In the past we have considered a robot especially built to penetrate ice overlying a mass of liquid—the so-called 'cryobot' (Horvath *et al.*, 1997). Others, more recently, are approaching the question of melting probes (Biele *et al.*, 2002). A somewhat more remote possibility is to build a corresponding submersible robot ('hydrobot') capable of bearing some experiments in its interior.

J. Seckbach et al. (eds.), Life in the Universe, 327–331.

2. Are there Other Environments in the Cosmos that are Favorable to Life's Origin?

In the Europa Ocean, or possibly on the iced surface itself, we are presented with the problem of deciding whether biology experiments should be planned in due course (Gatta and Chela-Flores, 2004), and which tests should be taken to the stage of feasibility studies. We discuss some possible biosignatures of biochemical nature (Bhattacherjee and Chela-Flores, 2004) and comment on their relevance in the search for evolutionary biosignatures. In the original theory of Darwin the possibility had been raised that local environments shape how organisms change with time through natural selection. In view of the evidence discussed earlier (Chela-Flores, 2003a), we assume that natural selection is the main driving force of evolution in the universe. For these reasons it is relevant to question whether local environments that were favorable for the emergence of life on the early Earth, were at all unique, occurring exclusively in our own solar system. Alternatively, we may question, as we do in the present paper, whether other environments fulfill conditions favorable to life's origin, either within our solar system, or in any of the planets, or satellites, in the multiple examples of solar systems known at present. In addition, we suppose that on such bodies steady conditions are preserved. By steady conditions it should be understood that the planet where life may evolve is bound to a star that lasts long enough: in other words, the time available for the origin and evolution of life should be sufficient to allow life itself to evolve before the solar system of the host planet, or satellite, reaches the final stages of stellar evolution, such as the red-giant and supernova phases. It is also assumed that major collisions of large meteorites with the world supporting life are infrequent after the solar system has passed through its early period of formation.

3. Evolutionary Convergence at the Cosmic Level

Cosmic evolutionary convergence may have some evidence at various levels, comets, meteorites and high red shift data. Firstly, hydrogen and helium make up almost the totality of the chemical species of the Universe. Only 2% of matter is of a different nature, of which approximately one half is made by the five additional biogenic elements (C, N, O, S, P). We know that nuclear synthesis is relevant for the generation of the elements of the Periodic Table beyond hydrogen and helium and, eventually, for the first appearance of life in solar systems. The elements synthesized in stellar interiors are needed for making the organic compounds that have been observed in the circumstellar, as well as in interstellar medium, in comets, and other small bodies. The same biogenic elements are also needed for the synthesis of biomolecules of life. Besides, the spontaneous generation of amino acids in the interstellar medium is suggested by general arguments based on biochemistry: the detection of amino acids in the room-temperature residue of an interstellar ice analogue that was ultraviolet-irradiated in a high vacuum has yielded 16 amino acids, some of which are also found in meteorites (Muñoz Caro *et al.*, 2002). There are factors, which contribute to the formation of habitable planets. Secondly, the Murchison meteorite may even play a role in the origin of life: According to chemical analyses, some amino acids have been found in several meteorites: in Murchison we find basic molecules for the origin of life such as lipids, nucleotides, and over 70 amino acids (Cronin and Chang,

1993). Most of the amino acids are not relevant to life on Earth and may be unique to meteorites. This remark demonstrates that those amino acids present in the meteorite, which also play the role of protein monomers, are indeed of extraterrestrial origin.). If the presence of biomolecules on the early Earth is due in part to the bombardment of interplanetary dust particles, or comets and meteorites, then the same phenomenon could have taken place in any of other solar systems. We shall not consider convergence at the molecular level (Doolittle, 1994), since it will be discussed separately in this volume (Akindahunsi, and Chela-Flores, 2004).

4. On the Inevitability of Biological Evolution

There is some evidence that once life originates provided sufficient (geologic) time is available, evolution is going to provide pressures to living organisms in every conceivable environment. This remark further advocates in favor of the hypothesis that once life appears at a microscopic level in a given planet or satellite, the eventual evolution of intelligent behavior is just a matter of time.

Cambrian fauna, such as lamp shells (inarticulate brachiopods) and primitive mollusks (Monoplacophora), were maintained during Silurian times by microorganisms that lived in hydrothermal vents (Little *et al.*, 1997). Taxonomic analysis of Cenozoic fossils suggests that shelly vent taxa are not ancestors of modern vent mollusks or brachiopods. We may conclude that modern vent taxa support the hypothesis that the vent environment is not a refuge for evolution. In fact, there is evidence that since the Paleozoic (e.g., the Silurian) and through the Mesozoic there has been movement of taxonomic groups in and out of the vent ecosystem through time—no single taxon has been unable to escape evolutionary pressures. Some independent support for the thesis of deep–water extinction has also been presented (Jacobs and Lindberg, 1998). Hence, these remarks rule out the possibility that these deep-sea environments are refuges against evolutionary pressures. In other words, the evidence so far does not support the idea that there might be environments, where ecosystems might escape biological evolution, not even at the very bottom of deep oceans. These remarks give some support to the hypothesis that any microorganism, in whatever environment on Earth, or elsewhere, would be inexorably subject to evolutionary pressures. As we have shown above, fossils from Silurian hydrothermal-vent fauna demonstrate that there has been extinction of species on these locations, which at first sight seem to be far removed from evolutionary forces. For these reasons we may ask whether over geologic time the most primitive cellular blueprint has inevitably bloomed into full eukaryogenesis and beyond the evolutionary pathway to organisms displaying intelligent behavior.

5. Discussion

The arguments presented in this paper militate in favor of planning experiments based on standard biology in solar system search for microorganisms, in view of both evolutionary convergence and universal Darwinism. Ever since the publication of *The origin of species*, it has been argued that the possible course of evolution may be dominated by either contingency or the gradual action of natural selection. Random gene changes accumulating over time may imply that the course of evolution is generally unpredictable over time. But some care is needed in this assertion: What is certainly unpredictable is the future of a given lineage. This is due to the strong role in shaping life's evolutionary pathways played by contingent factors, such as extinction of species due to asteroid collisions with a given inhabited world, or other calamities.

However, the main issue is the inevitability of the appearance of biological features, such as vision, locomotion, nervous systems, brains and, consequently intelligent behavior. We have argued that contingency does not contradict a certain degree of predictability of the eventual biological properties that are likely to evolve. We should underline "biological property", as opposed to a "lineage", which is clearly a strongly dependent on contingency. However, from what we have explained above, such limited predictability is nevertheless relevant to aspects of the question of life in the solar system, as well as evolution of intelligent behavior in the universe. We have argued that fossils of hydrothermal vent fauna militate in favor of the possibility that once life appears at a microscopic level, the eventual evolution of intelligent behavior is just a matter of the environment surviving over geologic time. Possible tests of evolutionary biomarkers with the techniques of molecular biology were considered.

6. References

Akindahunsi, A. A. and Chela-Flores, J. (2004) On the question of convergent evolution in biochemistry, in this volume.

Aretxaga, R. (2004) Astrobiology and Biocentrism, in this volume.

Bhattacherjee, A. B and Chela-Flores, J. (2004) Search for bacterial waste as a possible signature of life on Europa, in this volume.

Biele, J., Ulamec, J.S., Garry, Sheridan, S., Morse, A.D., Barber, S., Wright, I. P. Tug, H. and Mock, T. (2002) Melting probes at Lake Vostok and Europa, *ESA SP* **518**, pp. 253–260.

Chela-Flores, J. (2001) *The New Science of Astrobiology From Genesis of the Living Cell to Evolution of Intelligent Behavior in the Universe.* Kluwer Academic Publishers: Dordrecht, The Netherlands, p. 161.

Chela-Flores, J. (2003a) Astrobiology's Last Frontiers: The distribution and destiny of Life in the Universe. In: J. Seckbach (ed.) *Origins: Genesis, Evolution and Diversity of Life*, volume 6 of the **COLE** book. Kluwer Academic Publishers, Dordrecht, The Netherlands. pp. xx–yy.

Chela-Flores, J. (2003b). Testing Evolutionary Convergence on Europa. International Journal of Astrobiology (Cambridge University Press), **in press.**

Cronin, J. R. and Chang, S. (1993) Organic matter in meteorites: Molecular and isotopic analyses of the Murchison meteorite, In: J.M. Greenberg, C.X. Mendoza-Gomez, and V. Pirronello, (eds.) *The chemistry of life's origins.* Dordrecht: Kluwer Academic Publishers, pp. 209–258.

Dawkins, R. (1983) Universal Darwinism, In: D.S. Bendall (ed.) *Evolution from molecules to men*, London, Cambridge University Press, pp. 403–425.

De Duve, C. (1995) *Vital Dust. Life as a cosmic imperative*, New York, Basic Books, A Division of Harper Collins Publishers, pp. 296–297.

Doolittle, R.F. (1994) Convergent evolution: the need to be explicit, *Trends Biochem. Sci.* **19**, pp. 15–18.

Gatta, R. S. and Chela-Flores, J. (2004) Application of molecular biology techniques in astrobiology, in this volume.

Horvath, J., Carsey, F., Cutts, J. Jones, J. Johnson, E., Landry, B., Lane, L., Lynch, G., Chela-Flores, J., Jeng, T-W. and Bradley, A. (1997), http://www.ictp.trieste.it/~chelaf/searching_for_ice.html

Jacobs, D.K. and Lindberg, D.R. (1998) Oxygen and evolutionary patterns in the sea: Onshore/offshore trends and recent recruitment of deep-sea faunas, *Proc. Natl. Acad. Sci. USA* **95**, pp. 9396–9401.

Little, C.T.S., Herrington, R.J., Maslennikov, V.V., Morris, N.J. and Zaykov, V.V. (1997) Silurian hydrothermal-vent community from the southern Urals, Russia, *Nature* **385**, pp. 146–148.

Muñoz Caro, G.M., Meierhenrich, U.J., Schutte, W.A., Barbier, B., Arcones Segovia, A., Rosenbauer, H.,Thiemann, W.H.P., Brack, A., and Greenberg, J.M. (2002) Amino acids from ultraviolet irradiation of interstellar ice analogues, *Nature* **416**, pp. 403–406.

Vicuña, R. and Serani-Merlo, A. (2004) Chance or Design in the Origin of Living Beings An epistemological point of view, in this volume.

Ward, P.D. and Brownlee, D. (2000) Rare Earth: Why Complex Life is Uncommon in the Universe. Copernicus, New York.

EVOLUTION OF LANGUAGE AS INNATE MENTAL FACULTY

K. TAHIR SHAH
D.E.E.I., Università di Trieste, Trieste, Italy

1. Introduction

This note summarizes our general line of research on the evolution of intelligence with a particular emphasis on human language and mathematics as innate cognitive faculties, capabilities, which set our species apart from non-human primates. At the psychological level of abstraction, human intelligence can be considered as a collection of interacting modules of such mental faculties as language, theory of Mind (ToM), meta-mind and others (Shah 1998). Modules at lower level can be considered as "pre-adaptations" in the evolutionary sense that these are required for the emergence of higher cognitive functions like inventing mathematics and generation and comprehension of language. We outline our "emergence-at-threshold" model of language in terms of other 'simpler' mental faculties such as ToM and unbounded generativity, namely FLN of Hauser-Chomsky-Fitch (2002). Using this model, we estimate that modern human language, FLB in Hauser-Chomsky-Fitch (HCF) sense, emerged around 40,000 years ago.

Our effort to model language evolution was stimulated by HCF and the following two questions posed by Christiansen & Kirby (2003):

1. *What are the necessary and sufficient pre-adaptations for language?*
2. *Can genetic and archaeological evidence converge on a timetable for the origins of language in hominids?*

2. The Model

In the astrobiological context, a research program on the origins and evolution of intelligence as defined above requires a three-fold effort. First, evolutionary convergence in the broad sense (ECB) is to be investigated. This includes not only the re-emergence of terrestrial intelligence 'if the tape is played again' but also emergence of extraterrestrial intelligence (Shah 2001) elsewhere in the universe. A study of ECB should take into account adaptation, exaptation, task-directed evolution due to organism-environment interaction, and emergent higher cognitive modules as a consequence of threshold behaviour of a collection of interacting pre-adaptations, i.e., lower level cognitive modules. Second, an understanding of the nature and evolution of complex cognitive systems and their underlying nervous systems is necessary. Third, we need to clarify the nature of cognitive reality—reality as perceived by a

J. Seckbach et al. (eds.), Life in the Universe, 333–334.

cognitive system. The basic assumptions of this line of research are continuity of evolution, modularity (Fodor 1983) and domain-specificity (see, e.g., Hirshfeld & Gelman 1994).

There seems to be consensus that some pre-adaptations became available to the hominid line prior to emergence of language (Christiansen & Kirby 2003). We should, therefore, take into account all such underlying pre-adaptations in order to explain the emergence of language. At present, this is debated as to what these may have been. This line of thinking confirms our view point presented in Shah (1998) in the context of general intelligence. Our hypothesis is that these pre-adaptations reached a threshold and began to interact with each other only in hominid lineage, but not in non-human primates or in other species, but they are shared with many other species and seems qualitatively to be the same. An outline of necessary cognitive modules as pre-adaptation is given as follows (details are to be found in their respective references).

Symbolic Representation (see, e.g., Fodor & Pylyshyn 1988): This capacity could be as old as 1.5 million years and it is highly probable that symbolic representation pre-dated all other cognitive modules given below.

Recursive Generativity and Systematicity (Bloom 1994, Pylyshyn 1884)
Meta-mind (Suddendorf 1999)
Theory of Mind (Baron-Cohen 1999)
Granularity (Shah 1998, see references therein)

The other necessary modules are working memory, long-term memory and perceptual input interface (sensory modalities). We have not investigated the sufficiency issue in terms of pre-adaptations. The schizophrenia hypothesis of language origin considers left hemisphere dominance as the critical change dating it to 50,000 years (Crow 2000). The existence of meta-mind dates to around 50,000 years as well (Suddendorf 1999). However, theory of mind is suggested to be no more than 40,000 years old (Baron-Cohen 1999). Since ToM is essential to pragmatics of language, FLB, the full human language originated no more than 40,000 years ago.

3. References

Baron-Cohen, S. (1999) The evolution of a theory of mind, In: M.C. Corballis and S.E.G. Lea (eds.)*The Decent of Mind, psychological perspectives on hominid evolution*, , Oxford University Press, pp. 261–277.
Bloom, P. (1994) Generativity within language and other cognitive domains, *Cognition* **51**, 177–189.
Christiansen, M.H. and Kirby, S. (2003) Language evolution, *TRENDS in Cognitive Sciences* **7**, 300–307.
Crow, Y.J. (2000) Schizophrenia as the price that Homo sapiens pays for language: a resolution of the central paradox in the origin of species, *Brain Research Interactive* **31**, 118–129.
Fodor, J.A. & Pylyshyn, Z.W. (1988) Connectionism and Cognitive Architecture, *Cognition,* **28**, 3–71.
Hauser, M.D., Chomsky, N. and Fitch, W.T. (2002) The Faculty of Language: What Is It, Who Has It, and How Did It Evolve? *Science* **298**, 1569–1579.
Hirshfeld, L.A., and Gelman, S. (eds.) (1994) *Mapping the mind*, Cambridge University Press, New York.
Pylyshyn, Z. W. (1984) *Computation and Cognition*, The MIT Press, Cambridge (Mass), USA.
Shah, K. T. (1998) Cognitive Universals: Abstract Psychology of Terrestrial and Extraterrestrial Intelligence, In: J. Chela-Flores and F. Raulin (eds.) *Exobiology: Matter,Energy and Information in the Origin and Evolution of Life in the Universe*, Kluwer Academic Press, pp. 161–164.
Shah, K. T. (2001), Testing Evolutionary Convergence: From RNA World to Intelligence, In: J. Chela-Flores and F. Raulin (eds.) *First Steps in the Origin of Life in the Universe*, Kluwer Academic Press, pp. 261–266.
Suddendorf, T. (1999), The rise of meta-mind, In: M.C. Corballis and S.E.G. Lea (eds.) *The Decent of Mind, psychological perspectives on hominid evolution*, Oxford University Press, pp. 218–261.

HOW ADVANCED IS ET?

PAOLO MUSSO
Pontifical University of the Holy Cross,
P. Sant'Apollinare 49, 00186 Roma (RM)–Italy

1. Age and Advancement

According to a very widespread commonplace, if any extraterrestrial civilization does exist, it is very likely to be much more advanced than ours. Surely, it is likely to be even one billion years *older* (Norris, 2000). But does it implies that it would be also one billion years more *advanced*? In reality, this equivalence requires a further assumption, i.e. that technological progress is an endless process. But technological progress is based on scientific one. In fact, without substantial advancements in science, we can only *improve the existing techniques* until the full exploitation of their possibilities, established by the laws of nature they are based on. So, an endless technological progress implies that scientific progress is endless, too.

2. The End of Science? The Horgan's Challenge

In 1996 appeared in the USA *The end of science*, by John Horgan (Horgan, 1996), radically challenging such a view. In fact, in his book Horgan maintained not only that science is not endless in principle, but that it is ending actually, so that we are already in the decreasing part of the curve. Horgan's argument is very simple: science is aimed to discovery the fundamental laws of nature; of course, if it cannot succeed, it will have to stop somewhere before reaching them; but if it *can* succeed, then it will discovery *all* the laws and then it will have to stop. Thus, in any case science have to stop, soon or later. Then, he showed many historical examples, all seeming to support his theory.

3. A Statistical Evaluation

While in principle Horgan is surely right, the most challenging part of his theory is the second one, i.e. that science on Earth is ending just now. In fact, if science *wouldn't* be endless, but would take a billion years to end actually, the previous equivalence would still hold. So, a first (and very rough) attempt of a statistical evaluation of this crucial point is here provided, based on the data coming from a popular handbook (Rivieccio, 2001). Both discoveries and inventions seem to follow a typical bell curve, which is already in its decreasing half, even though the second is less sharply defined (fig. 1). Moreover, shifting the discovery curve

J. Seckbach et al. (eds.), Life in the Universe, 335–337.

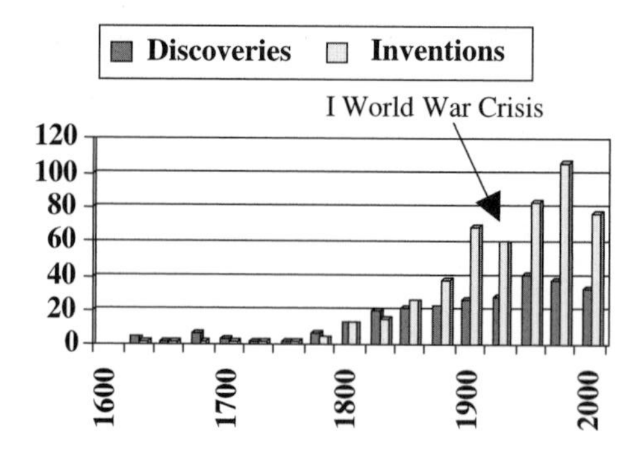

Figure 1.

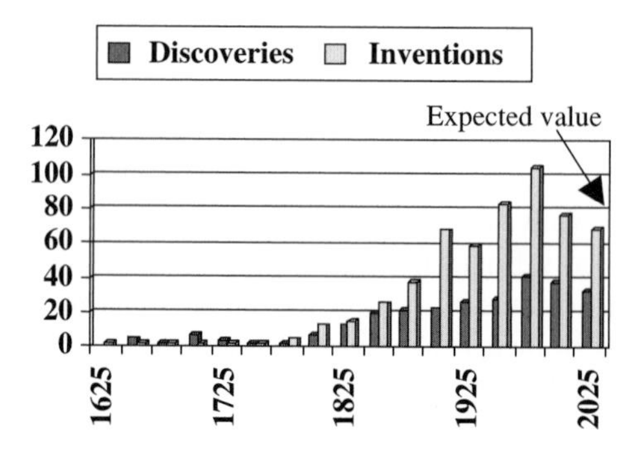

Figure 2.

towards the right of 25 years (just a generation, that is the time presumably needed to learn how to apply scientific discoveries to technology) we see that their fitness becomes almost perfect, the only exception possibly being due to the big crisis which followed the First World War (fig. 2).

4. Conclusions

1. Despite the increasing amount of new technological products, the number of *substantial* technological novelties seems to be decreasing, and it is likely to be decreasing at least in the next 25 years, too. If so, Horgan could be right.
2. A possible inversion of this trend seems to strictly depend on a new deal of discoveries in fundamental science, whose number, on the contrary, is regularly decreasing from about 1950.
3. According to the *present* trend, the age of substantial technological innovations on Earth is likely to end in about two centuries. Notice that such a conclusion fits very well with the widespread opinion that, at least in a well defined field, i.e. radioastronomy, at present we have already reached the half of the maximum possible degree of efficiency (Oliver, 1971). If so, an extraterrestrial civilization one billion years older than ours might be *only two centuries more advanced.*

4. Since we are still very far from reaching the capability for interstellar travels, and only two centuries more of (decreasing) progress may not be enough, this may be an impossible goal. If so, SETI could really be our only chance to get in touch with any possible extraterrestrial civilization.
5. Further and more professional studies are needed to better clarify this issue.

5. References

Norris, R. (2000) How Old is ET?, In: A. Tough (ed.) (2000) *When SETI Succeeds: The Impact of High-Information Contact*, Foundation For the Future, Washington, pp. 103–105.

Horgan, J. (1996) *The End of Science*, New York; trad. it. 1998, *La fine della scienza*, Adelphi, Milano.

Rivieccio, G. (2001) *Dizionario delle scoperte scientifiche e delle invenzioni*, Rizzoli, Milano.

Oliver, B.M. (1971, reprint 1996) *Project Cyclops*, SETI League & SETI Institute.

XIII. Epistemological and Historical Aspects of Astrobiology

CHANCE OR DESIGN IN THE ORIGIN OF LIVING BEINGS
An Epistemological Point of View

RAFAEL VICUÑA[1] AND ALEJANDRO SERANI-MERLO[2]
[1] *Facultad de Ciencias Biológicas, Pontificia Universidad Católica de Chile. Casilla 114-D, Santiago, Chile, and Millenium Institute of Fundamental and Applied Biology;* [2]*Facultad de Medicina, Universidad de los Andes. San Carlos de Apoquindo 2200, Santiago, Chile.*

1. The Rational Understanding of Living Beings

The judgment that the natural world is composed by living and non-living entities is one of the most ancient and spontaneous intuitions of humankind. Natural science, which is the critical, systematic and methodical deepening of these first and firm intuitions, consequently divides itself into biology, with its wide cohort of disciplines, and physico-chemical sciences, encompassing the majority of divisions.

The intellectual understanding and the formal conceptual expression of the uniqueness of living beings has challenged biologists and natural philosophers since the days of the Greek naturalists. Henceforth, the origin of living beings has been viewed as being tightly linked to the definition and explanation of the essence of life (Maturana *et al*, 1974).

If living beings are no more than complex material devices, organized fortuitously by the free interaction of blind mechanical forces and whose survival depends on its fitness to the environment—as the ancient Greek atomists first sustained—, then the explanation of the origin of life must lie on strictly mechanical explanations. If, on the other hand, living beings are unique natural entities, fundamentally different from non-living beings—as Plato and Aristotle affirmed—, then the explanation of their origin must include other causes in addition to mechanical ones. These non-mechanical causes were not conceived by these philosophers as forces in the vitalist way, but as real aspects or levels of explanation in natural beings.

In the ancient Greek non-creationist cosmological context, the explanation about the origin of life reverted to chance in the materialistic view, or to some kind of formal and final causality in the non-materialist view. In spite of the many nuances that have been introduced to the problem through the centuries, the basic opposing philosophical views seem to remain virtually untouched. Some authors, in the vein of Oparin and others, adhering explicitly to philosophically materialistic views, have sought grounding for their theoretical speculations in the experimental evidence provided by modern science. On the other hand—and also claiming to base their speculations in data provided by contemporary biological knowledge—an increasing number of authors try to substantiate that the origin and diversification of living beings would have been impossible without the recourse to

J. Seckbach et al. (eds.), Life in the Universe, 341–344.

other kinds of causes (non-reductionism), to the intervention of an intentional intelligent cause (intelligent design) or even to the direct intervention of God (creationism).

Herein, we propose an epistemological clarification that could facilitate interdisciplinary discussions and open the way to conciliate opposing philosophical views by recognizing the fragments of truth contained within each stance.

2. Epistemological Considerations

Every scientific discipline, whatever the subject it covers, relates to facts and with the rational understanding of those facts (Maritain, 1983). As Maritain sustained, the notion of fact is analogical. Therefore, evidence in any science, including philosophy, mathematics, the natural sciences and the social sciences, must possess its own facts from where to proceed in its way of conceptualization and reasoning. In the realm of natural experimental sciences, facts are defined in a quite restrictive manner as those perceptible empirical phenomena related to the particular field of study (Simon, 1999). Even if empirical facts are entirely referred to sensible perception, they are formally expressed in judgements.

Starting from this empirical phenomenology, a variety of possible objectives become possible. When a student of biology decides to devote himself to systematics, morphology, biochemistry, physiology or ethology, what he is in fact doing is choosing an option for a particular and restricted point of view. Every biologist working in subdisciplines of experimental biology accepts other disciplines as legitimate but distinct intellectual avenues to the understanding of living phenomena.

Experimental biologists don't need an explicit formal definition of living beings to carry out their work. However, it is plain, from an epistemological point of view, that for every biologist the affirmation of the existence of living beings as original and distinguishable entities in the physical world, is an absolute pre-requisite for the existence of its corresponding discipline. This constitutes a fact which every experimental scientist in any branch of biology has to take for granted (Meyerson, 1951).

This also applies to the definition of living beings. It is implicitly obvious for every biologist that if living entities exist they must have a proper definition, whatever it could be. From a philosophical point of view, this is one of the first and most evident facts on which all human understanding lies: everything that exists is something and the adequate conceptual expression of this "somethingness" is what we call its definition.

Although biologists do not need to formally define the distinct existence of living beings, two precisions must be made:

1. Even if biologists don't need a formal definition of living beings in order to conduct their work, they absolutely require at least an implicit recognition of the original existence of their subject matter and a general or working definition of this same study subject.
2. This kind of fundamental questioning does not leave scientists indifferent in regards to their reflective thinking and interpretation of experimental work. Moreover, it seems to be in the natural order of things that as an experimental biologist grows older, the more intrigued they become in relating the questions answered and proposed in their discipline to a unifying theory of scientific doctrine.

What kind of questions are these, whose answers are in some manner presupposed by the natural sciences, but that cannot be answered formally within their own conceptual frame? On which grounds can these questions be adequately answered? In which manner scientific developments of the natural experimental sciences contribute to a better understanding of the distinct nature of living beings and of the causes of their origin? Are, in some way, the natural experimental sciences continuingly answering these fundamental questions in their own particular manner?

We propose that the adequate definition of the uniqueness of living beings and the inevitable following questions, namely the how and why of the origin of this uniqueness, cannot be univocally answered. This definition must be approached in different but related ways.

Experimental biologists, who must in practice take for granted the existence and philosophical definition of living beings, have the task of defining living beings from a strictly experimental stance. Each time that biochemists, for example, describe the chemical composition of a particular organism and elucidate the metabolic pathways that produce and maintain those entities, they are defining in very concrete terms a living organism from a biochemical viewpoint. The epistemological question, dealing with the existence of other possible views, either in the experimental or the theoretical realms, is not a matter that can be biochemically approached. In other words, a biochemist is unable to answer a question that cannot be rigorously defined in biochemical terms.

3. Epistemological Proposal

We propose that in discussions concerning the origin of life, at least three distinct epistemologically arguments are applied within the same discussion.

First of all, experimental scientific questions are relevant for the discussion on the origin of life. Studies on the physico-chemical properties of water, the biochemical characteristics of ribozymes, or the genetic code of the mitochondria are some examples of these pertinent scientific matters that can be rightly approached on experimental terms.

Secondly, there are questions whose answers are out of the reach of experimental demonstration and that must be ultimately resolved in philosophical terms. Examples of these could be: Is the origin of living beings the result of the interaction of purely physicochemical forces that coincided by chance in some way under fortuitous circumstances? Or are living beings the result of an intentional design either immanent in natural processes or transcendent to the physical world? Is there any conceivable manner in which both answers could be legitimately combined?

Finally, there are historical or paleontological questions that are out of the reach of actual experimentation: Did the first living beings emerge from a primordial soup or from a primitive solid substrate of clay? Did eukaryotes emerge from prokaryotes? What exactly happened in the pre-Cambrian to Cambrian transition? Such singular historical biological facts, which are not deducible from verified general natural laws and that occurred under non-reproducible conditions, cannot be directly studied by current scientific methods. For example, in investigating the physical transformation that a subset of living beings underwent during Cambrian transition, we are confronting an historical fact that can only be approached by historical or paleontological speculation.

A biochemist can speculate on philosophical or historical questions that are outside the strict realm of biochemistry and provide experimental evidence for this speculation. A biochemist is free to speculate about the origin of living beings in time and place, and on the mutations they have subsequently undergone. However, a classical biochemical approach cannot wholly answer philosophical or historical questions. Biochemistry cannot have biochemical answers to non-biochemical questions, even if they are posed in biochemical terms.

We advocate in this proposal that paleontological speculations, based on actual experimental data are not necessarily contradictory with philosophical reasoning. Indeed, what appears contradictory are the mechanistic versus non mechanistic philosophical statements. We propose, that even on philosophical grounds, chance and design are not necessarily contradictory. There are numerous examples in which human design considers or even relies on the role of chance.

If correct in our affirmation, it would be legitimate for a biologist, -while thinking 'paleontologically'—, to reason on empirical basis, while at the same time, thinking as a philosopher, reason on intentional design. Only when a biologist/biochemist also affirms *as a philosopher* the truth of an empirical view, a final/intentional design position can be regarded as contradictory to his statements.

4. References

Maritain, J. (1983) *Les degrees du savoir, n°18 Éclaircissements sur la notion de fait.* Fribourg-Paris, Éditions Universitaires de Fribourg Suisse/ Éditions Saint Paul, pp. 361–365.

Maturana H.R., Varela, F.R. and Uribe, R. (1974) Autopoiesis: The organization of the living, its characterization and a model, *Bio Systems* **5**, pp. 187–196.

Meyerson, (1951) E. *Identité et Realité*, Paris, pp. 327–384.

Simon, I. (1999) *Maritain's Philosophy of the Sciences.* In A. Simon *Philosopher at work*, Lanham/Maryland, Rowman & Littlefield Publishers Inc., pp. 21–40.

ASTROBIOLOGY AND BIOCENTRISM

ROBERTO ARETXAGA
Philosophy Department, School of Philosophy and Educational Sciences
University of Deusto, Bilbao, Spain

1. Philosophical and Environmental Biocentrism

The term "biocentrism" is polysemic in as far as it has, at least, three different meanings. One of them is to be found in the field of Philosophy, another in the Environmental Sciences, and a third interpretation is also provided in the area of astrobiology. In the field of philosophy, the term "biocentrism" is used to describe that ethical theory which denies that human beings occupy a privileged position with respect to other living creatures, as well as humankind's centrality as a source of universal values. Life at large is taken as the only source and holder of any value by biocentrism, which implies that humanity is displaced from its central position, and so biocentrism is anti-anthropocentric. This is the usage the term is given in the "deep ecology" and conservation movement, based on the theories developed by Aldo Leopold and Paul W. Taylor. The second use of "biocentrism" is opposed to that of "functionalism". In this sense, these two designations refer to opposing views in the study and management of the environment, which, in turn, have generated two distinct scientific disciplines: population ecology and system ecology, respectively. Bearing this difference in mind, biocentrism is best characterized as focusing on organisms and taking the "biota" as its basic component. Besides, biocentrism relies on natural selection as its explanatory paradigm and defends biodiversity.

Functionalism, on the other hand, conceives both organisms and the abiotic component as a whole (holistically), that is, not just a mere addition of the parts. It has also made of the flowing of matter and energy its main object of analysis, and of the laws of thermodynamics its explanatory paradigm. Functionalism favors ecodiversity over biodiversity, and maintains that it is the preservation of the flowing of matter and energy typical of any ecosystem that guarantees the survival of its organisms.

Before I move on to discuss the astrobiological sense of the term "biocentrism", I will consider some of the implications that the above mentioned usage of the term:

(a) The philosophical conception of the term "biocentrism" brings up a relevant issue for astrobiology; since this science assumes the existence of a common ancestor and the evolutionary theory, it would seem natural to align it with the biocentristic—anti-anthropocentric—position in the debate. However, under no circumstances could astrobiology ignore the fact that human culture represents a real peculiarity among the different forms of life and adaptation on our planet. If life is found on other planets, this fact would broaden the horizon of the philosophical debate started by ethical biocentrism, a horizon

J. Seckbach et al. (eds.), Life in the Universe, 345–348.

that could then be opened to hitherto unheard-of ideas and views. One possible solution to the underlying philosophical dilemma may come from the distinction between "strongly anthropocentric" and "weakly anthropocentric" (Norton, 1984) proposed by Bryan G. Norton.

(b) We may assume that the environmental conception of the term "biocentrism" has implications for astrobiology, too. Thus, since it takes an interest in the origin and distribution of life, and it focuses on the study of microscopic life and its exchanges with the environment, astrobiology seems to require a functional approach. But as it is also concerned with the evolution and destiny of life, and especially with multi-cellular intelligent organisms, the biocentric approach would be the most adequate.

The conflict between biocentrism and functionalism in the ecological sciences can be solved by emphasizing the complementary nature of both paradigms, and by arguing that the use of one or the other depends exclusively on the spatial and temporal parameters to be considered in each case. Thus, while the functional approach offers a better understanding in the study and management of large ecosystems with small organisms; the biocentric one, is rather more suitable for the analysis of ecosystems of a smaller size but with bigger organisms. With respect to astrobiology, its multi/inter-disciplinary character, and the diversity of its subject matter, also seem to encourage the integration of both approaches, depending on the spatial and temporal parameter that is necessary in each case.

2. Astrobiological Biocentrism

The phrase "astrobiological biocentrism" (AB) or "astrobiological conception of the term biocentrism" refers to biocentrism in the sense in which Chela-Flores understands it when he defines it as "the belief that life has occurred only on Earth" (Chela-Flores, 1998), or "the doctrine which defends the singularity of the biological evolution that has taken place on Earth, from a bacteria to human beings" (Chela-Flores, 2003).

Chela-Flores' definition of the term "biocentrism" is likely to generate a philosophically relevant question. As a "belief", biocentrism does not constitute in itself a rationally-based and articulated system of ideas, but rather an inner conviction of an individual or a community which, admittedly or not, consciously or not, justifiably or not organizes and governs their thoughts and deeds. From this point of view, the term "biocentrism" may be analyzed in the light of Husserl's "life's world" (*Lebenswelt*) and Ortega y Gasset's "belief" (*creencia*). As a "doctrine", however, biocentrism is a theory—on the same footing as that built by J. Monod and others—which takes part in what T. S. Kuhn calls a "scientific paradigm". In this sense, the term "biocentrism" would belong in the category that Ortega y Gasset refers to as "idea" (*idea*). However, in both cases biocentrism is one of the explanatory keys in our contemporary way of knowing, understanding, evaluating and explaining the universe, life and humankind. Consequently, biocentrism also becomes one of the main pillars of our contemporary "conception of the world" (*Weltanschauung*, Dilthey), and it is this aspect precisely that needs to be highlighted when talking about AB, which comprises both meanings—that of "belief" and that of "idea". With all this in mind, I will now dwell upon the relation between AB and astrobiology as a science.

The issue of the existence of extraterrestrial life and of the plurality of worlds was already raised in ancient times (Dick, 1984), which means that astrobiology has its roots in an age-old human quest. In any case, as a contemporary science, astrobiology is indebted to 20th-century theories, techniques and methods that have revolutionized the way in which human beings have access to and present "the real"—physical, or living. Moreover, astrobiology is likely to engage all the other disciplines of human knowledge in its investigations (Aretxaga, 2003).

Copernicus and Galileo put an end to geocentrism. Darwin laid the foundations to leave behind anthropocentrism. In my opinion, what is really important about each of their scientific contributions in astronomy and biology is that they caused a radical change in our conception of the universe, of man and of man's role in this world, something that became apparent in profound cultural and sociological transformations.

With regard to AB, it is still, as geocentrism and anthropocentrism were in their time, just one of the pillars of our civilization, since up to now there is no strong evidence for the existence of other life forms in different planets. But it is also well-known that the lack of any evidence of the existence of extraterrestrial life does not necessarily entail its absence. Astrobiological discoveries do not only support this hypothesis, but also begin to undermine the foundations of biocentrism as a scientific theory. As a result, such knowledge has historic relevance, since they constitute the basis to prove empirically the falsity of biocentrism. This fact allows us to nourish hopes that we are facing a future—and perhaps not a very remote one—of scientific contributions which, like Galileo's or Darwin's, will go down in history, not just on account of its scientific and technological significance, but above all due to its new and revolutionary consequences for all the other aspects that constitute the different human cultures and societies (for instance, philosophy, art, religion, politics and literature).

In view of what has been said above, it seems reasonable to consider biocentrism as an obstacle for human progress (Chela-Flores, 2001) because, similar to geocentrism and anthropocentrism in their heyday, at present, biocentrism would seem a hindrance to mankind's development of a more truthful image of itself and, therefore, to a finer understanding of its real place in the world, and of the new type of responsibilities that accompany this change. Furthermore, if as a general rule a reliable knowledge contributes to raising the levels of adaptability, learning the truth about biocentrism will ensure the survival of the human race.

Taking the above arguments under consideration, the need to discover and analyze the role of biocentrism seems both inescapable and responsible, since it is one of the elements shaping the numerous and complex aspects that constitute human cultures and societies. This task leads to a better understanding of the character and depth of the changes and implications that the eventual decline of biocentrism would involve. This, in turn, makes it easier for the complementary task of investigating alternative models designed to approach future problems with more flexibility and effectiveness. In this context, and although it is not the business of humanists, but rather astrobiologists to demonstrate the falsity of biocentrism, philosophers and humanists do have to exercise and to encourage thought processes that help mankind as a whole to understand and take in the implications that the effects of an eventual success of astrobiology in its quest for life, present or past, outside planet earth.

In this particular area, some invaluable contributions have been made by the SETI Institute concerning extraterrestrial intelligence (Billingham *et al.*, 1994; Tough, 2000). Considering everything that has been stated so far, there is little doubt of the necessity to strengthen and promote cooperation between astrobiologists and humanists.

3. Discussion and Conclusion

The existence of three different conceptions of the term "biocentrism" has important implications for astrobiology. Thus, the philosophical and environmental conceptions have ethical and methodological consequences, respectively. In what concerns the conception, theories, methods and techniques of which astrobiology makes use can be said to offer us the historic opportunity of experimentally solving the question of whether we are alone or not in the universe or, the relation existing between our own evolution and that of other forms of life that may have developed somewhere else in the universe (Chela-Flores, 2001). The current state of the art suggests AB makes no reference to reality, but only represents an unjustified belief and a scientific theory based on partly out-dated knowledge. Thus, speaking of an incipient crisis of biocentrism brought about by the new astrobiological contributions does not seem hasty. Given the important role played by AB in the shaping of our culture and society, the possibility of its demise as a belief and as a theory would cause not only profound scientific changes, but also, and perhaps more importantly, cultural and social ones.

Astrobiology then, far from being a field of specialization only open to scientists, should also hold great interest for humanists since this theory may compel humankind to readjust their own perceptions as a race and to question their place in the universe, which will eventually contribute to their progress. This insight implies a responsible and efficient practice of reflection and investigation that requires, in turn, increasing cooperation between astrobiology and the humanities that should draw closer in order to make the aforesaid progress evident in all the dimensions that constitute the different human cultures and societies. To conclude, and in an attempt to avoid problems of terminology, I would recommend that the term "biogeocentrism", which has already been employed by Chela-Flores himself occasionally, be used to refer to what I have called here "astrobiological biocentrism".

4. References

Aretxaga, R. (2003) La ciencia astrobiológica. Un nuevo reto para el humanismo del siglo XXI. *Humanismo para el siglo XXI. Congreso Internacional (Bilbao, marzo 2003)*. Proceedings (CD-Rom), University of Deusto, Bilbao.

Billingham, J., Heyns, R., Milne, D., Doyle, S., Klein, M., Heilbron, J., Ashkenazi, M., Michaud, M., Lutz, J. and Shostak, S. (eds.) (1994) *Social Implications of the Detection of an Extraterrestrial Civilization*, SETI Press, SETI Institute, California.

Chela-Flores, J. (1998) Search for the Ascent of Microbial Life towards Intelligence in the Outer Solar System. In: R. Colombo, G. Giorello and E. Sindoni (eds.) *Origin of the life in the universe*. Edizioni New Press, Como, pp. 143–157.

Chela-Flores, J. (2001) La astrobiología, un marco para la discusión de la relación hombre-universo. Principia (Universidad Centro Occidental L. Alvarado, Barquisimeto, Venezuela) 18, pp. 12–18.

Chela-Flores, J. (2003) Marco cultural de la astrobiología. Letras de Deusto (University of Deusto, Bilbao, Spain) N° 98, Vol. XXXIII, January–March, pp. 199–215.

Norton, B. G. (1984) Environmental Ethics and Weak Anthropocentrism, Environmental Ethics, 6, pp. 131–148.

Dick, S. J. (1984) *Plurality of Worlds: The Origins of the Extraterrestrial Life Debate from Democritus to Kant.* Cambridge University Press.

Tough, A. (ed.) (2000) *When SETI Succeeds: The Impact of High-Information Contact*, Foundation for the Future, Washington, USA.

ANALYSIS OF THE WORKS OF THE GERMAN NATURALIST ERNST HAECKEL (1834–1919) ON THE ORIGIN OF LIFE

FLORENCE RAULIN-CERCEAU
Centre Alexandre Koyré (CNRS-EHESS-MNHN-UMR 8560)
Muséum national d'Histoire naturelle, 57 rue Cuvier—F-75005 Paris—France

1. Introduction

The German naturalist Ernst Haeckel (1834–1919) was probably the most prolific writer on the subject of the origin of life in the late 19th century. He was well-known for his monistic philosophy (monism) in which all phenomena stood in causal relation and all historical developments were strictly continuous (Kamminga, 1980). All of Haeckel's writings stressed the fundamental unity of the living and the non-living world. Haeckel was one of the most ardent protagonists of Darwin in Germany and his monism was clearly inspired by the theory of evolution. Haeckel transformed the doctrine of evolution into a monistic system in which an abiogenic origin of life was a major part (Farley, 1977).

Haeckel's opinion about the origin of life was directly dependent on his philosophy and on his view of the nature of life. His description of the organization of the simplest organisms (*Monera*), as made of only one substance, led him to adopt the general idea of *abiogenesis* (living material coming from inorganic substance). Regarding a genuine *abiogenesis*, this idea became a logical postulate of the theory of biological evolution and of the concept of the unity of nature. Therefore, Haeckel supported the thesis of *archigonia* (genuine *abiogenesis*), a past process that would have occurred during the early evolution of the planet.

However, partly because of the extreme simplicity of his *Monera*, Haeckel also believed in a continuous spontaneous generation for very primitive biological material in accordance with a process of present-day-*abiogenesis* taking place in the deep sea beds.

In this paper, we have analyzed the following writings of Haeckel:

- *Generelle Morphologie der Organismen* (General Morphology of Organisms) (1866)
- *Natürliche Schöpfungsgeschichte* (Natural History of Creation) (1868)
- *Die Welträthsel* (The Riddles of the Universe) (1899)
- *Die Lebenswunder* (The Wonders of Life) (1904)

J. Seckbach et al. (eds.), Life in the Universe, 349–352.

2. The Unity of Nature

Haeckel explained the whole living world by means of his monistic philosophy: a unitary notion of nature in which matter was supposed to evolve from a primitive cosmic condition to a complex biological state: a vast, uniform, uninterrupted and eternal process of development occurring throughout all nature. According to Haeckel, this idea was itself the fruit of the positive results of the sciences.

From Haeckel, the unity of nature (organic and inorganic) was supported by a general doctrine of development. This doctrine was directly dependent on the theory of biological evolution and on the mechanism of natural selection (this last mechanism gave an important nonteleological explanation for this development) (Farley, 1977). The main points resulting from the monistic philosophy are the following ones:

- There is no fundamental distinction between the living and the non-living world.
- There is continuity between them.
- The same "natural" laws rule the mineral and the organic worlds.
- The origin of life links together the mineral and the organic worlds.

3. What is Life?

The consequences of the monistic philosophy of the unity of nature led Haeckel to defend precise point of views about the nature of life. Haeckel deduced from the experiments of organic chemistry that all elements found in living organisms were also present in the inorganic domain and that carbon (in combination with H, O, N, P, S) was the element essential for building proteins (Haeckel described his carbon theory in *General Morphology of Organisms*). He upheld the idea that life depended on the properties of carbon. Proteins (or protoplasmic substances constituted mainly by carbon) were the basic components of life. Life was the result of physicochemical processes based on the properties of proteins. Life amounted to physicochemical actions but living organisms represented a specific state of matter: life resulted from combinations formed by carbon and these combinations, in interaction with water, led to a specific state, neither solid nor liquid, but a semi-fluid or viscid state.

The theory of the unity of nature included the concept that life was not a qualitatively new principle of evolution and then, that life didn't display any emerging properties. According to Haeckel, explaining the phenomenon of life was no more complicated than explaining the physical properties of the inorganic matter. In *The wonders of Life*, Haeckel went even further: the theory of the unity of nature implied that life could just as well has been found in any organic as inorganic substances.

4. The Origin of Life

In Haeckel's books, the term "spontaneous generation" holds a specific meaning: "spontaneous generation" inferred in fact the idea of a continuity between the organic and the inorganic worlds, and, in Haeckel's view, this term was equivalent to "*archigonia*" (or a genuine *abiogenesis*). About *archigonia*, Haeckel wrote: "under conditions quite different from those today, the spontaneous generation which now is perhaps no longer possible, may have taken place" (*Natural History of Creation, vol 1, p. 342*). In short, and according to Haeckel, studying the origin of life consists in two major points. The first point is: studying the simplest organisms. In *General Morphology of Organisms* and in *Natural History of Creation*, Haeckel claimed that the most primitive organisms (such as the *Monera*) were "undifferentiated, anucleate, naked pieces of albuminous jelly, far simpler in construction than a simple cell" (Farley, 1977). He concluded that the first living organisms on Earth were such "homogeneous, structureless, formless lumps of protein or *Monera*, similar to a *Protoamoeba*." (*General Morphology of Organisms, p. 179*, quoted in Kamminga, 1980). The second point is: studying the natural cosmology of the Earth with the help of the global concept of mechanical causes and evolutionary processes.

Haeckel wondered about the spontaneous generation from inorganic matter in the past. Where do the first ancestral organisms come from? Haeckel answered this question with the help of the following hypotheses and theories: the theory of descent—including the idea of adaptation and spontaneous generation (Lamarck), the notion of heredity and natural selection (Darwin), the principles of cosmogony ruled by mechanical causes (theory of evolution applied to the universe and to the celestial bodies, from Kant, Laplace, Herschel). Haeckel reached the following conclusions:

- The characteristic environmental conditions for the primitive Earth were perpetual rains, lower temperatures, and a specific electric state of the atmosphere.
- The major steps leading to the formation of living organisms were the first appearance of liquid water on the surface and the presence of a huge amount of carbon as carbonic acid mixed with the atmosphere.
- Life started on the Earth at a well-defined period in the past.

But nevertheless, Haeckel regretted deeply that no experiment regarding *archigonia* had yet been attempted in laboratory.

5. Criticism of Haeckel's Theory

Many criticisms were put forward about Haeckel's theories, all of these theories depending on his *Monism*. One of the most scathing comments has come from Carl Semper, professor of zoology and comparative anatomy in Germany. According to Semper, neither the theory of *archigonia*, nor the theory of origin of life, expressed with the help of carbon theory, could be checked with observational experiments. Moreover, Semper considered that the protoplasmic theory—as well as the carbon theory—was a simplified theory in comparison with what could be the explanation of life. Hence, the question of the origin of life seemed to be reduced to a mechanical and physical problem and the emergence of the first organisms was considered as an epiphenomenon of matter's evolution.

Haeckel defended himself against the first criticism in claiming that the problem of the origin of life—as well as the whole evolutionary biology—was an historical science that couldn't follow the model of the exact sciences (such as Physics or Mathematics).

Concerning the second criticism, Haeckel thought that a mechanical explanation did not necessarily involve a full reduction to physical processes. He criticized the so-called exact method in embryology on the grounds that it attempted to reduce complex historical processes to simple physical phenomena. In fact, Haeckel included historical explanations in terms of evolutionary development under the general heading of mechanical explanations (Kamminga, 1980).

6. Conclusion

Regarding the problem of the origin of life, Haeckel could be considered as a forerunner in many points: in order to explain the origin of life, he clearly proposed a past process (*archigonia*) occurring in a completely different environment from the present one. His theory contained the premises of the idea of chemical evolution, since the suggested steps leading to life are integrated into a broad evolutionary process including geology and chemistry. The origin of life is included in the history of the Earth, and therefore, very long spans of time are suggested to favor the transition from inert to living matter. From the epistemological point of view, he asserted that the problem of the origin of life was an historical problem that couldn't be solved in the same way as (for instance) problems of Physics. However, perhaps we could reproach Haeckel for having:

- underestimated the complexity of living matter and therefore having considered the transition from inert to living matter to be much too easy.
- kept the solution of a continuous and present-day spontaneous generation since this transition would be so easy and natural.
- used too much theory instead of practice, and finally having founded a philosophy more than a scientific theory.

7. References

Farley, J. (1977) *Spontaneous Generation Controversy from Descartes to Oparin*, The Johns Hopkins University Press, Baltimore and London.

Girard, J. (1874) *Les explorations sous-marines*, Savy, Paris.

Haeckel, E. (1866) *Generelle Morphologie des Organismen*, G. Reimer, Berlin.

Haeckel, E. (1874) *Histoire de la création des êtres organisés d'après les lois naturelles, Conférences scientifiques sur la doctrine de l'évolution en général et celle de Darwin, Goethe et Lamarck en particulier*, trad. C. Letourneau, C. Reinwald, Paris.

Haeckel, E. (1902) *Les énigmes de l'univers*, trad. C. Bos, Schleicher Frères, Paris.

Haeckel, E. (1907) *Les merveilles de la vie : études de philosophie biologique pour servir de complément aux "Enigmes de l'univers"*, Schleicher Frères, Paris.

Kamminga, H. (1980) *Studies in the History on the Origin of Life from 1860*, Thesis for the degree of Doctor of Philosophy, Department of History and Philosophy of Science, Chelsea College, University of London.

Raulin-Cerceau, F. (2004) Historical Review on the Origin of Life and Astrobiology, In: J. Seckbach (ed.) *Origins: Genesis, Evolution and the Diversity of Life*, Kluwer Academic Publishers, Dordrecht, The Netherlands, in press.

A REEXAMINATION OF ALFONSO HERRERA'S SULFOCYANIC THEORY ON THE ORIGIN OF LIFE

E. SILVA[1], L., PEREZGASGA[2], A., LAZCANO[1], and A. NEGRÓN-MENDOZA[3]
[1]Facultad de Ciencias, UNAM Apdo. Postal 70-407 Cd. Universitaria, 04510 México, D.F., MÉXICO, [2]Instituto de Biotecnología, UNAM, Apdo. Postal 510-3 Cuernavaca, Mor., 62250 Mexico and [3]Instituto de Ciencias Nucleares, UNAM Apdo. Postal 70-543 Cd. Universitaria 04510 México, D.F., México.

1. Introduction

Based on Pflüger's proposal on the role of CN-containing derivatives in biological catalysis (Pflüger, 1875), Alfonso L. Herrera developed in the late 1920's the sulfocyanic theory of the origin of life (Herrera 1942). According to this idea, the physical structure of cellular plasma was derived from sulfur- and CN-containing compounds that formed a molecular matrix within which the primordial fixation of CO_2 took place via its reduction to H_2CO.

As described in his extensive bibliography (Beltrán, 1968), Herrera achieved the formation of microscopic structures, which he claimed were comparable to cells, due to their growth, motility, and osmotic properties. He promptly divided then into two major groups: (a) *colpoids*, which were produced when olive oil, gasoline, and other complex molecules were used; and (b) *sulphobes*, which resulted from the mixture of NH_4SCN and H_2CO (Herrera, 1942). After many trials, Herrera found that the best starting material for the formation of his sulphobes was ammonium thiocyanate, which he dissolved in formaline and spread in thin layers until evaporation. According to Herrera, the reactions of these precursors gave rise not only to several kinds of cell-like microstructures, but also to (a) starch; (b) two uncharacterized amino acids; (c) globules of red, green and yellow pigments; and (d) what he described as a "proteinoid condensation product" (Herrerra 1942).

As shown by photocopies of some of his laboratory notes (available upon request), by 1933 Herrera was convinced that he had achieved the synthesis of glycine, cysteine and cystine. The formation of these compounds, which Herrera synthesized using formaldehyde and ammonium thiocyanate as starting materials, were based on the glycine synthesis from formaldehyde and potassium cyanide reported by Klages (1903) and Ling and Nanji (1922). Although Herrera (1942) mentioned the synthesis of "starch, and at least two amino acids", he did not list them nor characterize the other products he obtained. Here we attempt to do so, based on the repetition of some of his experiments and the use of modern analytical tools.

J. Seckbach et al. (eds.), Life in the Universe, 353–356.

2. Materials and Methods

We first mixed 20 ml of 37% formaldehyde (0.05 M final) with 36 g of ammonium chloride (0.067 M final) and put this solution into a three-mouth flask that was kept on ice. The mixture was stirred all the time with a mechanical stirrer. We then prepared a 0.055 M ammonium thiocyanate solution that was poured for 30 min with a separator funnel. When half of the thiocyanate solution was added, 25 ml of glacial acetic acid was dropped and the solution was stirred for two more hours. The temperature was always kept below 10^0 C. A precipitate was formed with yellow and white crystals. The mixture was filtered and the crystals were air-dried. By a fractionated precipitation, we obtained two more precipitates (the second one contained yellow crystals, and the third one yellow and white crystals that were separated). Based on their infrared spectra, we decided to work with the precipitate 1 and precipitate 3 white crystals. These were analyzed by infrared spectroscopy, a size-exclusion separation method, and HPLC, using a Beckman Instruments HPLC series chromatograph. The column used was a C_{18} with a particle size of 5 μ. The analyses were done at room temperature at a wavelength of 340 nm. The retention time of various amino acids was determined using a Beckman reference mixture. The fractions of precipitates 1 and 3 were hydrolyzed with sequential grade HCl (Pierce), acetic acid for HPLC and water HPLC grade, derivatized with ortho-ophtaldehyde (Lindroth and Moppen, 1979) and analyzed for the presence of amino acids (Ladrón de Guevara *et al.*, 1985). In order to identify the amino acids that could be present in the reaction mixture, the white crystals of precipitates 1 (in two runs of 500 mg each) and 90 mg of precipitate 3 were first purified by a size exclusion separation method using a SP-sephadex 25 column.

3. Results

The HPLC chromatograms shown in Figure 1 indicate the presence of several compounds whose retention times correspond to glycine, alanine, cysteine, and methionine. Of these, only cysteine and methionine are sulfur containing amino acids.

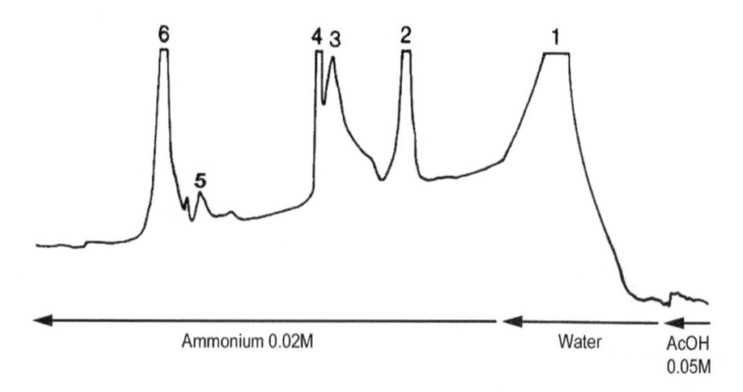

Figure 1. Size exclusion separation of products from precipitate 1, using a SP-Sephadex cm-25 column. Peak 1 corresponds to the thiocyanate ion that did not react. Peaks 2 to 6 were analyzed separately for amino acid content. HPLC identification of 1, glycine; 2, alanine; 3, cysteine; 4, methionine. Amino acids were identified by their retention times as described in the text.

Although we analyzed standard samples of other sulfur-containing amino acids, such as taurine and cystine, the peaks obtained in the chromatograms do not correspond to their retention times. Other unidentified peaks in the chromatogram could correspond to non-proteinic amino acids. The amino acid formation yield was low and was determined by the electronic integration of the peak area (Perezgasga et al., 2004). The same amino acids were observed in both the hydrolyzed and non-hydrolyzed fractions, although the yield was higher in the former one.

4. Discussion and Conclusions

Like many of his contemporaries, Herrera was convinced that the first living beings had been autotrophes. The popularity of Pflüger's (1875) ideas on biological catalysis had led Herrera to believe that CN derivatives played an essential role in biochemical processes and, hence, that cyanogen and its derivatives must have been present at the origin of life. Led by the schemes which suggested at the time that H_2CO was a central intermediate in the photosynthetic fixation of CO_2, Herrera attempted the laboratory formation of an autotrophic protoplasm by mixing the precursors he thought were essential for a minimal living system. Thus, it was not because of foresight that he employed compounds which are nowadays recognized as potential components of the prebiotic environment. Instead, he should be recognized as a careful worker, whose deep knowledge of the major theories of his contemporaries led him to study the origins of life within the framework of his times.

Since Herrera was following the reactions described by Klages (1903) and Ling and Nanji (1922), he was actually performing a variation of the Strecker synthesis, in which ammonium thiocyanate could take the place of NH_4CN, and where the hydrolysis of the nitrile was achieved by boiling with a highly concentrated solution of barium hydroxide (Ling and Nanji, 1922). Our results suggest that in the experiments which we have performed following the procedures described by Herrera, four amino acids were synthesized: glycine, alanine, cysteine and methionine, with a total yield of 2%. Alanine was the most abundant amino acid in all samples analyzed.

The preliminary results presented here suggest that a variation of the Strecker synthesis involving formaldehyde and ammonium thiocyanate could lead to low yields of amino acids, including sulfur-containing amino acids. Alternatively, it is possible that the amino acids reported here, are the outcome of an hydrolysis of oligomers that could be formed by the self-condensation of SCN, by a mechanism equivalent to that described by Ferris *et al.* (1978).

Although the starting materials used by Herrera were determined by his autotrophic hypothesis on the origin of cells, our results show that his experiments may provide insights on the abiotic synthesis of sulfur-containing amino acids within the framework of a heterotrophic emergence of life.

5. Acknowledgements

We are indebted to Arturo Becerra for technical assistance. A PAPIIT grant (ES 116601) to A.N. and DGAPA-UNAM PAPIIT IN 111003-3 to A.L. support from CONACyT grant (I36264-N) was provided to L.P.

6. References

Beltran, E. (1968) Alfonso L. Herrera (1868–1968): Primera figura de la Biología Mexicana. *Revista de la Sociedad Mexicana de Historia Natural.* **29**, 37–91.

Ferris, J. P., Joshi, P. D., Edelson, E. H., and Lawless, J. G. (1978) HCN: a plausible source of purines, pyrimidines and amino acids on the primitive Earth. *J. Mol. Evol.* **11**, 293–311.

Herrera, A.L. (1942) A new theory of the origin and nature of life *Science* **96**, 14.

Klages, A. (1903) Ueber das methilamino-acetonitril. *Berichte der deutschen chemischen Gesellschaft* **36**, 1506.

Ladrón de Guevara, O., Estrada, G., Antonio, S., Alvarado, X, Guereca, L., Zamudio, F. and Bolívar, F. (1985) Identification and isolation of human insulin A and B chains by high-performance liquid chromatography. *J. Chromatogr.* **329**, 428.

Lindroth, P. and Moppen K. (1979) High performance liquid chromatographic determination of subpicomole amounts of amino acids by precolumn fluorescence derivation with O-ophtaldehyde. *Anal. Chem.* **51**, 1667–1674.

Ling, A.R. and Nanji, D.R. (1922) The synthesis of glycine from formaldehyde. *Biochem. J.* **16**, 702.

Pflüger, E. (1875) Ueber die physiologische verbrennung in den ledendigen organismen. *Arch. Gesam.Physiol.* **10**, 641–644.

Perezgasga L., Silva E., Lazcano A. and Negron-Mendoza A. (2004) A reexamination of Alfonso Herrera's sulfocyanic theory on the origin of life. *Journal of Astrobiology* (in press).

DETERMINISM AND THE PROTEINOID THEORY

ARISTOTEL PAPPELIS[1] and PETER R. BAHN[2]
[1]Department of Plant Biology, Southern Illinois University, Carbondale, Illinois 62901 USA; [2]Bahn Biotechnology Co., RR2, Box 239A, Mount Vernon, Illinois 62864 USA.

1. Introduction

The discovery of proteinoids (thermal proteins; branched proteins) in the 1950s and protocells there from, filled the gap between physical/chemical and biological evolution. The ease in finding amino acids in nature for the autocatalytic synthesis of branched proteins (self-ordering) and the ease of finding water in which these macromolecules would form protocells (self-assembly) made the proteinoid theory appealing for those interested in finding a model for the synthesis of living cells (Fox, 1988) and those seeking a candidate system for the origin and early evolution of life (Pappelis, *et al.*, 2001). We worked with Sidney W. Fox and his associates. Together with Fox, we moved his "model" into biology as the branched protein-first paradigm to explain the origin of chemical and biological life. We inferred the pathways in the early evolution of protolife to provide new opportunities for experimental verification.

2. Scientific Determinism

"The night sky of my youth has been explored to the outermost distance and earliest beginnings of the universe (de Duve, 2002, p. vii):" i.e., big bang → elementary particles → atoms → compounds → life [= Cosmic evolution (Chaisson, 2001)]. Life emerges as chemical life (branched proteins) that yields biological life (branched protein protocells) (Fox, 1988; Pappelis, *et al.*, 2001). We infer (believe): the Universe (energy, matter, time, etc.) is real; Cosmic laws, discovered by scientists, govern the evolutionary activities of the Universe; life obeys Cosmic laws; and, the collective minds of our species can comprehend the Universe (scientism; scientific determinism).

3. Philosophical and Theological Views That Conflict With Scientific Determinism

Some theists and deists believe that the Natural Laws ["the skeleton of the Universe" (Trefil, 2003) that existed prior to the Universe and extended into its time] are "God given". Some

J. Seckbach et al. (eds.), Life in the Universe, 357–360.

wrote: God acted, created the Universe: "'And the Earth brought forth' life. (Gen. 1:12)" [sic] (Schroeder, 1998, pp. 5, 85).

Thus, this origin (and evolution) of life—implies Natural Laws at work. Others wrote: An impersonal God would not direct events in the Universe. "The Earth itself had within it the (necessary) special properties (self-organization and/or catalysts) to encourage the emergence of life. There is no biblical mention of a special creation for the origin of life (Schroeder, 1998, pp. 29, 85)." (See also, Armstrong, 1993, p. 355.)

Religious writings are seen by many as inspirational rather than documents for literal interpretation. However, deistic (impersonalized God) thinking is now seen in the Anthropic Principle: "Large numbers of apparent coincidences existed between things (Universal constants) and persuaded many a philosopher, theologian and scientist of the past that none of this (Natural Laws) was an accident. The Universe was designed with an end in view. This end involved the existence of life—perhaps even ourselves—and the plainness of the evidence for such design meant that there had to be a Designer (Barrow, 2002, p. 157)." Agnostics say that following this type of logic is going beyond the data (insufficient to prove the existence of God). Atheists claim there is no logic for such a belief since God does not exist in their minds (Armstrong, 1993).

Many humans retain the Determiner-view (personalized God; caring Creator of Heaven and Earth and all things visible and invisible) (based on the argument from design). The belief includes: the "something" that preceded the Universe was/is God.

Scientists generally limit their beliefs to events in the Universe and some claim:—strong evidence (data) to support a belief in Cosmic evolution does not prove nor disprove the existence of a conscious Determiner (Creator; God). Fundamentalism and creationism exhibit religious preferences not considered to be science (Armstrong, 1993, Armstrong, 2000).

4. Discussion

"Evolution (of the Cosmos) is a synthesis of determinism and chance, and this synthesis makes it a creative process.—Formation of the precursor molecules of life's origin provides—an illustration of the limited role of chance in Nature"—"—the soupy organic matter trapped in this (Miller's) chemical evolutionary experiment always yields the same kinds and proportions of amino-acid-rich, protein-like compounds—and in approximately the same relative abundances found embedding meteoric rocks and composing living organisms (Chaisson, 2001, p. 37)." Chaisson continues by describing other sources of these compounds (boiling, sulfurous pools; hot, mineral-laden deep-sea volcanic vents; interstellar space via comets or meteorites). We could suggest improvements in his review but the point is that if the reactants were "forming and reforming into larger molecules by chance alone, that is, in the hypothetical case of no forces at all, the products would be a hopeless mess comprising billions of possibilities and would likely vary each time.—But the results show no such chemical diversity, nor does life at the biochemical level. Specificity and reproducibility are hallmarks of this classic (Miller-type) experiment, ensuring that bonding of the amino acids does not occur entirely at random (yielding molecular variations), but that certain associations were advantaged while others were excluded (yielding molecular selection).—Selective assembly seems more of a rule than an exception—(Chaisson, 2001, p. 38)."

Stereo-electronic forces guide and bond small molecules into larger ones appropriate to protolife (amino acids undergoing self-ordering reactions → branched proteins undergoing self-assembly reactions → protocellular life) (Fox, 1988; Pappelis, *et al.*, 2001). The natural agents of order tend to tame chance. Molecular variations are important—"but natural (molecular) selection is a decidedly deterministic reaction that directs evolutionary change (Chaisson, 2001)." Patterns of valid and reliable continuity are clearly evident in the past 500 years of data obtained by human, scientific studies of the various levels of the Cosmos. Life histories (of chemical and cellular life) are now understood to be real, functioning structures of evolving processes.

Scientists can take samples of the Cosmos for studies (using statistical methods for design and analysis) and periodically stop to summarize their findings (develop theories and models). They formulate their future using concepts that arise from their predictive (valid and reliable) data. They accept that they cannot know all of the cosmic (moment-to-moment emergence of sub-atomic particles in the earliest seconds of the Universe) and biological past (99+% of the species that emerged according to Darwinian evolution are extinct). In this view, long-cherished religious and philosophical tenets have been replaced by scientific findings. Yet, many questions remain. (For example, following Stanley Miller's presentation at this conference, Antonio Lazcano asked him about the chemical origin of life—"If it wasn't DNA; and, if it wasn't RNA; what was it?" Miller replied—"I don't know." Our answer would be branched proteins.)

Modern concepts in Cosmic evolution provided the insights needed to replace older explanations (possibly 150,000 years old) about starry night skies. Similarly, the branched protein-first paradigm provides continuity within Cosmic evolution: i.e., amino acids → branched proteins (chemical life) → biological life. Therefore, we propose that it is most appropriate to partition our thinking (and research goals) about: the origins of chemical and biological life; the origin of protocells and their early evolution; and, the evolution of protocells and the emergence of prokaryotic and eukaryotic life.

Science is a way of solving problems; its methods apply to what is assumed to be the reality of the Universe. Questions concerning morals, value judgments, social issues, and attitudes cannot be answered using the science process skills. Science is self-correcting. The change in thinking about the sky and Earth was a new paradigm. Similarly, the branched protein-first paradigm, developed by Fox (1988) and associates (Pappelis, *et al.*, 2001), has changed our thinking about the origin of life and its early evolution on Earth and in the Universe. We believe that protolife could not have been discovered except by synthesis. Further, we do not propose that a deity will "fill the gaps" in the creation and evolution theories. Rather, we believe that good science will.

People understand that science and scientists cannot presently answer all of the problems of our time. Scientists believe that they cannot prove nor disprove the soundness of a belief in God or Gods. Thus, they prefer to assume that science merely provides one type of knowledge—the kind that has predictive value in what is often called the "real" Universe (even though most scientists prefer to "assume" that the Universe is real). The religious people have faith in their beliefs although there is no data to say their knowledge is either valid or reliable. Armstrong (1993) writes that many people around the world believe science has replaced (or will replace) religion.

5. Conclusion

We conclude: life is a self-sustained chemical system [amino acids from space via micrometeorites, from hydro-thermal systems, etc. → self-ordering with conformational changes → branched proteins = chemical life → + water → colloidal suspension → phase shifting and self-assembly of branched proteins to form wall-membrane vesicles = protocells = biological life] capable of undergoing Darwinian evolution (Pappelis, *et al.*, 2001). We infer that the branched proteins that self-assemble into protocells are at the same time multizymic structural elements of the wall-membrane boundary involved in protometabolism. As protocells enlarge, they form buds ("genetically" active). Branched proteins can be induced to form protocells in water (4–60 °C), degenerate (heat to 100 °C), and reform after drying or after storage in water (cycles). Microencapsulated macromolecules (branched proteins, etc.) and the branched proteins lining the inner surface of the wall-membrane could account for the synthesis sites of linear oligo-peptides, oligonucleotides of RNA, and, finally, oligonucleotides of DNA (reviewed by Fox, 1988). Since branched proteins might be in the "glass state" in meteoritic material in cold space, they may represent a kind of space-traveling life (litho-cosmozoa) (Pappelis, *et al.*, 2001). The branched protein-first paradigm bridges the gap between chemical and biological evolution and functions in the evolution of protocells.

6. Acknowledgements

Our memories of Sidney W. Fox have inspired us throughout this work. Aristotel Pappelis dedicates his efforts to the memory of Constantine Zahariades, Gerald and Peter Andrews, and Reverend Father Nicholas Karras, his first teacher of religion and philosophy.

7. References

Armstrong, K. (1993) *A History of God: The 4000-Year Quest of Judaism, Christianity and Islam.* Ballantine Books, NY.

Armstrong, K. (2000) *The Battle for God: A History of Fundamentalism.* Ballantine Books, NY.

Barrow, J. D. (2002) *The Constants of Nature: From Alpha to Omega—The Numbers That Encode the Deepest Secrets of the Universe.* Pantheon Books, NY.

Chaisson, E. J. (2001) *Cosmic Evolution: The Rise of Complexity in Nature.* Harvard University Press, Cambridge, MA, (pp. 37–41).

de Duve, C. (2002) *Life Evolving: Molecules, Mind, and Meaning.* Oxford University Press, NY.

Fox, S. W. (1988) *The Emergence of Life: Darwinian Evolution from the Inside.* Basic Books, NY.

Pappelis, A., Bahn, P., Grubbs, R., Bozzola, J., and Cohen, P. (2001) From inanimate macro-molecules to the animate protocell: In search of thermal protein phase-shifting, in J. Chela-Flores, T. Owen, and F. Raulin (eds.), *First Steps in the Origin of Life in the Universe*, Kluwer Academic Publishers, Dordrecht, pp. 65–68.

Schroeder, G. L. (1998) *The Science of God: The Convergence of Scientific and Biblical Wisdom.* Broadway Books, NY. (pp. 5, 29, 85).

Trefil, J. (2003) *The Nature of Science: An A-Z Guide to the Laws & Principles Governing Our Universe.* Houghton Mifflin, NY.

GLIMPSES OF TRIESTE CONFERENCES ON CHEMICAL EVOLUTION AND ORIGIN OF LIFE

A Pictorial Overview

MOHINDRA S. CHADHA
Beach House, Juhu, Mumbai 400 049, INDIA
chadhams@mail.com

A re-capitulation of Trieste Conferences, on Chemical Evolution and Origin of Life from their inception in 1992 is presented. An attempt is made to highlight the Scientific, Educational and Cultural aspects (through pictures taken at the six conferences held until 2000). Participation in the First Trieste Conference held at International Centre for Theoretical Physics (ICTP) in Oct 1992 and now at the Seventh Conference and all the others in between has provided the author with a perspective, which is being shared.

The first Trieste Conference, which was the brainchild of Prof. Abdus Salam and Prof. Cyril Ponnamperuma, was held under the joint sponsorship of the International Atomic Energy Agency and the United Nations Educational Scientific and Cultural Organisation and witnessed active participation by both the founders of this unique Series of Conferences. Prof. Julian Chela-Flores, who has been the backbone of this activity eversince, assisted them in their efforts. This inter-disciplinary international conference was entitled Chemical Evolution: Origin of Life and defined the scope and objectives of this novel effort.

The second Trieste Conference (Oct. 1993) was entitled Chemical Evolution: Self-organisation of the Macromolecules of Life and was dedicated to Cyril Ponnamperuma on his 70th birthday. The focal theme of self-organisation was sub-divided into chemical aspects, geophysical aspects, biochemical aspects, biophysical aspects and chirality. At this conference many of the erstwhile collaborators of Cyril Ponnamperuma participated and the proceedings of the conference were edited by Julian Chela-Flores and Cyril Ponnamperuma's former collaborators namely: Mohindra S. Chadha, Alicia Negron-Mendoza and Tairo Oshima.

The third Trieste Conference (Sept. 1994) was entitled Chemical Evolution: Structures and Model of the First Cell and was designated as the Alexander Ivanovich Oparin 100th Anniversary Conference. This was the best-attended conference of the three Trieste conferences till then and the presentations on planetary, extraterrestrial and interstellar conditions broke a lot of new ground. The topics covered the beginning of cellular organisation (the early paleonotoligical record, physical, chemical and biological aspects of the origin and structure of membrane and origin and structure of the cell).

The fourth Trieste Conference (Sept. 1995) was the first conference of the series, which was held after the untimely and sudden demise of Cyril Ponnamperuma, the co-founder of this series of conferences. He was involved in the early part of the preparation of this

J. Seckbach et al. (eds.), Life in the Universe, 361–363.

conference but unfortunately was not to live to witness its deliberations. Such is Life! The Conference was entitled Chemical Evolution: Physics of the Origin and Evolution. The Conference was aptly dedicated to the memory of Cyril Ponnamperuma. There were general overview presentations by John Oro on Cosmic Evolution and by Sidney Fox entitled Experimental Retracement of Terrestrial Origin of An Excitable Cell: Was it Predictable? The sessions covered a broad range of topics e.g. Origins; From Geophysics to Prebiotic Chemistry; Physicochemical Aspects; Biophysical Aspects—General problems & Biomolecular Chirality; Evolutionary Aspects; Information Theory: Communication and Instrumentation in Exobiology and Mars Exploration.

In a special session, homage was paid to Cyril Ponnamperuma and excerpts of letters received from various academies, institutions and admirers of Cyril Ponnamperuma from different parts of the Globe were read out. A touching note received from his wife (Valli) and daughter (Roshini) was also shared.

The fifth Trieste Conference (Sept. 1997) was entitled Exobiology, Matter, Energy and Information in the Origin and Evolution of Life in the Universe and was dedicated to the memory of Abdus Salam, the co-founder of the Trieste series of Conferences and Director of ICTP, who had unfortunately passed away after prolonged illness. This conference had as many as 12 sponsors and was largely attended. The highlights were talks by: Chela-Flores entitled: Abdus Salam—From Fundamental Interactions to the Origin of Life. The Abdus Salam lecture by John Oro entitled: Cosmocological Evolution—A Unifying and Creative Process in the Universe. The Cyril Ponnamperuma lecture by Frank Drake, namely The Search for Intelligent Life in the Universe. The opening lecture: The Theory of Common Descent by Richard D. Keynes and a Public Lecture by Paul Davies entitled: Are We Alone in the Universe (?).

The sessions at this Conference dealt with Matter in the Origin; Energy from Inert to Living matter; Information; Early Evolution; Exobiology on Mars and Europa; The Interstellar Medium, Comets and Chemical Evolution; Exobiology on Titan; Extrasolar Planets and Search for Extraterrestrial Intelligence.

This Conference was unique in focussing on Exobiology in general and on Comets, Planets and the Interstellar Medium in particular.

The sixth Trieste Conference (Sept. 2000) was entitled First Steps in the Origin of Life in the Universe and was dedicated to Giordano Bruno whose intuitive concepts have relevance to current enquiries related to the First Steps in the Origin of Life in the Universe. This conference had as many as 11 sponsors. The conference featured a special lecture by Stanley Miller who presented the possibility of Peptide Nucleic Acids as A Possible Primordial Genetic Polymer. The Abdus Salam lecture entitled: Physics and Life was delivered by Paul Davies and the Cyril Ponnamperuma lecture by J. William Schopf entitled: Solutions to Darwin's Dilemma: Discovery of the Missing Precambrian Records of Life. All the three presentations were quite scintillating.

A new feature of the sixth Trieste Conference was inclusion of a section entitled: Historical Aspects which featured work of Sidney Fox (by Aristotle Pappalis), Theories on Origins of Life between 1860–1900 (by F. Raulin-Carceau) and Reminiscences: Pont-a-Mousson-1970 to Trieste-2000 (by Mohindra Chadha).

The other main topics of the conference dealt with Life without Starlight and questions asked were: Can Life Originate in the absence of Starlight? The other important topics discussed were: Is there indigeneous material on Mars? The search for prebiotic and biological indicators in the satellites of the outer solar system; New Approaches to the Detection of Extraterrestrial Radio Signals and the Implications. In a Pre-Dinner lecture entitled New Paradigms Of Seti, Frank Drake brought the audience up to date in this challenging and exciting venture of scientific endeavour.

The Sixth Trieste conference had all the embellishments of the previous conferences as envisioned by the two founders (Abdus Salam and Cyril Ponnamperuma). The Keynote addresses given by leading researchers in their respective fields were authoritative. Many young scientists from both the industrialised countries and those from developing countries, coming from different parts of the Globe enriched themselves greatly from their participation in this conference as they did in the earlier ones.

The last Six Trieste Conferences and now the Seventh Conference held to honor Stanley Miller for his seminal discovery some 5 decades back have brought together scientists (both experimental and theoretical), philosophers and theologians of repute on the same platform. In all, nearly 600 participants have benefited from the Trieste Conferences. Unfortunately some of the luminaries who participated in the earlier conferences have departed from this world and a special mention of their contributions has been made.

The intellectual flavor of the discussions and many debates has been of a high order and the future for these inter-disciplinary international deliberations seems to be very bright. Our thanks are due to Julian Chela-Flores, Tobias Owen and Francois Raulin who have continued to carry the torch lit by Cyril Ponnamperuma and Abdus Salam still further. The sponsors and the organising committee also deserve applause. This tribute to the activities of over a decade of Trieste Conferences on Chemical Evolution and Origin of Life is presented through a Pictorial Overview.

LIST OF PARTICIPANTS IN LITV CONFERENCE TRIESTE SEPT. 2003

ABRAMSON Guillermo
Centro Atomico Bariloche
Instituto Balseiro
8400 Bariloche
Rio Negro
Argentina

ACHARYYA Kinsuk
Centre for Space Physics
P-61, Southend Garden, Garia
Kolkata 700084
India

ADEBOWALE Kayode Oyebode
University of Ibadan
Department of Chemistry
P.O. Box 22788
U.I.P.O.
Ibadan
Nigeria

AFOLABI Tahjudeen Adeniyi
Dept. of Chemistry
University of Ado Ekiti
Ekiti State
Ado Ekiti
Nigeria

AKINDAHUNSI Afolabi Akintunde
Federal University of Technology
Biochemistry Department
P.M.B. 704
Ondo State
Akure
Nigeria

AREXAGA Roberto
Universidad de Deusto
Departamento de Filosofia
Avenida de las Universidades s/n
48080 Bilbao
Spain

ASSAOUI Fatna
Universite' Mohammed V
Laboratoire de Physiques Des Hautes
Energies
Faculty of Sciences
Av. Ibn Battota
B.P. 1014
Rabat
Morocco

ATENEKENG KAHOU Guy Antoine
Univeriste de Yaounde I
Faculte Des Sciences
B.P. 812
Yaounde
Republic of Cameroon

BAHN Peter
Bahn Biotechnology Co.
RR2 Box 239A
Mt. Vernon 62864 IL
United States of America

BALTSCHEFFSKY Herrick
University of Stockholm
Arrhenius Laboratory
Department of Biochemistry
S-106 91 Stockholm
Sweden

BALTSCHEFFSKY Margareta
University of Stockholm
Arrhenius Laboratory
Department of Biochemistry
S-106 91 Stockholm
Sweden

BECERRA Arturo
Universidad Nacional Autonoma
de Mexico
Facultad de Ciencias
Apdo. Postal 70 407
Cd. Universitaria
D.F.
04510 Mexico City
Mexico

BERNARD Jean-Michel
Université Paris Xii (Val-De-Marne)Lisa
61 Ave. Gal de Gaulle
94010 Creteil Cedex
France

BHATTACHARJEE Aranya
Universita' di Pisa
Dipartimento di Fisica
Via Buonarroti. 2
56127 Pisa
Italy

BIONDI Elisa
Universita' degli Studi di Firenze
Dipartimento di Biologia Animale
e Genetica
Via Romana 17/19
50125 Firenze
Italy

BOICE Anand Erik
Indiana University
Department of Geological Sciences
1001 East 10th St.
Bloomington 47405-1403
Indiana
United States of America

BONACCORSI Rosalba
Universita' degli Studi di Trieste
D.Sci.Geo.Ambientali.Marine
V. E. Weiss 2
34127 Trieste
Italy

BRACK Andre
Centre National de La Recherche
Scientifique
Centre de Biophysique-Moleculaire
Rue Charles Sandron
45071 Orleans Cedex 2
France

BRILLI Matteo
Universita' degli Studi di Firenze
Dipartimento di Biologia Animale
e Genetica
Via Romana 17/19
50125 Firenze
Italy

BUHSE Thomas Werner
Centro de Investigaciones Quimicas
Universidad Autonoma del Estado
de Morelos
Av. Universidad 1001
Col. Chamilpa
62210 Cuernavaca
Morelos
Mexico

CANTONI Roberto
Universita' di Napoli 'Federico Ii'
Facolta' di Scienze
Dipartimento di Fisica
Via Cinthia
80100 Napoli
Italy

CARNERUP Anna
Australian National University
Research School of Physical Sciences
D. of Applied Mathematics
Gpo Box 4
ACT 2601 Canberra
Australia

CETINKAYA Berkan
Ege University
Institute of Nuclear Sciences
Bornova
35100 Izmir
Turkey

CHADHA Mohindra S.
Bhabha Atomic Research Centre
Bio-Organic Division
400 085 Bombay-Mumbai
India

CHAKRABARTI Sandip Kumar
S.N. Bose National Centre For Basic
Sciences
Theoretical Astrophysics
Jd Block. Sector-111

Salt Lake
700098 Calcutta
India

CIPRIANI FITA Roberto
Universidad Simon Bolivar
Depto Estudios Ambientales
Apartado 89.000
Baruta
Caracas 1080A
Venezuela

COLLIS William
Strada Sottopiazzo 18
14056 Asti
Italy

COSMOVICI Cristiano
Consiglio Nazionale Delle Ricerche
Istituto di Fisica Dello Spazio
Interplanetario
Via Fosso del Cavaliere 100
Area di Ricerca Roma-Tor Vergata
00133 Roma
Italy

COYNE George Vincent
Specola Vaticana
Citta' del Vaticano
00120 Roma
Italy

CRISMA Marco
Consiglio Nazionale delle Ricerche
Istituto di Chimica Biomolecolare
Via Marzolo 1
35131 Padova
Italy

DE SOUZA BARROS Fernando
Universidade Federal Do Rio
de Janeiro
Instituto de Fisica
Cidade Universitaria. Ct Bl.A 4
Caixa Postal 68528
Ilha Do Fundao
21945-970 Rio de Janeiro
Brazil

DELAYE Luis
Universidad Nacional Autonoma de Mexico
Facultad de Ciencias
Apdo. Postal 70 407
Cd. Universitaria
D.F.
04510 Mexico City
Mexico

DICK Steven J.
U.S. Naval Observatory
3450 Massachussetts Ave. NW
Washington 20392-5420
DC
United States of America

DRAKE Frank
Seti Institute
2035 Landings Drive
94043 CA Mountain View
United States of America

DRAKE Leila
Seti Institute
2035 Landings Drive
94043 CA Mountain View
United States of America

DRAKE Nadia
Seti Institute
2035 Landings Drive
94043 CA Mountain View
United States of America

FANI Renato
Universita' degli Studi di Firenze
Dipartimento di Biologia Animale
e Genetica
Via Romana 17/19
50125 Firenze
Italy

FOUPOUAGNIGNI Mama
Universite de Yaounde 1
Ecole Normale Superieure
Departement de Mathematiques
B.P. 47
Yaounde
Republic of Cameroon

FRANCHI Marco
Universita' degli Studi di Firenze
Dipartimento di Biologia Animale
e Genetica
Via Romana 17/19
50125 Firenze
Italy

FRASER Donald Gordon
University of Oxford
Department of Earth Sciences
Parks Road
OX1 3PR Oxford
United Kingdom

FRAY Nicolas
Universite' de Paris Xii (Val-De-Marne)
Avenue Du General de Gaulle
94010 Creteil Cedex
France

FUSI Luca
Universita' degli Studi di Firenze
Dipartimento di Biologia Animale
e Genetica
Via Romana 17/19
50125 Firenze
Italy

GALLARDO Jose
Pontificia Universidad Catolica de Chile
Facultad de Fisica
Campus San Joaquin
Av. V. Mackenna 4860
(22) Santiago
Chile

GALLORI Enzo
Universita' degli Studi di Firenze
Dipartimento di Biologia Animale
e Genetica
Via Romana 17/19
50125 Firenze
Italy

GATTA Riccardo S.
Via Biasoletto, 27
34142 Trieste
Italy

GENTA Giancarlo
Politecnico di Torino
Dipartimento di Meccanica
Corso Duca degli Abruzzi 24
10129 Torino
Italy

GIRARDI Leo
Osservatorio Astronomico di Trieste
Via G.B. Tiepolo 11
34131 Trieste
Italy

GRYMES Rosalind
NASA Astrobiology Institute
Ames Research Center MS240-1
Moffett Field 94035-1000
CA
United States of America

GUIMARAES Romeu Cardoso
Universidade Federal de Minas Gerais
Instituto de Ciencias Biologicas
Depto. de Biologia Geral
Mg
31270-901 Belo Horizonte
Brazil

IGWEBUIKE Udensi Maduabuchi
University of Nigeria
Dept. of Veterinary Anatomy
Nsukka
Nigeria

JOHNSON Torrence
California Institute of Technology
Jet Propulsion Laboratory
Department of Physics
4800 Oak Grove Drive
Ca 91109
Pasadena
United States of America

KAMALUDDIN . . .
University of Roorkee
Department of Chemistry
247 667 (U.P.) Roorkee
India

KRITSKIY Mikhail S.
Russian Academy of Sciences
A.N. Bach Institute of Biochemistry
Leninsky Prospekt 33
117071 Moscow
Russian Federation

LANCET Doron
the Weizmann Institute of Science
Crown Human Genome Center
Dept. of Molecular Genetics
76100 Rehovot
Israel

LAZCANO Antonio
Universidad Nacional Autonoma
de Mexico
Facultad de Ciencias
Apdo. Postal 70 407
Cd. Universitaria
D.F.
04510 Mexico City
Mexico

LEGER Alain
Universitè Paris-Sud
IAS
Batiment 121
F-91405 Orsay
France

MACCONE Claudio
International Academy of Astronautics
"Lunar Farside Radio Lab" Cosmic Study
Via Martorelli 43
10155 Torino
Italy

MASSON Philippe
Departement des Sciences de la Terre
FRE 2566 CNRS-UPS (ORSAYTERRE)
Universite Paris-Sud
Bat. 509
F-91405 Orsay Cedex
France

MATTEUCCI Francesca
Universita' di Trieste
Dipartimento di Astronomia
Via G.B. Tiepolo. 11
34131 Trieste
Italy

MAYOR Michel
Observatoire de Geneve
Chemin Des Maillettes 51
CH-1290 Sauverny
Switzerland

MEIERHENRICH Uwe
University of Bremen
Leobenerstrasse
D-28359 Bremen
Germany

MEJIA CARMONA Diego Fernando
Universidad del Valle
Departamento de Bioquimica
Sede San Fernando Piso 5
Cali
Colombia

MERAZKA Fatiha
Ecole Nationale Polytechnique
D. Elect. & Computer Eng.
10 Av. Hassen Badi
B.P. 182
Al Harrach
Algiers
Algeria

MESSEROTTI Mauro
Osservatorio Astronomico
di Trieste
Succursale di Basovizza
Loc. Basovizza. 302
34012 Trieste
Italy

MEYER Michael
National Aeronautics and Space
Administration (Nasa)
Nasa Headquarters
Code SE
300 E. Street S.W.
20546 Washington D.C.
United States of America

MICHEAU Jean-Claude
Université P. Sabatier
Laboratoire IMRCP
118 Route de Narbonne
F-31062 Toulouse
France

MILLER Stanley L.
University of California At San Diego
Chemistry & Biochemistry
9500 Gilman Drive
CA 92093-0371 La Jolla
United States of America

MINNITI Dante
Pontificia Universidad Catolica
de Chile
Facultad de Fisica
Dpto. de Astronomia y Astrofisica
Vicuna Mackenna 4860
Santiago
Chile

MINNITI NOGUERAS Alicia
Pontificia Universidad Catolica de Chile
Facultad de Biologia
Alameda 340
Santiago
Chile

MOLARO Paolo
Osservatorio Astronomico di Trieste
Via G.B. Tiepolo 11
34131 Trieste
Italy

MONTEBUGNOLI Stelio CNR
Radiotelescopio
Via Fiorentina
40060 Villafontana (BO)
Italy

MOORBATH Stephen
University of Oxford
Department of Earth Sciences
Parks Road
OX1 3PR Oxford
United Kingdom

MUSSO Paolo
Universita' della Santa Croce
Dipartimento di Filosofia
Piazza S. Apollinari 49
00186 Roma
Italy

NAHAL Arashmid
Institute For Advanced Studies in
Basic Sciences (IASBS)
P.O. Box 45195-159
Gava Zang
Zanjan
Islamic Republic of Iran

NDOUNDAM Rene
University of Yaounde
Faculty of Science
Dept. of Computer Science
B. P. 812
Yaounde
Republic of Cameroon

NEGRON-MENDOZA Alicia
Universidad Nacional Autonoma
de Mexico
Instituto de Ciencias Nucleares
Apdo. Postal 70-543
Circuito Exterior. C.U.
Df
04510 Mexico City
Mexico

OBOH Ganiyu
Federal University of Technology
Biochemistry Department
P.M.B. 704
Ondo State
Akure
Nigeria

OWEN Tobias
Institute For Astronomy
2680 Woodlawn Drive
Hawaii
96822 Honolulu
United States of America

PAPPELIS Aristotel
Southern Illinois University At Carbondale
Illinois
62901-6509 Carbondale
United States of America

PENSADO DIAZ Hector Omar
Instituto de Ciencias Avanzadas
Edificio Tlacolulan I Depto 201
Ignacio Allende 93
Xalapa 91097
Ver.
Mexico

PEREZ DE VLADAR Harold Paul
Instituto de Estudios Avanzados—IDEA
Centro de Biotecnologia
Apartado Postal 17606
Parque Central
Caracas 1015-A
Venezuela

PEREZ-MERCADER Juan
Centro de Astrobiologia
(Nasa Astrobiology Institute)
Ctra. de Ajalvir Km.4
Torrejon de Ardoz
28850 Madrid
Spain

PLATTS Nicholas Simon
Carnegie Institution of Washington
Geophysical Laboratory
5251 Broad Branch Road. N.W.
20015 Washington
United States of America

PUY Denis
Observatoire de Geneve
Chemin Des Maillettes 51
CH-1290 Sauverny
Switzerland

RAMIREZ {JIMENEZ} Sandra Ignacia
Centro de Investigaciones Quimicas
Universidad Autonoma del
Estado de Morelos
Av. Universidad 1001
Col. Chamilpa
62210 Cuernavaca
Morelos
Mexico

RAMOS-BERNAL Sergio
Universidad Nacional Autonoma
de Mexico
Instituto de Ciencias Nucleares
Apdo. Postal 70-543
Circuito Exterior. C.U.
Df
04510 Mexico City
Mexico A39

RAULIN Francois
Universites Paris 7 and 12. of Cnrs
Lisa
Faculte' Des Science Et Tech.
61 Ave. General de Gaulle
94010 Creteil
France

RIGHINI Simona
Radiotelescopi di Medicina
AIA Cavicchio
Via della Fiorentina
40060 Villafontana (Bologna)
Italy

ROEDERER Juan Gualterio
University of Alaska
Geophysical Institute
P.O. Box 757320
99775-0800 Alaska
Fairbanks
United States of America

SALAWU Sule Ola
Federal University of Technology
Biochemistry Department
P.M.B. 704
Ondo State
Akure
Nigeria

SCAPPINI Flavio
Consiglio Nazionale Delle

Ricerche (C.N.R.)
Istituto di Spettroscopia Molecolare (I.S.M.)
Via Gobetti 101
40129 Bologna
Italy

SCHWEHM Gerhard
European Space Agency (Esa)
Estec
Space Science Department
Postbus 299
2200 AG Noordwijk
Netherlands

SECKBACH Joseph
the Hebrew University of Jerusalem
Jerusalem
Israel

SHAH Tahir K.
Universita degli Studi di Trieste
D.E.E.I.
Via Valerio 10
34100 Trieste
Italy

SHIL Pratip
University of Pune
Biophysics Laboratory
Department of Physics
411 007 Pune
India

SI LAKHAL Bahia
Universite de Blida
Institut Des Sciences Exactes
Departement de Physique
B.P. 270
Soumaa
Blida
Algeria

SIEFERT Janet
Rice University
Department of Statistics, MS 138
P.O. Box 1892
Houston 77251-1892
Texas
United States of America

SIMAKOV Mikhail
Russian Academy of Sciences
Institute of Cytology
Tikhoretsky Av. 4
194064 St. Petersburg
Russian Federation

SIMON Istvan
Hungarian Academy of Sciences
Institute of Enzymology
Pob 7
H-1518 Budapest
Hungary

SINGER Emily
New Scientist
151 Wardour St.
London
United Kingdom

SINGLETON. JR. Robert
Los Alamos National Laboratory
X-7, MSF699
Los Alamos 87545
New Mexico
United States of America

STAN-LOTTER Helga
Universitat Salzburg
Institut Fur Allgemeine Biologie
Hellbrunner Strasse 34
5020 Salzburg
Austria

TEWARI Vinod Chandra
Wadia Institute of Himalayan
Geology
33 General Mahadeo Singh Road
P.O. Box 74
248001 Dehra Dun
India

TORRES AROCHE Leonel Alberto
Centre For Clinical Research
34 No. 4501 E/45 Y 47
Reparto Kohly
11300 Havana
Cuba

TURNBULL Margaret C.
University of Arizona
Steward Observatory
933 N. Cherry Ave.
Tucson 85716
Arizona
United States of America

UNAK (DARCAN) Perihan
Ege University
Institute of Nuclear Sciences
Bornova
35100 Izmir
Turkey

VAN DUNNE Hein Johan Francois
Kluwer Academic Publishers
P.O. Box 17
3300 AA Dordrecht
Netherlands

VAN ZUILEN Mark
CRPG-CNRS
15 Rue Notre Dame des Pauvres
BP 20
54501 Vandoeuvre les Nancy
France

VANCE Steven
Institut fuer Planetologie
Wilhelm-Klemms Strasse 10
Munster 48149
Germany

VIDYASAGAR Pandit Bhalchandra
University of Pune
Department of Physics
Ganeshkhind
411 007 Pune
India

VIEYRA Adalberto
Universidade Federal Do
Rio de Janeiro
Instituto de Biofisica
Ibccf-Ccs-Ilha Do Fundao
22000 Rio de Janeiro
Brazil

VLADILO Giovanni
Osservatorio Astronomico
di Trieste
Via G.B. Tiepolo 11
34131 Trieste
Italy

WANG Wenqing
Beijing University
{University of Peking}
Dept. of Technical Physics
100871 Beijing
People's Republic of China

WARD Peter D.
University of Washington
Astrobiology Program
Department of Biology
P.O. Box 357242
Seattle 98195-7242
WA
United States of America

WESTALL Frances
Centre National de La Recherche
Scientifique
Centre de
Biophysique-Moleculaire
Rue Charles Sandron
45071 Orleans Cedex 2
France

INDEX

INDEX OF AUTHORS